An Introduction to
Environmental Chemistry

An Introduction to Environmental Chemistry

SECOND EDITION

J.E. Andrews, P. Brimblecombe, T.D. Jickells, P.S. Liss and B. Reid

School of Environmental Sciences
University of East Anglia
United Kingdom

Blackwell
Publishing

350 Main Street, Malden, MA 02148-5020, USA
108 Cowley Road, Oxford OX4 1JF, UK
550 Swanston Street, Carlton, Victoria 3053, Australia

The right of J.E. Andrews, P. Brimblecombe, T.D. Jickells, P.S. Liss and B. Reid to be
identified as the Authors of this Work has been asserted in accordance with
the UK Copyright, Designs, and Patents Act 1988.

First published 1996 by Blackwell Science Ltd
Second edition 2004

Library of Congress Cataloging-in-Publication Data

An introduction to environmental chemistry / J.E. Andrews . . . [et al.]. – 2nd ed.
    p.   cm.
Includes bibliographical references and index.
    ISBN 0-632-05905-2 (pbk.: alk. paper)
    1. Environmental geochemistry.   I. Andrews, J.E. (Julian E.)

QE516.4.I57 2004
551.9 – dc21   2003002757

A catalogue record for this title is available from the British Library.

Set in $9\frac{1}{2}$/12 pt Janson
by SNP Best-set Typesetter Ltd., Hong Kong
Printed and bound in the United Kingdom
by TJ International Ltd, Padstow, Cornwall

For further information on
Blackwell Publishing, visit our website:
http://www.blackwellpublishing.com

Contents

5 The chemistry of continental waters 141

6 The oceans 181

Boxes

Preface to the Second Edition

In revision of this book we have tried to respond to constructive criticism from reviewers and students who have used the book and at the same time have pruned and grafted various sections where our own experience as teachers has prompted change. Not least, of course, science has moved on in the eight years since we prepared the first edition, so we have had to make some substantial changes to keep up with these developments, especially in the area of global change.

We have tried to retain the ethos of the first edition, using concise and clear examples of processes that emphasize the chemistry involved. We have also tried to highlight how the chemistry, processes or compounds interlink between the chapters and sections, so that no compartment of environmental science is viewed in isolation.

The substantial changes include more emphasis on organic chemistry, soils, contaminants in continental water and remediation of contaminated land. To do this effectively, the terrestrial environments chapter from the first edition has been split into two chapters dealing broadly with solids and water. We have re-organized the box structure of the book and have placed some of the original box material, augmented by new sections, to form a new chapter outlining some of the basic chemical principles that underpin most sections of the book.

Much of the new material has been prepared by Brian Reid, who, in 1999, joined us in the School of Environmental Sciences at the University of East Anglia. Brian has very much strengthened the organic chemistry dimension of the book and we are very pleased to welcome him to the team of authors.

Julian Andrews, Peter Brimblecombe, Tim Jickells, Peter Liss and Brian Reid
University of East Anglia, Norwich, UK

Preface to the
First Edition

During the 1980s and 1990s environmental issues have attracted a great deal of scientific, political and media attention. Global and regional-scale issues have received much attention, for example, carbon dioxide (CO_2) emissions linked with global warming, and the depletion of stratospheric ozone by chlorofluorocarbons (CFCs). Local issues, however, have been treated no less seriously, because their effects are more obvious and immediate. The contamination of water supplies by landfill leachate and the build up of radon gas in domestic dwellings are no longer the property of a few idiosyncratic specialists but the concern of a wide spectrum of the population. It is noteworthy that many of these issues involve understanding chemical reactions and this makes environmental chemistry a particularly important and topical discipline.

We decided the time was right for a new elementary text on environmental chemistry, mainly for students and other readers with little or no previous chemical background. Our aim has been to introduce some of the fundamental chemical principles which are used in studies of environmental chemistry and to illustrate how these apply in various cases, ranging from the global to the local scale. We see no clear boundary between the environmental chemistry of human issues (CO_2 emissions, CFCs, etc.) and the environmental geochemistry of the Earth. A strong theme of this book is the importance of understanding how natural geochemical processes operate and have operated over a variety of timescales. Such an understanding provides baseline information against which the effects of human perturbations of chemical processes can be quantified. We have not attempted to be exhaustive in our coverage but have chosen themes which highlight underlying chemical principles.

We have some experience of teaching environmental chemistry to both chemists and non-chemists through our first-year course in Environmental

Chemistry, part of our undergraduate degree in Environmental Sciences at the University of East Anglia. For 14 years we used the text by R.W. Raiswell, P. Brimblecombe, D.L. Dent and P.S. Liss, *Environmental Chemistry*, an earlier University of East Anglia collaborative effort published by Edward Arnold in 1980. The book has served well but is now dated, in part because of the many recent exciting discoveries in environmental chemistry and also partly because the emphasis of the subject has swung toward human concerns and timescales. We have, however, styled parts of the new book on its 'older cousin', particularly where the previous book worked well for our students.

In places the coverage of the present book goes beyond our first-year course and leads on towards honours-year courses. We hope that the material covered will be suitable for other introductory university and college courses in environmental science, earth sciences and geography. It may also be suitable for some courses in life and chemical sciences.

Julian Andrews, Peter Brimblecombe, Tim Jickells and Peter Liss
University of East Anglia, Norwich, UK

Acknowledgements

We would like to thank the following friends and colleagues who have helped us with various aspects of the preparation of this book: Tim Atkinson, Rachel Cave, Tony Greenaway, Robin Haynes, Kevin Hiscock, Alan Kendall, Gill Malin, John McArthur, Rachel Mills, Willard Pinnock, Annika Swindell and Elvin Thurston. Special thanks are due to Nicola McArdle for permission to use some of her sulphur isotope data.

We have used data or modified tables and figures from various sources, which are quoted in the captions. We thank the various authors and publishers for permission to use this material, which has come from the following sources.

Books

Andres, R.J., Marland, G., Boden, T. & Bischof, S. (2000) In *The Carbon Cycle*, ed. by Schimel, D.S. & Wigley, T.M.L., pp. 53–62. Cambridge University Press, Cambridge.

Baker, E.T., German, C.R. & Elderfield, H. (1995) In *Seafloor Hydrothermal Systems: Physical, Chemical, Biological and Geological Interactions*, ed. by Humphris, S.E., Zierenberg, R.A., Mullineaux, L.S. & Thomson, R.E., pp. 47–71. Geophysical Monograph 91. American Geophysical Union, Washington, DC.

Baird, C. (1995) *Environmental Chemistry*, W.H. Freeman, New York.

Berner, K.B. & Berner, R.A. (1987) *The Global Water Cycle*. Prentice Hall, Englewood Cliffs, New Jersey.

Berner, R.A. (1980) *Early Diagenesis*. Princeton University Press, Princeton.

Birkeland, P.W. (1974) *Pedology, Weathering, and Geomorphological Research*. Oxford University Press, New York.

Brady, N.C. & Weil, R.R. (2002) *The Nature and Properties of Soils*, 13th edn. Prentice Hall, New Jersey.

Brimblecombe, P. (1986) *Air Composition and Chemistry*. Cambridge University Press, Cambridge.

Brimblecombe, P., Hammer, C., Rodhe, H., Ryaboshapko, A. & Boutron, C.F. (1989) In *Evolution of the Global Biogeochemical Sulphur Cycle*, ed. by Brimblecombe, P. & Lein, A. Yu, pp. 77–121, Wiley, Chichester.

Broecker, W.S. & Peng, T-H. (1982) *Tracers in the Sea*. Eldigio Press, New York.

Burton, J.D. & Liss, P.S. (1976) *Estuarine Chemistry.* Academic Press, London.

Chester, R. (2000) *Marine Geochemistry*, 2nd edn. Blackwell Science, Oxford.

Crawford, N.C. (1984) In *Sinkholes: Their Geology, Engineering and Environmental Impact*, ed. by Beck, B.F., pp. 297–304, Balkema, Rotterdam.

Davies, T.A. & Gorsline, D.S. (1976) In *Chemical Oceanography*, vol. 5, ed. by Riley, J.P & Chester, R., pp. 1–80, Academic Press, London.

Department of the Environment National Water Council (1984) *Standing Technical Advisory Committee on Water Quality, Fourth Biennial Report, February 1981-March 1983*, Standing Technical Committee Report No. 37. HMSO, London.

Drever, J.I., Li, Y-H. & Maynard, J.B. (1988) In *Chemical Cycles and the Evolution of the Earth*, ed. by Gregor, C.B., Gregor, C.B., Garrels, R.M., Mackenzie, F.T. & Maynard, J.B., pp. 17–53, Wiley, New York.

Garrels, R.M. & Christ, C.L. (1965) *Solutions, Minerals and Equilibria*, Harper and Rowe, New York.

Garrels, R.M., Mackenzie, F.T. & Hunt, C. (1975) *Chemical Cycles and the Global Environment.* Kaufmann, Los Altos.

Gill, R. (1996) *Chemical Fundamentals of Geology*, 2nd edn. Chapman & Hall, London.

Houghton, R.A. (2000) In *The Carbon Cycle*, ed. by Wigley, T.M.L. & Schimel, D.S., pp. 63–76. Cambridge University Press, Cambridge.

IPCC (1990) *Climate Change: The IPCC Scientific Assessment*, ed. by Houghton, J.T., Jenkins, G.J. & Ephramus, J.J. Cambridge University Press, Cambridge.

IPCC (1995) *Climate Change 1994. Reports of Working Groups I and III Intergovernmental Panel on Climate Change (IPCC).* Cambridge University Press, Cambridge.

IPCC (2001) *Climate Change 2001: The Scientific Basis. Contribution of Working Group I to the Third Assessment Report, Intergovernmental Panel on Climate Change.* Cambridge University Press, Cambridge.

Killops, S.D. & Killops, V.J. (1993) *An Introduction to Organic Geochemistry.* Longman Scientific and Technical, Harlow, Essex.

Koblentz-Mishke, O.J., Volkovinsky, V.V. & Kabanova, J.G. (1970) In *Scientific Exploration of the South Pacific*, ed. by Wooster, W.S., pp. 183–193, National Academy of Sciences, Washington, DC.

Krauskopf, K.B. & Bird D.K. (1995) *Introduction to Geochemistry*, 3rd edn. McGraw-Hill, New York.

Lister, C.R.B. (1982) In *The Dynamic Environment of the Ocean Floor*, ed. by Fanning, K.A. & Manheim, F.T., pp. 441–470. D.C. Heath, Lexington, Massachusetts.

Marland, G., Boden, T. & Andres, R.J. (2002) Global, regional and national CO_2 emissions. In *Trends: A Compendium of Data on Global Change.* Carbon Dioxide Information Analysis Center, Oak Ridge National Laboratory, US Department of Energy, Oak Ridge Tennessee.

McKie, D. & McKie, C. (1974) *Crystalline Solids.* Nelson, London.

Moss, B. (1988) *Ecology of Freshwaters.* Blackwell Scientific Publications, Oxford.

Polynov, B.B. (1937) *The Cycle of Weathering.* Murby, London.

Raiswell, R.W., Brimblecombe, P, Dent, D.L. & Liss, P.S. (1980) *Environmental Chemistry.* Edward Arnold, London.

Schaug, J. *et al.* (1987) *Co-operative Programme for Monitoring and Evaluation of the Long Range Transport of Air Pollutants in Europe (EMEP).* Summary Report of the Norwegian Institute for Air Research, Oslo.

Scoffin, T.P. (1987) *An Introduction to Carbonate Sediments and Rocks.* Blackie, Glasgow.

Sherman, G.D. (1952) In *Problems in Clay and Laterite Genesis*, p. 154. American Institute of Mining, Metallurgical, Petroleum Engineers, New York.

Spedding, D.J. (1974) *Air Pollution.* Oxford University Press, Oxford.

Spiedel, D.H. & Agnew, A.F. (1982) *The Natural Geochemistry of our Environment.* Perseus Books, New York.

Strakhov, N.M. (1967) *Principles of Lithogenesis*, vol. 1. Oliver & Boyd, London.

Svedrup, H., Johnson, M.W. & Fleming, R.H. (1941) *The Oceans.* Prentice Hall, Englewood Cliffs, New Jersey.

Taylor, R.S. & McLennan, S.M. (1985) *The Continental Crust: Its Composition and Evolution.* Blackwell Scientific Publications, Oxford.

Todd, D.K. (1980) *Groundwater Hydrology*, 2nd edn. John Wiley, New York.

Von Damm, K.L. (1995) In *Seafloor Hydrothermal Systems: Physical, Chemical, Biological and Geological Interactions*, ed. by Humphris, S.E., Zierenberg, R.A., Mullineaux, L.S. & Thomson, R.E., pp. 222–247, Geophysical Monograph 91. American Geophysical Union, Washington, DC.

Wood, L. (1982) *The Restoration of the Tidal Thames.* Adam Hilger, Bristol.

Articles

Boyd, P.W. and 34 others (2000) *Nature* **407**, 695–702, Macmillan, London.

Boyle, E.A., Collier, R., Dengler, A.T., Edmond, J.M., Ng, A.C. & Stallard, R.R. (1974) *Geochimica Cosmochimica Acta* **38**, 1719–1728, Pergamon, Oxford.

Bruland, K.W. (1980) *Earth and Planetary Science Letters* **47**, 176–198, Elsevier, Amsterdam.

Buddemeier, R.W., Gattuso, J-P. & Kleypas, J.A. (1998) *LOICZ Newsletter* No 4.

Coffey, M., Dehairs, F., Collette, O., Luther, G., Church, T. & Jickells, T. (1997) *Estuarine Coastal Shelf Science* **45**, 113–121, Academic Press, London.

Crane, A. & Liss, P.S. (1985) *New Scientist* **108** (1483), 50–54, IPC Magazines, London.

Duce, R.A. *et al.* (1991) *Global Biogeochemical Cycles* **5**, 193–259, American Geophysical Union, Washington, DC.

Edwards, A. (1973) *Journal of Hydrology* **18**, 219–242, Elsevier, Amsterdam.

Elderfield, H. & Schultz, A. (1996) *Annual Review of Earth and Planetary Sciences* **24**, 191–224, Annual Reviews Inc., Palo Alto, California.

Fell, N. & Liss, P.S. (1993) *New Scientist* **139** (1887), 34–38, IPC Magazines, London.

Fichez, R., Jickells, T.D. & Edmonds, H.M. (1992) *Estuarine Coastal Shelf Science* **35**, 577–592, Academic Press, London.

Fonselius, S. (1981) *Marine Pollution Bulletin*, **12**, 187–194, Pergamon, Oxford.

Galloway, J.N., Schofield, C.L., Peters, N.E., Hendrey, G.R. & Altwicker, E.R. (1983) *Canadian Journal of Fisheries and Aquatic Science* **40**, 799–806, NRC Research Press, Ottawa.

Gibbs, R.J. (1970) *Science* **170**, 1088–1090, American Association for the Advancement of Science, Washington DC.

Gieskes, J.M. & Lawrence, J.R. (1981) *Geochimica Cosmochimica Acta* **45**, 1687–1703, Pergamon, Oxford.

Heany, S.I., Smyly, W.J. & Talling, J.F. (1986) *Internationale Revue der Gestanten Hydrobiologie* **71**, 441–494, Wiley–VCH Verlag, Berlin.

Herlihy, A.T., Kaufmann, P.R., Mitch, M.E. & Brown, D.G. (1990) *Water, Air and Soil Pollution* **50**, 91–107, Kluwer, Amsterdam.

Husar, R.B., Prospero, J.M. & Stowe, L.C. (1997) *Journal of Geophysical Research*, **D 102**, 16889–16909, American Geophysical Union.

Kado, D., Baross, J. & Alt, J. (1995) In *Seafloor Hydrothermal Systems: Physical, Chemical, Biological and Geological Interactions*, ed. by Humphris, S.E., Zierenberg, R.A., Mullineaux, L.S. & Thomson, R.E., pp. 446–466, Geophysical Monograph 91. American Geophysical Union, Washington, DC.

Kimmel, G.E. & Braids, O.C. (1980) *US Geological Survey Professional Paper 1085*, 38 pp. US Government Printing Office, Washington, DC.

Legrand, M., Feniet-Saigne, C., Saltzman, E.S., Germain, C., Barkov., N.I. & Petrov, N. (1991) *Nature*, **350**, 144–146, Macmillan, London.

Likens, G., Wright, R.F., Galloway, J.N. & Butler, T.J. (1979) *Scientific American* **241**(4), 39–17, Scientific American Inc., New York.

Livingstone, D.A. (1963) Chemical composition of rivers and lakes. *US Geological Survey Professional Paper*. US Government Printing Office, Washington, DC.

Macintyre, I.G. & Reid, R.P. (1992) *Journal of Sedimentary Petrology* **62**, 1095–1097, Society for Sedimentary Geology, Tulsa.

Manabe, S. & Wetherald, R.T. (1980) *Journal of the Atmospheric Sciences* **37**, 99–118, American Meteorological Society, Boston.

Martin, J.H., Gordon, R.M., Fitzwater, S. & Broenkow, W.H. (1989) *Deep Sea Research*, **36**, 649–680, Pergamon, Oxford.

Martin, J.-M. & Whitfield, M. (1983) In *Trace Metals in Sea Water*, ed. by Wong C.S., Boyle, E., Bruland, K.W., Burton, J.D. & Goldberg, E.D., Plenum Press, New York.

Martin, R.T., Bailey, S.W., Eberl, D.D. *et al.* (1991) *Clays and Clay Minerals* **39**, 333–335, Clay Minerals Society, Bloomington.

Maynard, J.B., Ritger, S.D. & Sutton, S.J. (1991) *Geology* **19**, 265–268, Geological Society of America, Boulder.

Meybeck, M. (1979) *Revue de Geologie Dynamique et de Geographie Physique* **21**(3), 215–246, Masson, Paris.

Michot, L.J. & Pinnavaia, T.J. (1991) *Clays and Clay Minerals* **39**, 634–641, Clay Minerals Society, Bloomington.

Millero, F.J. & Pierrot, D. (1998) *Aquatic Geochemistry* **4**, 153–199, Kluwer, Amsterdam.

Milliman, J.D. & Meade, R.H. (1983) *Journal of Geology*, **91**, 1–21, University of Chicago Press, Chicago.

Nehring, D. (1981) *Marine Pollution Bulletin* **12**, 194–198, Pergamon, Oxford.

Nesbitt, H.W. & Young, G.M. (1982) *Nature* **299**, 715–717, Macmillan, London.

Nesbitt, H.W. & Young, G.M. (1984) *Geochimica Cosmochimica Acta* **48**, 1523–1534, Pergamon, Oxford.

Orians, K.J. & Bruland, K.W. (1986) *Earth Planetary Science Letters* **78**, 397–410, Elsevier, Amsterdam.

Petit, J.R. and 18 others (1999) *Nature* **399**, 429–436, Macmillan, London.

Savoie, D.L. *et al.* (1993) *Journal of Atmospheric Chemistry* **17**, 95–122. Kluwer, Amsterdam.

Shen, G.T. & Boyle, E.A. (1987) *Earth Planetary Science Letters* **82**, 289–304, Elsevier, Amsterdam.

Sohrin, Y., Isshiki, K. & Kuwamoto, T. (1987) *Marine Chemistry* **22**, 95–103, Elsevier, Amsterdam.

Spilhaus, A.F. (1942) *Geographical Review* **32**, 431–435, American Geographical Society, New York.

Stallard, R.F. & Edmond J.M. (1983) *Journal of Geophysical Research* **88**, 9671–9688, American Geophysical Union, Washington, DC.

Stommel, H. (1958) *Deep Sea Research* **5**, 80–82, Pergamon, London.

United States Environmental Protection Agency. Superfund Innovative Technology Evaluation (SITE). EPA/540/S5-91/009. Pilot-scale demonstrationof a slurry-phase biological reactor for creosote-contaminated soil. Project Summary. United States Environmental Protection Agency, Cincinnati.

Watson, A.J., Bakker, D.C.E., Ridgwell, A.J., Boyd, P.W. & Law, C.S. (2000) *Nature*, **407**, 730–733, Macmillan, London.

Wedepohl, K.H. (1995) *Geochimica Cosmochimica Acta* **59**, 1217–1232, Pergamon, Oxford.

Woof, C. & Jackson, E. (1988) *Field Studies* **7**, 159–187, Field Studies Council, Shrewsbury.

Symbols and Abbreviations

Multiples and submultiples

Symbol	Name	Equivalent
T	tera	10^{12}
G	giga	10^{9}
M	mega	10^{6}
k	kilo	10^{3}
d	deci	10^{-1}
c	centi	10^{-2}
m	milli	10^{-3}
μ	micro	10^{-6}
n	nano	10^{-9}
p	pico	10^{-12}

Chemical symbols

Symbol	Description	Units
a	activity	$mol\,l^{-1}$
c	concentration	$mol\,l^{-1}$
eq	equivalents	$eq\,l^{-1}$
I	ionic strength	$mol\,l^{-1}$
IAP	ion activity product	$mol^{n}\,l^{-n}$
K	equilibrium constant	$mol^{n}\,l^{-n}$
K'	first dissociation constant	$mol^{n}\,l^{-n}$
K_a	equilibrium constant for acid	$mol^{n}\,l^{-n}$

K_b	equilibrium constant for base	$mol^n l^{-n}$
K_H	Henry's law constant	$mol l^{-1} atm$
K_{sp}	solubility product	$mol^n l^{-n}$
K_w	equilibrium constant for water	$mol^2 l^{-2}$
mol	mole (amount of substance – see Section 2.5)	
p	partial pressure	atm

General symbols and abbreviations

Symbol	Description
A	total amount of gas in atmosphere
(aq)	aqueous species
atm	atmosphere (pressure)
ATP	adenosine triphosphate
B[a]P	benzo[a]pyrene
°C	degrees Celsius (temperature)
CCD	calcite compensation depth
CCN	cloud condensation nuclei
CDT	Canyon Diablo troilite
CEC	cation exchange capacity
CFC	chlorofluorocarbon
CIA	chemical index of alteration
D	deuterium
DDT	2,2-bis-(p-chlorophenyl)-1,1,1-trichloroethane
DIC	dissolved inorganic carbon
DIP	dissolved inorganic phosphorus
DMS	dimethyl sulphide
DMSP	beta-dimethylsulphoniopropionate
DNA	deoxyribonucleic acid
DSi	dissolved silicon
$E°$	standard electrode potential (V)
e^-	electron
Eh	redox potential (V)
EPA	Environmental Protection Agency
F	flux
FACE	free-air CO_2 enrichment
FAO	Food and Agriculture Organization
Fs	furans
G	Gibbs free energy $(kJ mol^{-1})$
g	gram (weight)
(g)	gas
GEOSECS	US geochemical ocean sections programme
GtC	gigatonnes expressed as carbon
H	scale height
H	enthalpy $(J mol^{-1})$

HCFCs	hydrochlorofluorocarbons
HCH	hexachlorocyclohexane
$h\nu$	photon of light
IAP	ion activity product
IGBP	International Geosphere–Biosphere Programme
IPCC	Intergovernmental Panel on Climate Change
J	joule (energy, quantity of heat)
K	kelvin (temperature)
l	litre (volume)
(l)	liquid
ln	natural logarithm
$\log_{10}$	base 10 logarithm
m	metre (length)
M	a third body
MSA	methanesulphonic acid
N	neutron number
n	an integer
NAPL	non-aqueous phase liquid
Pa	pascal (pressure)
PAH	polycyclic aromatic hydrocarbon
PAN	peroxyacetylnitrate
PCBs	polychlorinated biphenyls
PCDD	polychlorinated dibenzo-*p*-dioxin
PCDF	polychlorinated dibenzo-*p*-furan
PCP	pentachlorophenol
PM	particulate matter
POP	persistent organic pollutant
ppb	parts per 10^9
ppm	parts per million
r	ionic radius
S	entropy ($J\,mol^{-1}\,K^{-1}$)
s	second (time)
(s)	solid
SOM	soil organic matter
SRB	sulphate reducing bacteria
SVOC	semi-volatile organic compound
T	absolute temperature (kelvin)
TBT	tributyl tin
TCA	tricarboxylic acid
TDIC	total dissolved inorganic carbon
UNESCO	United Nations Educational, Scientific and Cultural Organisation
USDA	United States Department of Agriculture
UV	ultraviolet (radiation)
V	volt (electrical potential)
V	volume

W	watt (power – Js^{-1})
WHO	World Health Organization
wt%	weight per cent
Z	atomic number
z	charge
$|z|$	charge ignoring sign

Greek symbols

α	alpha particle (radiation)
γ	activity coefficient
γ	gamma particle (radiation)
δ	stable isotope notation (Box 7.2)
$\delta-$	partial negative charge
$\delta+$	partial positive charge
τ	change in sum of residence time
Ω	degree of saturation

Constants

F	Faraday constant ($6.02 \times 10^{23}\,e^-$)
R	gas constant ($8.314\,Jmol^{-1}K^{-1}$)

Introduction

1

1.1 What is environmental chemistry?

It is probably true to say that the term environmental chemistry has no precise definition. It means different things to different people. We are not about to offer a new definition. It is clear that environmental chemists are playing their part in the big environmental issues—stratospheric ozone (O_3) depletion, global warming and the like. Similarly, the role of environmental chemistry in regional-scale and local problems—for example, the effects of acid rain or contamination of water resources—is well established. This brief discussion illustrates the clear link in our minds between environmental chemistry and human beings. For many people, 'environmental chemistry' is implicitly linked to 'pollution'. We hope this book demonstrates that such a view is limited and shows that 'environmental chemistry' has a much wider scope.

Terms like *contamination* and *pollution* have little meaning without a frame of reference for comparison. How can we hope to understand the behaviour and impacts of chemical contaminants without understanding how natural chemical systems work? For many years a relatively small group of scientists has been steadily unravelling how the chemical systems of the Earth work, both today and in the geological past. The discussions in this book draw on a small fraction of this material. Our aim is to demonstrate the various scales, rates and types of natural chemical processes that occur on Earth. We also attempt to show the actual or possible effects that humans may have on natural chemical systems. The importance of human influences is usually most clear when direct comparison with the unperturbed, natural systems is possible.

This book deals mainly with the Earth as it is today, or as it has been over the last few million years, with the chemistry of water on the planet's surface a recurrent theme. This theme emphasizes the link between natural chemical systems and organisms, not least humans, since water is the key compound in sustaining life itself. We will start by explaining how the main components of the near-surface Earth—the crust, oceans and atmosphere—originated and how their broad chemical composition evolved. Since all chemical compounds are built from atoms of individual elements (Box 1.1), we begin with the origin of these fundamental chemical components.

1.2 In the beginning

It is believed that the universe began at a single instant in an enormous explosion, often called the *big bang*. Astronomers still find evidence of this explosion in the movement of galaxies and the microwave background radiation once associated with the primeval fireball. In the first fractions of a second after the big bang, the amount of matter and radiation, at a ratio of about 1 in 10^8, was fixed. Minutes later the relative abundances of hydrogen (H), deuterium (D) and helium (He) were determined. Heavier elements had to await the formation and processing of these gases within stars. Elements as heavy as iron (Fe) can be made in the cores of stars, while stars which end their lives as explosive supernovae can produce much heavier elements.

Hydrogen and helium are the most abundant elements in the universe, relics of the earliest moments in element production. However, it is the stellar production process that led to the characteristic cosmic abundance of the elements (Fig. 1.1). Lithium (Li), beryllium (Be) and boron (B) are not very stable in stellar interiors, hence the low abundance of these light elements in the universe. Carbon (C), nitrogen (N) and oxygen (O) are formed in an efficient cyclic process in stars that leads to their relatively high abundance. Silicon (Si) is rather resistant to photodissociation (destruction by light) in stars, so it is also abundant and dominates the rocky world we see about us.

1.3 Origin and evolution of the Earth

The planets of our solar system probably formed from a disc-shaped cloud of hot gases, the remnants of a stellar supernova. Condensing vapours formed solids that coalesced into small bodies (planetesimals), and accretion of these built the dense inner planets (Mercury to Mars). The larger outer planets, being more distant from the sun, are composed of lower-density gases, which condensed at much cooler temperatures.

As the early Earth accreted to something like its present mass some 4.5 billion years ago, it heated up, mainly due to the radioactive decay of unstable isotopes (Box 1.1) and partly by trapping kinetic energy from planetesimal impacts. This heating melted iron and nickel (Ni) and their high densities allowed them to sink to the centre of the planet, forming the core. Subsequent cooling allowed

Box 1.1 Elements, atoms and isotopes

Elements are made from atoms—the smallest particle of an element that can take part in chemical reactions. Atoms have three main components: protons, neutrons and electrons. Protons are positively charged, with a mass similar to that of the hydrogen atom. Neutrons are uncharged and of equal mass to protons. Electrons are about 1/1836 the mass of protons, with a negative charge of equal value to the (positive) charge of protons.

Atoms are electrically neutral because they have an equal number (Z) of protons and electrons. Z is known as the atomic number and it characterizes the chemical properties of the element.

The atomic weight of an atom is defined by its mass number and most of the mass is present in the nucleus.

Mass number
= number of protons (Z) + number of neutrons (N)

eqn. 1

Equation 1 shows that the mass of an element can be changed by altering the number of neutrons. This does not affect the chemical properties of the element (which are determined by Z). Atoms of an element which differ in mass (i.e. N) are called isotopes. For example, all carbon atoms have a Z number of 6, but mass numbers of 12, 13 and 14, written:

^{12}C, ^{13}C, ^{14}C (isotopes of carbon)

In general, when the number of protons and neutrons in the nucleus are almost the same (i.e. differ by one or two), the isotopes are stable. As Z and N numbers become more dissimilar, isotopes tend to be unstable and break down by radioactive decay (usually liberating heat) to a more stable isotope. Unstable isotopes are called radioactive isotopes (see Section 2.8).

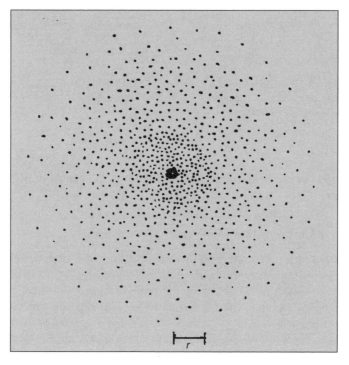

Fig. 1 Representation of the hydrogen atom. The dots represent the position of the electron with respect to the nucleus. The electron moves in a wave motion. It has no fixed position relative to the nucleus, but the probability of finding the electron at a given radius (the Bohr radius, r) can be calculated; $r = 5.3 \times 10^{-5}\,\mu m$ for hydrogen.

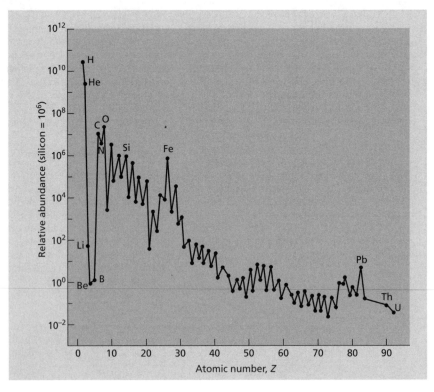

Fig. 1.1 The cosmic abundance of elements. The relative abundance of elements (vertical axis) is defined as the number of atoms of each element per 10^6 atoms of silicon and is plotted on a logarithmic scale.

solidification of the remaining material into the mantle of $MgFeSiO_3$ composition (Fig. 1.2).

1.3.1 Formation of the crust and atmosphere

The crust, hydrosphere and atmosphere formed mainly by release of materials from within the upper mantle of the early Earth. Today, ocean crust forms at mid-ocean ridges, accompanied by the release of gases and small amounts of water. Similar processes probably accounted for crustal production on the early Earth, forming a shell of rock less than 0.0001% of the volume of the whole planet (Fig. 1.2). The composition of this shell, which makes up the continents and ocean crust, has evolved over time, essentially distilling elements from the mantle by partial melting at about 100 km depth. The average chemical composition of the present crust (Fig. 1.3) shows that oxygen is the most abundant element, combined in various ways with silicon, aluminium (Al) and other elements to form silicate minerals.

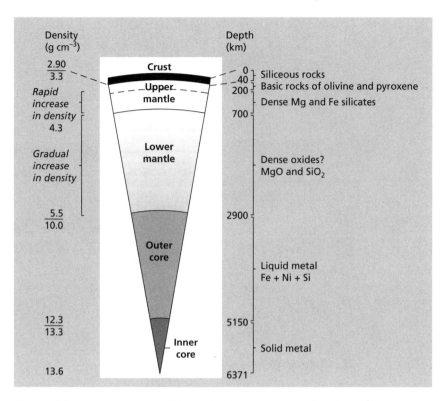

Fig. 1.2 Schematic cross-section of the Earth. Silica is concentrated in the crust relative to the mantle. After Raiswell *et al.* (1980).

Various lines of evidence suggest that volatile elements escaped (degassed) from the mantle by volcanic eruptions associated with crust building. Some of these gases were retained to form the atmosphere once surface temperatures were cool enough and gravitational attraction was strong enough. The primitive atmosphere was probably composed of carbon dioxide (CO_2) and nitrogen gas (N_2) with some hydrogen and water vapour. Evolution towards the modern oxidizing atmosphere did not occur until life began to develop.

1.3.2 The hydrosphere

Water, in its three phases, liquid water, ice and water vapour, is highly abundant at the Earth's surface, having a volume of 1.4 billion km^3. Nearly all of this water (>97%) is stored in the oceans, while most of the rest forms the polar ice-caps and glaciers (Table 1.1). Continental freshwaters represent less than 1% of the total volume, and most of this is groundwater. The atmosphere contains comparatively little water (as vapour) (Table 1.1). Collectively, these reservoirs of water are called the *hydrosphere*.

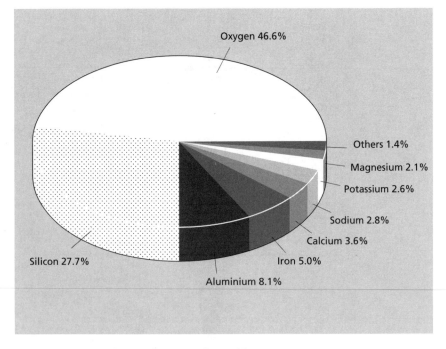

Fig. 1.3 Percentage of major elements in the Earth's crust.

Table 1.1 Inventory of water at the Earth's surface. After Speidel and Agnew (1982).

Reservoir	Volume (10^6 km³)	Percentage of total
Oceans	1350	97.41
Ice-caps and glaciers	27.5	1.98
Groundwater	8.2	0.59
Inland seas (saline)	0.1	0.007
Freshwater lakes	0.1	0.007
Soil moisture	0.07	0.005
Atmosphere*	0.013	0.001
Rivers	0.0017	0.0001
Biosphere	0.0011	0.00008
Total	1385.9	100

*As liquid equivalent of water vapour.

The source of water for the formation of the hydrosphere is problematical. Some meteorites contain up to 20% water in bonded hydroxyl (OH) groups, while bombardment of the proto-Earth by comets rich in water vapour is another possible source. Whatever the origin, once the Earth's surface cooled to 100°C, water vapour, degassing from the mantle, was able to condense. Mineralogical evidence suggests water was present on the Earth's surface by 4.4 billion years

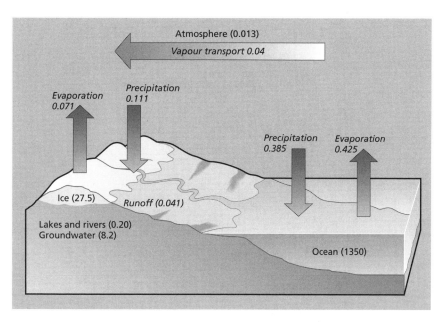

Fig. 1.4 Schematic diagram of the hydrological cycle. Numbers in parentheses are reservoir inventories ($10^6 \, km^3$). Fluxes are in $10^6 \, km^3 \, yr^{-1}$. After Speidel and Agnew (1982).

ago, soon after accretion, and we know from the existence of sedimentary rocks laid down in water that the oceans had formed by at least 3.8 billion years ago.

Very little water vapour escapes from the atmosphere to space because, at about 15 km height, the low temperature causes the vapour to condense and fall to lower levels. It is also thought that very little water degasses from the mantle today. These observations suggest that, after the main phase of degassing, the total volume of water at the Earth's surface changed little over geological time.

Cycling between reservoirs in the hydrosphere is known as the *hydrological cycle* (shown schematically in Fig. 1.4). Although the volume of water vapour contained in the atmosphere is small, water is constantly moving through this reservoir. Water evaporates from the oceans and land surface and is transported within air masses. Despite a short residence time (see Section 3.3) in the atmosphere, typically 10 days, the average transport distance is about 1000 km. The water vapour is then returned to either the oceans or the continents as snow or rain. Most rain falling on the continents seeps into sediments and porous or fractured rock to form groundwater; the rest flows on the surface as rivers, or re-evaporates to the atmosphere. Since the total mass of water in the hydrosphere is relatively constant over time, evaporation and precipitation must balance for the Earth as a whole, despite locally large differences between wet and arid regions.

The rapid transport of water vapour in the atmosphere is driven by incoming solar radiation. Almost all the radiation that reaches the crust is used to evaporate liquid water to form atmospheric water vapour. The energy used in this transformation, which is then held in the vapour, is called *latent heat*. Most of the

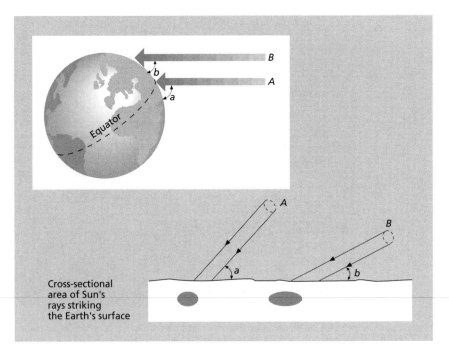

Fig. 1.5 Variation in relative amounts of solar radiation (energy per unit area) with latitude. Equal amounts of energy *A* and *B* are spread over a larger area at higher latitude, resulting in reduced intensity of radiation.

remaining radiation is absorbed into the crust with decreasing efficiency with increasing latitude, mainly because of the Earth's spherical shape. Solar rays hit the Earth's surface at 90 degrees at the equator, but at decreasing angles with increasing latitude, approaching 0 degrees at the poles. Thus, a similar amount of radiation is spread over a larger area at higher latitudes compared with the equator (Fig. 1.5). The variation of incoming radiation with latitude is not balanced by an opposite effect for radiation leaving the Earth, so the result is an overall radiation imbalance. The poles, however, do not get progressively colder and the equator warmer, because heat moves poleward in warm ocean currents and there is poleward movement of warm air and latent heat (water vapour).

1.3.3 The origin of life and evolution of the atmosphere

We do not know which chance events brought about the synthesis of organic molecules or the assembly of metabolizing, self-replicating structures we call organisms, but we can guess at some of the requirements and constraints. In the 1950s there was considerable optimism that the discovery of deoxyribonucleic acid (DNA) and the laboratory synthesis of likely primitive biomolecules from experimental atmospheres rich in methane (CH_4) and ammonia (NH_3) indicated a clear picture for the origin of life. However, it now seems more likely that the synthesis of biologically important molecules occurred in restricted, specialized environments, such as the surfaces of clay minerals, or in submarine volcanic vents.

Best guesses suggest that life began in the oceans some 4.2–3.8 billion years ago, but there is no fossil record. The oldest known fossils are bacteria, some 3.5 billion years old. In rocks of this age there is fossil evidence of quite advanced metabolisms which utilized solar energy to synthesize organic material. The very earliest of autotrophic (self-feeding) reactions were probably based on sulphur (S), supplied from volcanic vents.

$$CO_{2(g)} + 2H_2S_{(g)} \rightarrow CH_2O_{(s)} + 2S_{(s)} + H_2O_{(l)} \qquad \text{eqn. 1.1}$$
$$\text{(organic}$$
$$\text{matter)}$$

However, by 3.5 billion years ago photochemical splitting of water, or photosynthesis was happening.

$$H_2O_{(l)} + CO_{2(g)} \rightarrow CH_2O_{(s)} + O_{2(g)} \qquad \text{eqn. 1.2}$$

(If you are unfamiliar with chemical reactions and notation, see Chapter 2.)

The production of oxygen during photosynthesis had a profound effect. Initially, the oxygen gas (O_2) was rapidly consumed, oxidizing reduced compounds and minerals. However, once the rate of supply exceeded consumption, O_2 began to build up in the atmosphere. The primitive biosphere, mortally threatened by its own poisonous byproduct (O_2), was forced to adapt to this change. It did so by evolving new biogeochemical metabolisms, those that today support the diversity of life on Earth. Gradually an atmosphere of modern composition evolved (see Table 3.1). In addition, oxygen in the stratosphere (see Chapter 3) underwent photochemical reactions, leading to the formation of ozone (O_3), protecting the Earth from ultraviolet radiation. This shield allowed higher organisms to colonize the continental land surfaces.

In recent decades a few scientists have argued that the Earth acts like a single living entity rather than a randomly driven geochemical system. There has been much philosophical debate about this issue, often called the Gaia hypothesis, and more recently, Gaia theory. This view, suggested by James Lovelock, argues that biology controls the habitability of the planet, making the atmosphere, oceans and terrestrial environment comfortable to sustain and develop life. There is little consensus about these Gaian notions, but the ideas of Lovelock and others have stimulated active debate about the role of organisms in mediating geochemical cycles. Many scientists use the term 'biogeochemical cycles', which acknowledges the role of organisms in influencing geochemical systems.

1.4 Human effects on biogeochemical cycles?

In discussing the chemistry of near-surface environments on Earth it is important to distinguish between different types of alteration to Earth systems caused by humans. Two main categories can be distinguished:

1 Addition to the environment of exotic chemicals as a result of new substances synthesized and manufactured by industry.

2 Change to natural cycles by the addition or subtraction of existing chemicals by normal cyclical and/or human-induced effects.

The first category of chemical change is probably easiest to understand. Some examples of substances which are found in the environment only as a result of human activities are given in Table 1.2 and include pesticides, such as 2,2-bis(*p*-chlorophenyl)-1,1,1-trichloroethane (DDT), which is broken down by bacteria in the soil to produce a number of other exotic compounds; polychlorinated biphenyls (PCBs), which have many industrial uses and are slow to degrade in the environment; tributyl tin (TBT), which is used in marine paints to inhibit organisms from settling on the hulls of ships; many drugs; some radionuclides; and a range of chlorofluorocarbon compounds (CFCs), which were developed for use as aerosol propellants, as refrigerants and in the manufacture of solid foams.

The list in Table 1.2 is by no means complete. It has been calculated that the chemical industry has synthesized several million different chemicals (mainly organic) never previously seen on Earth. Although only a small fraction of these chemicals are manufactured in commercial quantities, it is estimated that approximately a third of the total production escapes to the environment.

The impact of these exotic substances on the environment is difficult to predict, since there are often no similar natural compounds whose behaviour is

Table 1.2 Examples of substances found in the environment only as a result of human activities.

Name	Formula	Use	Environmental impact
DDT (2,2-bis (*p*-chlorophenyl)-1,1,1-trichloroethane	(structure: chlorophenyl groups with H–C–CCl$_3$)	Pesticide	Unselective poison, concentrates up food chain
PCBs (polychlorinated biphenyls)	biphenyl with x x x x / x x x x (x are possible chlorine positions)	Dielectric in transformers; hydraulic fluids and many other uses	Resistant to breakdown carcinogens
TBT (tributyl tin)	$(CH_3(CH_2)_3)_3Sn$	Antifouling agent in marine paints	Affects sexual reproduction of shellfish
CFCs (chlorofluorocarbons)	e.g. F-11, CCl_3F	Aerosol propellant, foam blower	Destruction of stratospheric ozone

understood. A new substance may be benign, but our lack of knowledge can lead to unforeseen and sometimes harmful consequences. For example, because of the chemical inertness of the CFCs, when they were first introduced it was assumed that they would be completely harmless in the environment. This was true in all environmental reservoirs except the upper layers of the atmosphere (stratosphere), where they were broken down by solar radiation. The breakdown products of CFCs led to destruction of ozone (O_3), which forms a natural barrier, protecting animal and plant life from harmful ultraviolet (UV) radiation coming from the sun (see Section 3.10).

The second category of chemical changes is concerned with natural or human-induced alterations to existing cycles. These types of changes are illustrated in Chapter 7 with the elements carbon and sulphur. The cycling of these elements has occurred throughout the 4.5 billion years of Earth history. Furthermore, the appearance of life on the planet had a profound influence on both cycles. As well as being affected by biology, the cycles of carbon and sulphur are also influenced by alterations in physical properties, such as temperature, which have varied substantially during Earth history—for example, between glacial and interglacial periods. It is also clear that changes in the cycles of carbon and sulphur can influence climate, by affecting variables such as cloud cover and temperature. In the last few hundred years, the activities of humans have perturbed both these and other natural cycles. Such anthropogenic changes to natural cycles essentially mimic and in some cases enhance or speed up what nature does anyway.

In contrast to the situation for exotic chemicals described earlier, changes to natural cycles should be easier to predict, since the process is one of enhancement of what already occurs, rather than addition of something completely new. Thus, knowledge of how a natural system works now and has done in the past should be helpful in predicting the effects of human-induced changes. However, we are often less able at such predictions than we would like to be, because of our ignorance of the past and present mode of operation of natural chemical cycles.

1.5 The structure of this book

In the following chapters we describe how components of the Earth's chemical systems operate. Chapter 2 is a 'toolbox' of fundamental concepts underpinning environmental chemistry. We do not expect all readers will need to pick up these 'tools', but they are available for those who need them. The emphasis in each of the following chapters is different, reflecting the wide range of chemical compositions and rates of reactions that occur in near-surface Earth environments. The modern atmosphere (see Chapter 3), where rates of reaction are rapid, is strongly influenced by human activities both at ground level, and way up in the stratosphere. In terrestrial environments (see Chapters 4 & 5), a huge range of solid and fluid processes interact. The emphasis here is on weathering processes and their influence on the chemical composition of sediments, soils and continental surface waters. Human influence in the contamination of soils and natural waters is also a strong theme. Terrestrial weathering links through to the oceans (see Chapter 6) as the major input of constituents to seawater. It soon becomes clear, however, that the chemical composition of this vast water reservoir is controlled by a host of other physical, biological and chemical processes. Chapter 7 examines environmental chemistry on a global scale, integrating information from earlier chapters and, in particular, focusing on the influence of humans on global chemical processes. The short-term carbon and sulphur cycles are examples of natural chemical cycles perturbed by human activities. Persistent organic pollutants (POPs) are used as examples of exotic chemicals that persist for years to decades in soils or sediments and for several days in the atmosphere. Their persistence has allowed them to be transported globally, often impacting environments remote from their place of manufacture and use. In all of these chapters we have chosen subjects and case studies that demonstrate the chemical principles involved. To help clarify our main themes we provide information boxes that describe, in simple terms, some of the laws, assumptions and techniques used by chemists.

1.6 Internet keywords

There is now a wealth of information available on the Internet (worldwide web, www). In an environmental chemistry context there are many thousands of sites that provide quality information. Information ranges from lecture notes and problems set by university and college staff, through society web pages, to pages managed by government institutions. These pages have the advantage of many excellent colour illustrations and photographs. The information can be used to consolidate on material covered in this book, or as way of starting to explore a subject in more depth. To help you find material on the Internet, at the end of each chapter we have included a list of keywords or phrases as input for search engines. We use keywords rather than specific site addresses as website addresses change rapidly and would soon become dated in a book. The keyword lists are not intended to be complete, but are based on the main themes discussed in each chapter. You will be able to adapt the keywords or think up your own. We have personally checked each of the keywords included in the lists and know they give sensible outcomes.

We do, however, ask you to take care in your Internet searches. Remember, unlike scientific books and papers, there has been no peer review of material. If you are unsure about the quality of information on a specific site do check with your course teachers. They will be able to advise you on the validity of information.

Finally, when using search engines we advise you to use a variety of search options. Advanced search options that search for exact word strings are better for finding specific factual sites, whereas wider, less-constrained searches, usually find more diverse sites. Be as specific as you can. For example, if you are interested in ion exchange in soils use the phrase 'ion exchange soil' rather than 'ion exchange'. This will help you home in to the subject of interest much more efficiently.

1.7 Further reading

Allegre, C. (1992) *From Stone to Star.* Harvard University Press, Cambridge, Massachusetts.
Broecker, W.S. (1985) *How to Build a Habitable Planet.* Lamont-Doherty Geological Observatory, Columbia University, Palisades, New York.
Lovelock, J. (1982) *Gaia: A New Look At Life on Earth.* Oxford University Press, Oxford, 157pp.
Lovelock, J. (1988) *The Ages of Gaia.* Oxford University Press, Oxford, 252pp.

1.8 Internet search keywords

big bang
formation chemical elements
elements stars
elements isotopes
differentiation Earth
origin atmosphere

origin hydrosphere
hydrological cycle
origin life earth
photosynthesis
Gaia theory

Environmental Chemist's Toolbox

2

2.1 About this chapter

Undergraduate students studying environmental science come from a wide variety of academic backgrounds. Some have quite advanced chemical knowledge, while others have almost none. Whatever your background, we want you to understand some of the chemical details encountered in environmental issues and problems. To do this you will need some basic understanding of fundamental chemistry. As a rule, we find most students like to learn a particular aspect of chemistry where they need it to understand a specific problem. Learning material for a specific application is much easier than wading through pages of what can seem rather dull or irrelevant facts. Consequently much of the basic chemistry is distributed throughout the book in boxes, sited where the concept is first needed to understand a term or process.

Some of the basic chemistry is, however, so fundamental—underpinning most sections of the book—that we describe it here in a dedicated chapter. We have laid out enough information for students with little or no chemistry background to get a foothold into the subject. You may only need to 'dip' into this material. We certainly don't expect you to read this chapter from beginning to end. Imagine the contents here as tools in a toolbox. Take out the tool (= facts, laws, etc.) you need to get the job (= understanding an aspect of environmental chemistry) done. Some of you will not need to read this chapter at all, and can move on to the more exciting parts of the book!

2.2 Order in the elements?

Most of the chemistry in this book revolves around elements and isotopes (see Box 1.1). It is therefore helpful to understand how the atomic number (Z) of an element, and its electron energy levels allow an element to be classified. The electron is the component of the atom used in bonding (Section 2.3). During bonding, electrons are either donated from one atom to another, or shared; in either case the electron is prised away from the atom. One way of ordering the elements is therefore to determine how easy it is to remove an electron from its atom. Chemists call the energy input required to detach the loosest electron from atoms, the ionization energy. As explained in Box 1.1, the number of positively charged components (protons, Z) in an atom is balanced by the same number of negatively charged electrons that form a 'cloud' around the nucleus. Although electrons do not follow precise orbits around the nucleus, they do occupy specific spatial domains called orbitals. We need only think in terms of layers of these orbitals. Those electrons in orbitals nearest the nucleus are tightly held by electrostatic attraction-forming core electrons that never take part in chemical reactions. Those further away from the nucleus are less tightly held and may be used in 'transactions' with other atoms. These loosely held electrons are known as valence electrons. Electrons normally occupy spaces available in the lowest energy orbitals such that energy dictates the electron distribution around the nucleus. The valence electrons reside in the highest occupied energy levels and are thus the easiest to remove. For example, the element sodium (Na) has a Z number of 11. This means that sodium has 11 electrons, 10 of which are core electrons, and one valence electron. It is this single valence electron that dictates the way sodium behaves in chemical reactions.

Plotting the expected first ionization energy—i.e. that required to detach the loosest valence electron from the atom—against atomic number (Fig. 2.1a), shows that as atomic number increases the energy required to detach valence electrons decreases from $Z = 1$ (H) to $Z = 20$ (Ca). In this diagram the increasing nuclear charge between hydrogen (H) and calcium (Ca) has been disregarded. The clear downward steps in energy mark large energy gaps where electrons occupy progressively higher energy orbitals further away from the nucleus. The steps in Fig. 2.1a predict a marked difference in atomic structure between helium (He) and lithium (Li), between neon (Ne) and sodium (Na) and between argon (Ar) and potassium (K). Although much simplified, this periodic repetition of the elements has long been used as the basis to tabulate the ordering of elements on a grid known as the Periodic Table (Fig. 2.2), first published in its modern form by Mendeleev in 1869.

If the ionization energy is corrected to account for nuclear charge (Fig. 2.1b) —because increasing nuclear charge makes electron removal more difficult—the energy pattern in each period becomes more like a ramp. Each 'period' begins with an element of conspicuously low ionization energy, the so-called alkali metals (Li, Na and K). Each of these elements readily lose their single valence electron to form singly charged or monovalent ions (Li^+, Na^+ and K^+). The periods of elements are depicted as 'rows' in the Periodic Table (Fig. 2.2), and when these rows are stacked on top of one another a series of 'columns' result (Fig. 2.2). Column Ia depicts the alkali metals. Moving up the energy ramps in Fig. 2.1b, the alkali metals are followed by the elements beryllium (Be), magnesium (Mg) and calcium (Ca), each with two, relatively easily removed valence

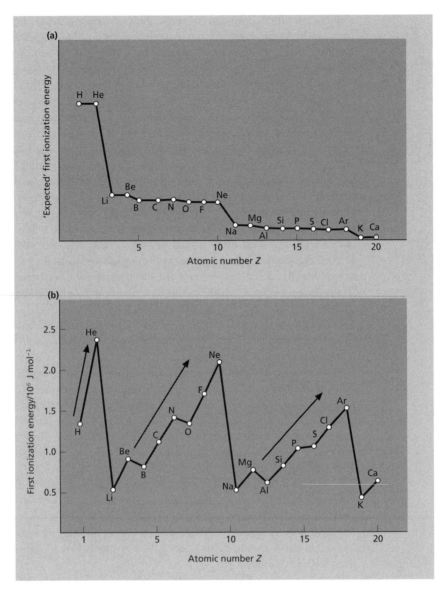

Fig. 2.1 (a) Expected first ionization energy plotted against atomic number (Z), up to $Z = 20$. This plot disregards the effects of increasing nuclear charge with Z. (b) Variation of measured first ionization energy with atomic number Z, up to $Z = 20$. The ramped profile (arrows) of increasing ionization energy following the abrupt drops reflects the increasing nuclear charge. After Gill (1996), with kind permission of Kluwer Academic Publishers.

electrons. These elements form doubly charged or divalent ions (Be^{2+}, Mg^{2+} and Ca^{2+}) and are known as alkali earth metals (column IIa in the Periodic Table). Continued progression up each energy ramp in Fig. 2.1b results in predictable patterns. For example, Mg is followed by aluminium (Al) which has three valence electrons, and then silicon with four valence electrons. Progressively more energy

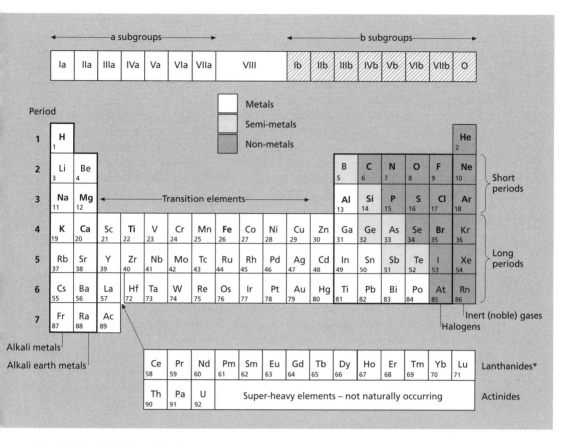

Fig. 2.2 Periodic Table of the elements and their Z numbers. Note that the periodic pattern is complicated by the transition metals between columns II and III. *La and the lanthanides are known as the rare earth elements (REE). The table has been constructed using conventional terminology and further details can be found in basic chemistry textbooks. Gill (1996) gives an accessible summary with a strong applied earth science stance. Elements in bold are those most abundant in environmental materials (see Fig. 2.3). After Gill (1996), with kind permission of Kluwer Academic Publishers.

is required to remove these electrons due to the increasing nuclear attraction. This means that aluminium will form trivalent cations whereas silicon typically will not: instead it shares its electrons in covalent bonds (Section 2.3), except in one special case (Section 2.3.2). At the top of each energy ramp are the elements He, Ne and Ar that cling tenaciously to all of their electrons. These elements have no valence electrons and therefore no significant chemical reactivity. These chemically inert elements are often called the inert or noble gases and form column O on the far right of the Periodic Table (Fig. 2.2).

Although the periodic pattern becomes more complicated above Z values of 20, the overall ordering persists. Complications arise in the so-called transition elements that occupy a position between columns II and III of the Periodic Table (Fig. 2.2). These elements have between one and three valence electrons. Importantly, however, the electrons in the orbital below the valence electrons have almost the same energy as the valence electrons themselves. In some compounds, usually depending on oxidation state (see Box 4.3), these lower orbital electrons act as additional valence electrons. For example, the element iron (Fe) exists in compounds in a reduced (Fe^{2+} or ferrous iron) and oxidized (Fe^{3+} or ferric iron) state. In general, the transition metals are less regular in their atomic properties when compared to the main groups, which also makes their behaviour more complicated to predict in nature.

It is clear from the discussion above, and by looking at the Periodic Table (Fig. 2.2) that some elements are classed as metals, some as semi-metals and some as non-metals. In each row of the Periodic Table the degree of metallic character decreases progressively from left to right, i.e. up the energy ramps of Fig. 2.1b. In essence this is because those elements with low ionization energy hold electrons loosely. In an applied electrical voltage these excited electrons will flow, conducting the electricity, whereas in non-metals there is a gap in the electron configuration that will not allow passage of excited electrons. In the case of semi-metals the gap in electron configuration is small enough that excited electrons can jump through, but only when activated by an external energy source. In effect the semi-metal flips between being an insulator (when not stimulated by external energy) and a conductor (when stimulated by external energy). Semi-metals such as silicon are also known as semi-conductors, and are used in various industrial applications to speed up electrical processes, most famously as the key component of the 'silicon chip' in computer microprocessors.

There have been many attempts to further classify the elements geologically and environmentally. In Fig. 2.3 we show the most abundant elements in four of the main environmental materials of the Earth. A glance at this figure shows that

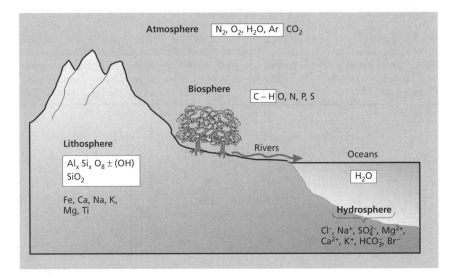

Fig. 2.3 Distribution of elements in the four main environmental materials, lithosphere, hydrosphere, atmosphere and biosphere. The elements are shown in their actual form as compounds, ions or molecules as appropriate. The main components of each material are shown in boxes, other major constituents are shown outside the boxes.

oxygen (O), and to a lesser extent hydrogen (H), are superabundant in most Earth surface materials such as air, water, organic matter and silicate minerals. In the lithosphere, silicon (Si) and aluminium (Al) are next most abundant forming the silicate minerals feldspar and quartz (see Chapter 4). In the hydrosphere it is the dissolved ions in seawater (see Chapter 6) that dominate the chemistry, particularly chloride (Cl^-) and sodium (Na^+), while the main atmospheric gases are nitrogen (N_2), oxygen (O_2), argon (Ar) and carbon dioxide (CO_2), along with water vapour (see Chapter 3). The organic matter of the biosphere is made principally of carbon and hydrogen bonded in various combinations (Section 2.7), along with lesser amounts of oxygen and the nutrient elements nitrogen (N) and phosphorus (P). Based on the information in this diagram it might be tempting to conclude that we need only understand the behaviour of these elements in nature to understand environmental chemistry. In fact the reverse is true. Paradoxically, it is often the elements present in trace amounts in the solids and fluids of the environment that tell us most about chemical processes.

2.3 Bonding

Many elements do not normally exist as atoms, but are bonded together to form molecules. The major components of air, nitrogen and oxygen for example, are present in the lower atmosphere as the molecules N_2 and O_2. By contrast, argon is rather unusual because as an inert element (or noble gas—Section 2.2) it is found uncombined as single argon atoms. Inert elements are exceptions and most substances in the environment are in the form of molecules.

2.3.1 Covalent bonds

Molecular bonds are formed from the electrostatic interactions between electrons and the nuclei of atoms. There are many different electronic arrangements that lead to bond formation and the type of bond formed influences the properties of the compound that results. It is the outermost electrons of an atom that are involved in bond formation. The archetypical chemical bond is the covalent bond and we can probably best imagine this as formed from outer electrons shared between two atoms. Take the example of two fluorine atoms that form the fluorine molecule:

$$:\ddot{F}. + :\ddot{F}. \rightleftharpoons :\ddot{F}\!:\!\ddot{F}: \qquad\qquad\qquad \text{eqn. 2.1}$$

In this representation of bonding, the electrons are shown by dots. In reality the bonding electrons are smeared out over the entire molecules, but their most probable position is between the nuclei. The bond is shown as the two electrons between the atoms. The bond is created from the two electrons shared between the atoms. In simple terms it can be argued that this arrangement of electrons achieves a structure similar to that of argon, i.e.:

$$:\ddot{Ar}:$$

Thus bond formation can be envisaged as a result of attaining noble-gas-type structures that have particularly stable configurations of electrons. Symbolically this covalent bond is written F–F. We can think of the bonding electrons, which tend to sit between the two nuclei, as shielding the repulsive forces of the protons in the nucleus.

Oxygen and nitrogen are a little different:

$$:\dot{O}. + :\dot{O}. \leftrightharpoons :O::O:$$ eqn. 2.2

For oxygen the argon-like structure requires two electrons from each atom and the double bond formed is symbolized O=O. For nitrogen we have:

$$:\dot{N}. + :\dot{N}. \leftrightharpoons :N:::N:$$ eqn. 2.3

symbolized $N \equiv N$ (a triple bond)

Gases in the atmosphere, water and organic compounds (Section 2.7) are typically formed with these kinds of covalent bonds.

2.3.2 Ionic bonding, ions and ionic solids

Unlike oxygen (O_2) and nitrogen (N_2), where individual atoms bond by sharing electrons, many crystalline inorganic materials bond by donating and accepting electrons. In fact, it can be argued that these structures have no bond at all, because the atoms entirely lose or gain electrons. This behaviour is usually referred to as ionic bonding. The classic example of an ionic solid is sodium chloride (NaCl):

$$Na. + :\ddot{C}l. \rightarrow Na^+ : \ddot{C}l^- :$$ eqn. 2.4

As in equations 2.1–2.3 the dots represent electrons.

The theory behind this behaviour is that elements with electronic structures close to those of inert (noble) gases lose or gain electrons to achieve a stable (inert) structure. In equation 2.4, sodium (Na; $Z = 11$) loses one electron to attain the electronic structure of neon (Ne; $Z = 10$), while chlorine (Cl; $Z = 17$) gains one electron to attain the electronic structure of argon (Ar; $Z = 18$). The compound NaCl is formed by the transfer of one electron from sodium to chlorine and the solid is bonded by the electrostatic attraction of the donated/received electron. The compound is electrically neutral.

Crystalline solids, for example NaCl, are easily dissolved in polar solvents such as water (see Box 4.1), which break down the ionic crystal into a solution of separate charged ions:

$$Na^+ : \ddot{C}l^- . \overset{H_2O}{\leftrightharpoons} Na^+_{(aq)} + :\ddot{C}l:^-_{(aq)}$$ eqn. 2.5

(Note: Most equations in this book will not show the electrons (·) on individual ions and atoms, only their charge (+/–).)

Positively charged atoms like Na^+ are known as cations, while negatively charged ions like Cl^- are called anions. Thus, metals whose atoms have one, two or three electrons more than an inert gas structure form monovalent (e.g. potassium, K^+), divalent (e.g. calcium, Ca^{2+}) or trivalent (e.g. aluminium, Al^{3+}) cations.

Similarly, non-metals whose atoms have one, two or three electrons less than an inert gas structure form monovalent (e.g. bromine, Br^-), divalent (e.g. sulphur, S^{2-}) and trivalent (e.g. nitrogen, N^{3-}) anions. In general, the addition or loss of more than three electrons is energetically unfavourable, and atoms requiring such transfers generally bond covalently (Section 2.3.1).

The silicon ion Si^{4+} is an interesting exception. The high charge and small ionic radius make this cation polarizing or electronegative (see Box 4.2), such that its bonds with the oxygen anion O^{2-} in silicate minerals (see Section 4.2) are distorted. This produces an appreciable degree of covalency in the Si–O bond.

2.4 Using chemical equations

The chemical principles discussed in this book are often illustrated using equations. It is useful to know a few of the ground rules chemists have adopted to construct these. Let us begin by looking at an equation depicting the process of rusting metallic iron:

$$4Fe_{(s)} + 3O_{2(g)} \rightarrow 2Fe_2O_{3(s)} \qquad \text{eqn. 2.6}$$

Firstly, the arrow shows that the reaction is favoured in one direction (we will demonstrate this later when discussing energy needed to drive reactions). Next we can see that the reaction balances, i.e. we have four atoms of iron and six atoms of oxygen on both sides of the equation. When chemical reactions take place, we neither gain nor lose atoms. Finally, the subscripted characters in brackets represent the status of the chemical species. In this book l = liquid, g = gas, s = solid and aq = an aqueous species, i.e. a component dissolved in water.

It is important to realize that these reactions are usually simplifications of the actual chemical transformations that occur in nature. In equation 2.6 we are representing rusted or oxidized iron as Fe_2O_3, the mineral haematite. In nature, rusted metal is a complex mixture of iron hydroxides and water molecules. So equation 2.6 summarizes a series of complicated reaction stages. It illustrates a product we might reasonably expect to form without necessarily depicting the stages of reaction or the complexity encountered in nature.

Many of the equations in this book are written with the reversible reaction sign (two-way half-arrows; e.g. eqn. 2.5). This shows that the reaction can proceed in either direction and this is fundamental to equilibrium-based chemistry (see Box 3.2). Reactions depicting dissolution of substances in water may or may not show the water molecule involved, but dissolution is implied by the (aq) status symbol. Equation 2.7, read from left to right, shows dissolution of rock salt (halite).

$$NaCl_{(s)} \rightleftharpoons Na^+_{(aq)} + Cl^-_{(aq)} \qquad \text{eqn. 2.7}$$

The reverse reaction (right to left) shows crystallization of salt from solution.

When writing chemical equations, the sum of charges on one side of the equation must balance the sum of charges on the other side. On the left side of equation 2.7, NaCl is an electrically neutral compound, whilst on the right side sodium and chloride each carry a single but opposite charge so that the charges cancel (neutralize) each other.

In this book the symbol 🌑, placed underneath the reaction arrow (e.g. see Box 4.16), is used to denote the presence of microorganisms involved in the reaction. We use the symbol where the presence of microorganisms is proven, although we should remember that in many other reactions microorganisms are often suspected to play a part.

2.5 Describing amounts of substances: the mole

Chemists have adopted a special unit of measurement called the mole (abbreviation mol) to describe the amount of substance. A mole is defined as 6.0221367×10^{23} molecules or atoms. This is chosen to be equivalent to its molecular weight in grams. Thus 1 mol of sodium, which has an atomic weight of 23 (if one were to be very accurate, it would be 22.9898), weighs 23 g and contains 6.0221367×10^{23} molecules. This special number of particles is called the Avogadro number, in honour of the Italian physicist Amedeo Avogadro.

The mole is used because it always refers to the same amount of substance, in terms of the number of molecules, regardless of mass of the atoms involved. Thus a mole of a light element like sodium, and a mole of a heavy element like uranium (U), contain the same numbers of atoms.

Units of concentration are, for this reason, also expressed in terms of the mole. Thus a concentration is given as the amount of substance per unit volume as mol dm^{-3} (the unit of molarity) or per unit weight as mol kg^{-1} (molality). The latter is now used frequently in chemistry because it has a number of advantages (such as it does not depend on temperature). However, at 25°C, in a dilute solution, mol kg^{-1}, mol dm^{-3} and mol l^{-1} are almost equivalent. Although mol l^{-1} is not an accepted SI unit it remains widely used in the environmental sciences, which is the reason we have decided to use it throughout this book.

2.6 Concentration and activity

When environmental chemists measure the amounts of chemical substances, for example in a water sample, they are usually measuring the concentration of that substance, for example the concentration of calcium (Ca) in the water. It is very easy to assume that the analysis has measured all of the free Ca^{2+} ions in the sample, but in fact it will almost certainly have measured all of the calcium-bearing dissolved species, called ion pairs, as well. Ions in solution are often sufficiently close to one another for electrostatic interactions to occur between oppositely charged species. These interactions reduce the availability of the free ion to participate in reactions, thereby reducing the effective concentration of the free ion. Collisions between oppositely charged ions also allow the transient formation of ion pairs, for example:

$$Na^+_{(aq)} + SO^{2-}_{4(aq)} \rightleftharpoons NaSO^-_{4(aq)} \qquad \text{eqn. 2.8}$$

The formation of these ion pairs (see Box 6.4) further reduces the effective concentration, and the frequency of collisions increases as the total amounts of chemical species in the solution increase. The effective concentration of an ion therefore becomes an important consideration in concentrated and complex solutions like seawater.

In order to predict accurately chemical reactions in a concentrated solution, we need to account for this reduction in effective concentration. This is done using a concentration term known as 'activity' that is independent of electrostatic interactions. Activity is the formal thermodynamic representation of concentration and it describes the component of concentration that is free to take part in chemical reactions. Activity is related to concentration by an activity coefficient (γ).

$$\text{activity} = \text{concentration} \times \gamma \qquad \text{eqn. 2.9}$$

Equation 2.9 shows that units of activity and concentration are proportional; in other words γ can be regarded as a proportionality constant. These constants, which vary between 0 and 1, can be calculated experimentally or theoretically and are quite well known for some natural solutions. Having said this, measuring γ in complex solutions like seawater has proved very difficult. Most importantly for our purposes, as solution strength approaches zero, γ approaches 1. In other words, in very dilute solutions (e.g. rainwater), activity and concentration are effectively the same.

In this book activity is expressed in units of $mol\,l^{-1}$ in the same way as concentration, but activity is denoted by the prefix 'a' in equations.

We should also note that all thermodynamic terms (e.g. equilibrium constants, see Box 3.2) are expressed as activity. Thus, measured concentrations of any chemical species should usually be converted to activities before comparison with thermodynamic data.

2.7 Organic molecules—structure and chemistry

Organic matter and organic compounds are integral components of all environmental reservoirs; it is therefore important to understand some of the basic facts about their structure and chemistry. Organic molecules contain carbon, hydrogen and often some other non-metallic elements such as oxygen, nitrogen, sulphur or halogens such as chlorine. Organic molecules are often complex structures, typically a skeleton of carbon atoms arranged in chains, branched chains or rings. It is more convenient to draw these complex structures as a simple picture (Fig. 2.4) rather than write the formula. The atoms in organic molecules are usually held together by covalent bonds, depicted as lines joining atoms (Section 2.3.1 & Fig. 2.4).

The simplest organic compounds contain only hydrogen atoms bonded to the carbon skeleton and are known as hydrocarbons, for example methane, the main

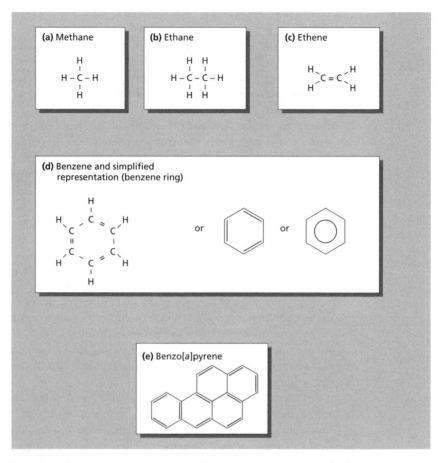

Fig. 2.4 Organic molecules. (a) Methane. (b) Ethane. (c) Ethene—note the double bond. (d) Benzene—the double bonds become delocalized so may be symbolized by a circle. It is conventional to omit the H atoms from pictorial representations of benzene. Compounds based on benzene are called aromatic compounds, for example (e) benzo[*a*]pyrene—a polycyclic aromatic hydrocarbon (PAH). The benzene ring is particularly stable, which enables it to be the building block of larger molecules such as this one. See also Figs 3.4, 4.28 and 4.34.

constituent of natural gas (Fig. 2.4a). When all the carbon atoms in a compound are joined by single bonds (Section 2.3.1) the structure is called aliphatic or saturated. Simple straight-chain aliphatic hydrocarbons are known as normal alkanes (n-alkanes), for example ethane (Fig. 2.4b). Molecules that contain one or more double-bonded (Section 2.3.1) carbon atoms are known as unsaturated molecules. Ethene is an example (Fig. 2.4c), but you are probably more familiar with the term unsaturated in relation to fats in foods. A particularly stable structure forms when double-bonded (Section 2.3.1) carbon atoms alternate with single bonded carbon atoms, i.e.:

$$C - C = C - C = C - C = C$$

It is easy to imagine this structure bent around upon itself into a ring or cycle, such that the first carbon atom in the chain also bonds with the sixth carbon atom (Fig. 2.4d). These six-carbon cyclic structures are very common and are known as aromatic compounds as many of them have a distinctive odour. Benzene is the simple hydrocarbon with this ring structure (Fig. 2.4d); it can be formally described as an unsaturated aromatic hydrocarbon. It is possible for stable aromatic rings of this type to be fused together into larger multiple ring structures. These complex but stable structures are known as polycyclic. The polycyclic aromatic hydrocarbons or PAHs (e.g. benzo[*a*]pyrene, Fig. 2.4e) are becoming quite familiar in well-publicized cases of urban air pollution (see Section 3.7). The presence of aliphatic or aromatic hydrocarbon groups in a molecule are often denoted by the symbol, R, when the specific identity of the molecule is unimportant (e.g. see eqns. 4.9 & 4.10).

To simplify the depiction of organic molecules it is usual not to label each carbon and hydrogen atom of the basic structure. Hydrogen atoms are usually omitted, while the position of each carbon atom in the 'skeleton' is indicated by the change in angle of the line drawing (Fig. 2.4d). Also, quite often, the bond system of aromatic rings is represented as a circle (Fig. 2.4d). This shows that the bonds are delocalized, i.e. the bonds between all the six carbon atoms are identical and intermediate in strength between single and double bonds.

2.7.1 Functional groups

Atoms of other elements, typically oxygen, nitrogen and sulphur, are incorporated into the basic hydrocarbon structures, usually as peripheral components known as functional groups (Table 2.1). Each functional group confers specific properties on the compound, and can be a major factor in determining the chemical behaviour of the compound. Functional groups include the hydroxyl (–OH), carboxyl (–COOH), amino (–NH$_2$) and nitro groups (–NO$_2$). The –OH and

Table 2.1 Important functional groups in environmental chemistry. Modifed from Killops and Killops (1993). Reproduced with kind permission of the authors.

Symbol	Group name	Resulting compound name
R—OH	Hydroxyl	Alcohol (R = aliphatic group)
		Phenol (R = aromatic group)
—C = O $\vert$ R	Carbonyl	Aldehyde (R = H)
		Ketone (R = aliphatic or aromatic group)
—C = O $\vert$ OH	Carboxyl	Carboxylic acid
—NH$_2$	Amino	Amine
—NO$_2$	Nitro	
—O—	Oxo	Ether*

*Ethers make bonds in biopolymers such as cellulose (see Box 4.11).

–COOH functional groups, for example, increase the polarity of the molecule (see Box 4.14), increasing its aqueous solubility. Organic molecules are often described collectively by the functional groups they contain, for example the carboxylic acids all contain –COOH, an example being ethanoic (acetic) acid:

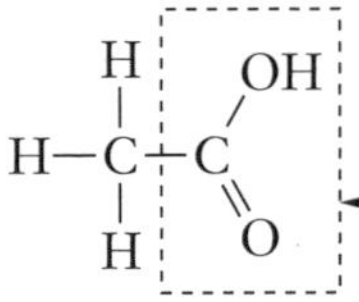

The word acid indicates that these substances act as a weak acid (see Box 3.3) in water, releasing H^+ by dissociation from the carboxyl group (see Box 4.5). Some molecules contain more than one functional group, for example the amino acids contain both –COOH and –NH_2:

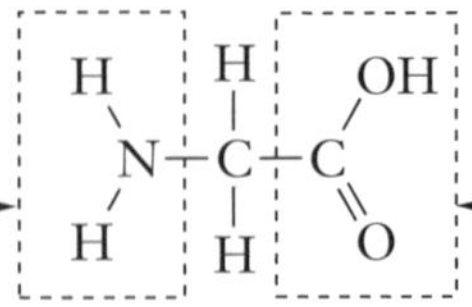

Amino acids also behave as weak acids due to the –COOH group, so they can release H^+ ions by dissociation, for example in the simple amino acid glycine:

$$NH_2CH_2COOH \rightarrow H^+ + NH_2CH_2COO^- \qquad \text{eqn. 2.10}$$
glycine

However, the amino functional group (–NH_2) is also able to accept an H^+ ion and act as a base (see Box 3.3), i.e.:

$$H^+ + NH_2CH_2COO^- \rightarrow {}^+NH_3CH_2COO^- \qquad \text{eqn. 2.11}$$

The amino acids thus have the unusual ability to form dipolar ions, denoted by the + sign on the amino group and the –sign on the COO. This property makes amino acids highly soluble in polar solvents like water or ethanol, each polar end attracted to the suitable solvent (see Box 4.14).

2.7.2 Representing organic matter in simple equations

Organic matter is implicated in many chemical reactions in natural environments. Organic matter is typically a mixture of materials derived from a number of different plants, which are themselves composed from a variety of complex biopolymers (see Box 4.10). The precise chemistry of organic matter in a soil or sediment is thus not usually known. This problem is usually avoided by using simple compounds such as carbohydrates to represent organic matter. Carbohydrates have the general formula $C_n(H_2O)_n$ (where n is an integer), which shows they contain only carbon, hydrogen and oxygen, the latter two elements in the same ratio as in water. In this book organic matter is often represented by the generic formula CH_2O. A specific example might be the sugar glucose, i.e.:

$$H-\overset{\overset{\displaystyle H}{|}}{\underset{\underset{\displaystyle OH}{|}}{C}}-\overset{\overset{\displaystyle H}{|}}{\underset{\underset{\displaystyle OH}{|}}{C}}-\overset{\overset{\displaystyle H}{|}}{\underset{\underset{\displaystyle OH}{|}}{C}}-\overset{\overset{\displaystyle OH}{|}}{\underset{\underset{\displaystyle H}{|}}{C}}-\overset{\overset{\displaystyle H}{|}}{\underset{\underset{\displaystyle OH}{|}}{C}}-\overset{\overset{\displaystyle H}{|}}{C}=O,$$ which gives the forumla $C_6H_{12}O_6$

Again, it is easy to see that glucose might form a six-carbon cyclic unit. However this molecule cannot be considered aromatic because it lacks the double bonds found in the benzene ring. Compounds like glucose that contain elements other than carbon in the ring structure are called heterocyclic.

Glucose

The drawing shows a three-dimensional representation of the molecule with, for example, some OHs sticking up from the ring and some hanging downward. Common sugars, for example sucrose—the main component of domestic sugar —are made of short chains of these six-carbon (C_6) units. In the case of sucrose just two C_6 units are present (disaccharide), whereas polymers such as cellulose (see Box 4.10) are made of much longer chains, giving rise to the term polysaccharide.

Use of the simple carbohydrate formula to represent organic matter has the advantage of simplifying reactions, but it also means the reactions are very approximate representations of the complexities found in nature (see e.g. Box 4.10).

2.8 Radioactivity of elements

Where the number of both protons and neutrons in an atom is known we are able to identify a specific isotope of a specific element and this is termed a nuclide. Some naturally occurring elements are radioactive and specific isotopes of these elements are called radionuclides. This term implies that their nuclei are unstable and spontaneously decay, transforming the nucleus into that of a different element. Radioactive decay is written in equations that look a little like those for chemical reactions, but they need to express the atomic mass of the elements involved and the type of rotation emitted. A number of modes of radioactive decay are possible, and here we outline some of the common ones. The decay of potassium (^{40}K) can be written:

$$^{40}K \rightarrow {}^{40}Ar + \gamma \qquad\qquad \text{eqn. 2.12}$$

In this transformation an electron of the potassium is captured by the nucleus, a proton within it is converted to a neutron and excess energy is lost as a γ

particle. Thus the Z number of the nucleus is decreased by 1 from K ($Z = 19$) to Ar ($Z = 18$), but the mass number is unchanged. This so-called gamma radiation (γ) is essentially a photon that carries a large amount of electromagnetic energy. This transformation is very important for the atmosphere as it produces the stable form (isotope) of argon which emanates from the potassium-containing rocks of the earth and accumulates in the atmosphere.

Unstable heavy elements with an excess of protons in the nucleus decay to produce radiation as an α particle (alpha decay), which is in fact a helium (He) nucleus, for example:

$$^{238}U \rightarrow {}^{234}Th + \alpha \qquad\qquad \text{eqn. 2.13}$$

As the α particle loses energy, it picks up electrons and eventually becomes ^{4}He in the atmosphere. As the helium nucleus contains two protons and two neutrons the nucleus Z number changes from that of U ($Z = 92$) to Th ($Z = 90$), while the mass number decreases by 4. Another source of helium is the alpha decay of radium (Ra):

$$^{226}Ra \rightarrow {}^{222}Rn + \alpha \qquad\qquad \text{eqn. 2.14}$$

which also produces the inert, but radioactive, gas radon (Rn) discussed in Box 4.13.

Other heavy elements with an excess of neutrons in the nucleus decay by transforming the neutron into a proton by ejecting an energized electron known as a negative beta particle (β^-). An example of beta decay is:

$$^{87}Rb \rightarrow {}^{87}Sr + \beta^- \qquad\qquad \text{eqn. 2.15}$$

As one neutron has been transformed into a proton the Z number in the nucleus is increased by 1 from Rb ($Z = 37$) to Sr ($Z = 38$), but the mass number is unaffected.

While many radionuclides are natural, human activities have produced either artificial radionuclides, or have greatly increased levels of otherwise natural ones. These anthropogenic radionuclides are produced by nuclear power generation (e.g. power stations, satellites and submarines), by reprocessing of nuclear waste or from nuclear weapons. For example, the atmospheric testing of nuclear weapons in the 1950s and 1960s vastly increased the concentrations of tritium (^{3}H), ^{14}C and ^{137}Cs (caesium) and dispersed them worldwide. Consequently, the fallout of these isotopes from the atmosphere also increased, producing a characteristic 'spike' increase in their flux to the surfaces of the oceans and the land. This sudden arrival of radionuclide has been used to trace movements of water masses and mixing rates in the oceans (see Box 7.1), while its burial in sediments (e.g. saltmarshes) can be used as a time marker.

Nuclear weapons testing was deliberate; however many other releases of radionuclides are accidental. These have included fires and spillages at nuclear reprocessing plants resulting in releases of an assortment of nuclides to the atmosphere and the oceans, including super-heavy elements from the actinide group of the Periodic Table (Fig. 2.2) such as plutonium (Pu). Similarly, accidental sinking of nuclear submarines has released radionuclides to the bottom waters of the oceans. Perhaps most infamous was the fire and explosion at the Ukranian Chernobyl power plant in 1986 which released a cocktail of radionuclides (e.g. ^{131}I (iodine), ^{134}Cs and ^{137}Cs) into the local area and the atmosphere. Most of the fallout at distance from the source was mediated by rainfall over parts of Europe, resulting in the contamination of upland pasture, where rainfall was heaviest, and ultimately of livestock and milk.

2.9 Finding more chemical tools in this book

Most of the other basic chemical 'tools' are dispersed in boxes elsewhere in the book, sited where the concept is first needed to understand a term or process. To help you find some of these more easily, Fig. 2.5 maps out the position of some of the key boxes. We have ordered these under three main headings: (i) system acidity and oxidation; (ii) water; and (iii) physical chemistry.

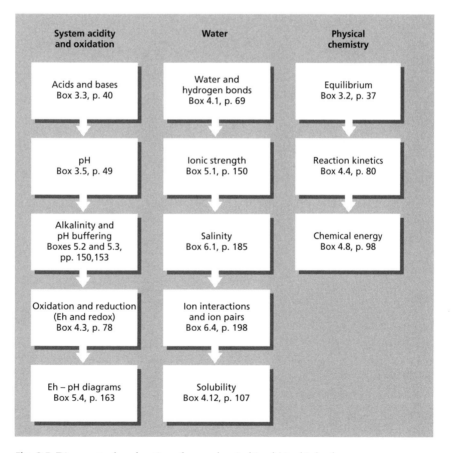

Fig. 2.5 Diagram to show location of more chemical 'tools' in this book.

2.10 Further reading

Atkins, P. (1995) *The Periodic Kingdom. A Journey into the Land of the Chemical Elements.*
   Phoenix, London.
Cox, P.A. (1995) *The Elements on Earth; Inorganic Chemistry in the Environment.* Oxford
   University Press, Oxford.
Gill, R. (1996) *Chemical Fundamentals of Geology*, 2nd edn. Chapman & Hall, London.

2.11 Internet search keywords

Periodic table
chemical bonding
covalent bonds
ionic bonds
balancing equations
mole Avogadro
concentration units chemistry

activity coefficient
basic hydrocarbon chemistry
benzene ring
aromatic hydrocarbon
functional groups
radioactive decay

The Atmosphere 3

3.1 Introduction

The atmosphere is in the news! Atmospheric chemistry has become a matter of public concern in the last two decades. While the complexities of modern science do not usually spark off great political and social debate, the changes in the atmosphere have evoked great interest. Heads of state have been forced to meetings in Stockholm, Montreal, Kyoto and Johannesburg and given their attention to the fate of our atmosphere. Television, which normally relegates scientific matters to off-peak hours, has shown skilfully created colourful images from remotely sensed measurements of the ozone (O_3) hole and huge emissions from forest fires of 1997 that have continued to raise concern into the current century. What has caused this interest in the atmosphere?

The atmosphere is the smallest of the Earth's geological reservoirs (Fig. 3.1). It is this limited size that makes the atmosphere potentially so vulnerable to contamination. Even the addition of a small amount of material can lead to significant changes in the way the atmosphere behaves.

We should note that the mixing time of the atmosphere is very rapid. Debris from a large accident, such as the one at the nuclear reactor at Chernobyl in 1986, can quickly be detected all over the globe. Pollutant particles from Europe and North America can be detected over China. This mixing, while distributing contaminants widely, dilutes them at the same time. By contrast, the spread of contaminants in the ocean is much slower and in the other reservoirs of the Earth takes place only over geological timescales of millions of years.

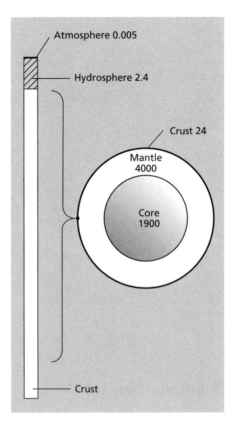

Fig. 3.1 Relative sizes of the major reservoirs of the Earth. Units, 10^{24} g.

3.2 Composition of the atmosphere

Bulk composition of the atmosphere is quite similar all over the Earth because of the high degree of mixing within the atmosphere. This mixing is driven in a horizontal sense by the rotation of the Earth. Vertical mixing is largely the product of heating of the surface of the Earth by incoming solar radiation. The oceans have a much slower mixing rate, but even this is sufficient to ensure a relatively constant bulk composition in much the same way as the atmosphere. However, some parts of the atmosphere are not so well mixed and here quite profound changes in bulk composition are found.

The lower atmosphere, which is termed the troposphere (Fig. 3.2), is well mixed by convection. Thunderstorms are the most apparent of the convective driving forces. Temperature declines with height in the troposphere (Fig. 3.2); solar energy heats the surface of the Earth and this in turn heats the directly over-lying air, causing the convective mixing. This is because the warmer air that is in contact with the surface of the Earth is lighter and tends to rise. However, at a height of some 15–25 km, the atmosphere is heated by the absorption of ultra-

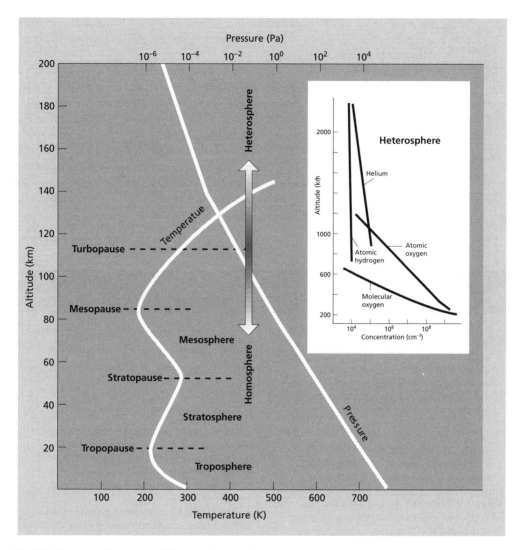

Fig. 3.2 The vertical structure of the atmosphere and associated temperature and pressure variation. Note the logarithmic scale for pressure. The inset shows gas concentration as a function of height in the heterosphere and illustrates the presence of lighter gases (hydrogen and helium) at greater heights.

violet radiation by oxygen (O_2) and O_3. The rise in temperature with height has the effect of giving the upper part of the atmosphere great stability against vertical mixing. This is because the heavy cold air at the bottom has no tendency to rise. This region of the atmosphere has air in distinct layers or strata and is thus called the stratosphere. The well-known O_3 layer forms at these altitudes. Despite this stability, even the stratosphere is well mixed compared with the atmosphere even higher up. Above about 120 km, turbulent mixing is so weak that individual gas molecules can separate under gravitational settling. Thus the relative concentrations (see Section 2.6) of atomic oxygen (O) and nitrogen (N) are greatest lower down, while the lighter hydrogen (H) and helium (He) are found to predominate higher up.

Figure 3.2 shows the various layers of the atmosphere. The part where gravitational settling occurs is usually termed the heterosphere, because of the varying composition. The better-mixed part of the atmosphere below is called the homosphere. Turbopause is the term given to the boundary that separates these two parts. The heterosphere is so high (hundreds of kilometres) that pressure is extremely low, as emphasized by the logarithmic scale in the figure.

In a mixture of gases like the troposphere, Dalton's law of partial pressure (Box 3.1) is obeyed. This means that individual gases in the atmosphere will decline in pressure at the same rate as the total pressure. This can all be conveniently represented by the barometric equation:

$$p_z = p_0 \exp(-z/H) \qquad \text{eqn. 3.1}$$

Box 3.1 Partial pressure

The total pressure of a mixture of gases is equal to the sum of the pressures of the individual components. The pressure–volume relationship of an ideal gas (i.e. a gas composed of atoms with negligible volume and which undergo perfectly elastic collisions with one another) is defined as:

$$pV = nRT \qquad \text{eqn. 1}$$

where p is the partial pressure, V the volume, n the number of moles of gas, R the gas constant and T the absolute temperature. Real gases behave like ideal gases at low pressure and we denote a mixture of gases (1, 2, 3) through the use of subscripts:

$$p_1 V = n_1 RT$$
$$p_2 V = n_2 RT$$
$$p_3 V = n_3 RT$$

Hence:

$$(p_1 + p_2 + p_3)V = (n_1 + n_2 + n_3)RT \qquad \text{eqn. 2}$$

or

$$p_T V = (n_1 + n_2 + n_3)RT \qquad \text{eqn. 3}$$

where p_T is the total pressure of the mixture. The implication that partial pressure p_i is a function of n_i means that the barometric law (Section 3.2):

$$p_z = p_0 \exp(-z/H) \qquad \text{eqn. 4}$$

can be rewritten:

$$n_z = n_0 \exp(-z/H) \qquad \text{eqn. 5}$$

or even:

$$c_z = c_0 \exp(-z/H) \qquad \text{eqn. 6}$$

where c is some unit of the amount of material per unit volume ($g\,m^{-3}$ or molecules cm^{-3}). The barometric law predicts that pressure and concentration of gases in the atmosphere decline at the same rate with height.

The relationship between partial pressure and gas-phase concentration explains why concentrations in the atmosphere are frequently expressed in parts per million (ppm) or parts per billion (ppb) (see Table 3.1). This is done on a volume basis so that 1 ppm means $1\,cm^3$ of a substance is present in $10^6\,cm^3$ of air. It also requires that there is one molecule of the substance present for every million molecules of air, or one mole of the substance present for every million moles of air. This ppm unit is thus a kind of mole ratio. It can be directly related to pressure through the law of partial pressure, so at one atmosphere (1 atm) pressure a gas present at a concentration of 1 ppm will have a pressure of $10^{-6}\,atm$.

Table 3.1 Bulk composition of unpolluted air. These are the components that provide the background medium in which atmospheric chemistry takes place. From Brimblecombe (1986).

Gas	Concentration
Nitrogen	78.084%
Oxygen	20.946%
Argon	0.934%
Water	0.5–4%
Carbon dioxide	360 ppm
Neon	18.18 ppm
Helium	5.24 ppm
Methane	1.7 ppm
Krypton	1.14 ppm
Hydrogen	0.5 ppm
Xenon	0.087 ppm

where p_z is the pressure at altitude z, p_0 the pressure at ground level and H, the scale height (about 8.4 km in the lower troposphere and a measure of the rate at which pressure falls with height). We can solve this equation and show that the pressure declines so rapidly in the lower atmosphere that it reaches 50% of its ground level value by 5.8 km. This is painfully obvious to people who have found themselves exhausted when trying to climb high mountains. We should note that if equation 3.1 is integrated over the troposphere it accounts for about 90% of all atmospheric gases. The rest are largely in the stratosphere and the low mass of the upper atmosphere reminds us that it will be sensitive to pollutants (Section 3.10). There is so little gas in the stratosphere that relatively small amounts of trace pollutants can have a big impact. Furthermore, pollutants will be held in relatively well-defined layers because of the restricted vertical mixing and this will prevent dispersal and dilution.

It is well known that the atmosphere consists mostly of nitrogen (N_2) and O_2, with a small percentage of argon (Ar). The concentrations of the major gases are listed in Table 3.1. Water (H_2O) is also an important gas, but its abundance varies a great deal. In the atmosphere as a whole, the concentration of water is dependent on temperature. Carbon dioxide (CO_2) has a much lower concentration, as do many other relatively inert (i.e. unreactive) trace gases. Apart from water, and to a lesser extent CO_2, most of these gases remain at fairly constant concentrations in the atmosphere.

Although the non-variant gases can hardly be said to be unimportant, the attention of atmospheric chemists usually focuses on the reactive trace gases. In the same way, much interest in the chemistry of seawater revolves around its trace components and not water itself or sodium chloride (NaCl), its main dissolved salt (see Chapter 6).

3.3 Steady state or equilibrium?

Let us look at an individual trace gas in the atmosphere. We will take methane (CH_4), not an especially reactive gas, as an illustration. It is present in the atmosphere at about 1.7 ppm (Box 3.1). Methane could be imagined to react with O_2 in the following way:

$$CH_{4(g)} + 2O_{2(g)} \rightarrow CO_{2(g)} + 2H_2O_{(g)}$$ eqn. 3.2

The reaction can be represented as an equilibrium situation (Box 3.2) and described by the conventional equation:

$$K = \frac{cCO_2 \cdot cH_2O^2}{cCH_4 \cdot cO_2^2}$$ eqn. 3.3

which can be written in terms of pressure (Box 3.1):

$$K = \frac{pCO_2 \cdot pH_2O^2}{pCH_4 \cdot pO_2^2}$$ eqn. 3.4

The equilibrium constant (K) is about 10^{140} (Box 3.2). This is an extremely large number, which suggests that the equilibrium position of this reaction lies very much to the right and that CH_4 should tend to be at low concentrations in the atmosphere. How low? We can calculate this by rearranging the equation and solving for CH_4. Oxygen, we can see from Table 3.1, has a concentration of about 21%, i.e. 0.21 atm, while CO_2 and H_2O have values of 0.000 36 and about 0.01 atm respectively. Substituting these into equation 3.4 and solving the equation gives an equilibrium concentration of 8×10^{-147} atm. This is very different from the value of 1.7×10^{-6} atm actually found present in air.

What has gone wrong? This simple calculation tells us that gases in the atmosphere are not necessarily in equilibrium. This does not mean that atmospheric composition is especially unstable, but just that it is not governed by chemical equilibrium. Many trace gases in the atmosphere are in steady state. Steady state describes the delicate balance between the input and output of the gas to the atmosphere. The notion of a balance between the source of a gas to the atmosphere and sinks for that gas is an extremely important one. The situation is often written in terms of the equation:

$$F_{in} = F_{out} = A/\tau$$ eqn. 3.5

where F_{in} and F_{out} are the fluxes in and out of the atmosphere, A is the total amount of the gas in the atmosphere and τ is the residence time of the gas.

To be in steady state the input term must equal the output term. Imagine the atmosphere as a leaky bucket into which a tap is pouring water. The bucket would fill for a while until the pressure rose and the leaks were rapid enough to match the inflow rate. At that point we could say that the system was in steady state.

Methane input into the atmosphere occurs at a rate of 500 Tg yr^{-1} (i.e. 500×10^9 kg yr^{-1}). We have seen that the atmosphere has CH_4 at a concentration of 1.7 ppm. The total atmospheric mass is 5.2×10^{18} kg. If we allow for the slight differences between the molecular mass of CH_4 and that of the atmosphere as a whole (i.e. 16/29), the total mass of CH_4 in the atmosphere can be estimated as 4.8×10^{12} kg. Substituting these values in equation 3.5 gives a residence time of

---

Box 3.2 Chemical equilibrium

Many chemical reactions occur in both directions such that the products are able to re-form the reactants. For instance, in rainfall chemistry, we account for the hydrolysis (i.e. reaction with water) of aqueous formaldehyde (HCHO) to methylene glycol ($H_2C(OH)_2$) according to the equation:

$$HCHO_{(aq)} + H_2O_{(l)} \rightarrow H_2C(OH)_{2(aq)} \qquad \text{eqn. 1}$$

but the reverse reaction also occurs:

$$H_2C(OH)_{2(aq)} \rightarrow HCHO_{(aq)} + H_2O_{(l)} \qquad \text{eqn. 2}$$

such that the system is maintained in dynamic equilibrium, symbolized by:

$$HCHO_{(aq)} + H_2O_{(l)} \rightleftharpoons H_2C(OH)_{2(aq)} \qquad \text{eqn. 3}$$

The relationship between the species at equilibrium is described in terms of the equation:

$$K = \frac{aH_2C(OH)_2}{aHCHO \cdot aH_2O} \qquad \text{eqn. 4}$$

where a denotes the activities of the entities involved in the reaction. Remember from Section 2.6 that activities are the formal thermodynamic representations of concentration. However, in dilute solutions activity and concentration are almost identical. Dilute solutions, such as rainwater, are almost pure water. The activity of pure substances is defined as unity, so in the case of rainwater the equation can be simplified:

$$K = \frac{aH_2C(OH)_2}{aHCHO} \qquad \text{eqn. 5}$$

K is known as the equilibrium constant and in this case it has the value 2000. An equilibrium constant greater than unity suggests that equilibrium lies to the right-hand side and the forward reaction is favoured. Equilibrium constants vary with temperature, but not with concentration if the concentrations have been correctly expressed in terms of activities.

The equilibrium relationship is often called the law of mass action and may be remembered by the fact that an equilibrium constant is the numerical product of the activity of the products of a reaction divided by the numerical product of the reactants, such that in general terms:

$$kA + lB \rightleftharpoons mC + nD \qquad \text{eqn. 6}$$

$$K = \frac{aC^m \cdot aD^n}{aA^k \cdot aB^l} \qquad \text{eqn.7}$$

It may be easier to grasp the notion of shifts in equilibrium in a qualitative way using the *Le Chatelier Principle*. This states that, if a system at equilibrium is perturbed, the system will react in such a way as to minimize this imposed change. Thus, looking at the formaldehyde equilibrium (eqn. 3), any increase in HCHO in solution would be lessened by the tendency of the reaction to shift to the right, producing more methylene glycol.

9.75 years. This represents the average lifetime of a CH_4 molecule in the atmosphere (at least, it would if the atmosphere was very well mixed).

Residence time is the fundamental quantity that describes systems in steady state. It is a very powerful concept that plays a central role in much of environmental chemistry. Compounds with long residence times can accumulate to relatively high concentrations compared with those with shorter ones. However, even though gases with short residence times are removed quickly, their high reactivity can yield reaction products that cause problems.

The famous atmospheric chemist C.E. Junge made an important observation about residence times and the variability of gases in the atmosphere. If a gas has

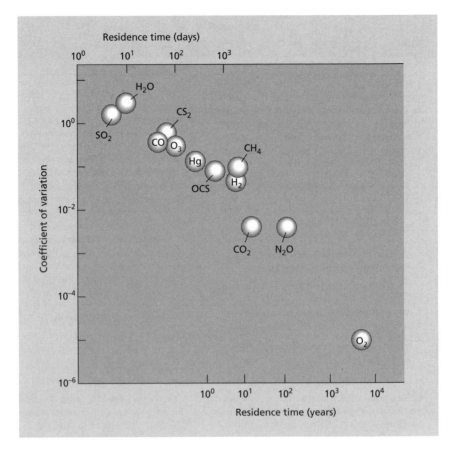

Fig. 3.3 Variability of trace and other components in the atmosphere as a function of residence time. Large coefficients of variation indicate higher variability. From Brimblecombe (1986). With kind permission of Cambridge University Press.

a long residence time, then it will have ample time to become well mixed in the atmosphere and thus would be expected to show great constancy in concentration all around the globe. This is the case and the results of measurements are illustrated in Fig. 3.3.

3.4 Natural sources

Since the atmosphere can be treated, on a large scale, as if it were in steady state, we have a model that views the atmosphere as having sources, a reservoir (i.e. the atmosphere itself) and removal processes, all in delicate balance. The sources need to be quite stable over the long term. If they are not, then the balance will shift. In terms of our earlier analogy, the level in the leaking bucket will change.

The best-known, and most worrying, example of such a shift is the increasing magnitude of the CO_2 source because of the consumption of vast amounts of fossil fuel by human activities. This has given rise to a continuing increase in the CO_2 concentration in the atmosphere. The predicted rise in temperature, due to the greenhouse effect, is explored in detail in Chapter 7.

There are many sources of trace components in the atmosphere, which can be divided into different categories, such as geochemical, biological and human or anthropogenic sources. Some of these sources are hard to categorize. Is a forest fire a geochemical, biological or human source—particularly if the forest was planted or the fire started through human activities? Although our definitions can become a little blurred, it is nevertheless useful to categorize sources.

3.4.1 Geochemical sources

Perhaps the largest geochemical sources are wind-blown dusts and sea sprays, which put huge amounts of solid material into the atmosphere (see also Chapter 6). The dust is largely soil from arid regions of the Earth. If this dust is fine enough, it can spread over large areas of the globe and is important in redistributing material. Often, however, the chemical effects of the dust in the atmosphere are not particularly evident, because dusts are not chemically very reactive. By contrast, wind-blown sea spray places a more reactive entity into the atmosphere as salt particles.

The salt particles from the oceans are hygroscopic and under humid conditions these tiny NaCl crystals attract water and form a concentrated solution droplet or aerosol. Ultimately, this process can take part in cloud formation. The droplets can also be a site for important chemical reactions in the atmosphere. If strong acids (Box 3.3) in the atmosphere, perhaps nitric acid (HNO_3) or sulphuric acid (H_2SO_4), dissolve in these small droplets, hydrogen chloride (HCl) can be formed. It is thought that this process is an important source of HCl in the atmosphere:

$$H_2SO_{4\,(in\ aerosol)} + NaCl_{(in\ aerosol)} \rightarrow HCl_{(g)} + NaHSO_{4\,(in\ aerosol)} \qquad \text{eqn. 3.6}$$

Incoming meteors also inject particles into the atmosphere. This is a very small source compared with wind-blown dust or forest fires, but meteors make their contribution to the upper parts of the atmosphere where the gas is at a low density. Here, a small contribution can be particularly significant and the metals ablated from incoming meteors enter a series of chemical reactions.

Volcanoes are a large source of dust and particularly powerful eruptions can push dust into the stratosphere. It has long been known that volcanic particles can change global temperature by blocking out sunlight. They can also perturb the chemistry at high altitudes. Along with the dust, volcanoes are huge sources of gases such as sulphur dioxide (SO_2), CO_2, HCl and hydrogen fluoride (HF). These gases can react in the stratosphere to provide a further source of particles, with H_2SO_4 being the most important particle produced indirectly from volcanoes.

It is important to realize that the volcanic source is a very discontinuous one, both in time and space. Large volcanic eruptions are infrequent. It may be that years pass without any really major eruptions and then suddenly more material

---

Box 3.3 Acids and bases

Acids and bases are an important class of chemical compounds, because they exert special control over reactions in water. Traditionally acids have been seen as compounds that dissociate to yield hydrogen ions in water:

$$HCl_{(aq)} \rightleftharpoons H^+_{(aq)} + Cl^-_{(aq)} \qquad \text{eqn. 1}$$

The definition of an acid has, however, been extended to cover a wider range of substances by considering electron transfer. For example, boric acid (H_3BO_3), which helps control the acidity* of seawater, gains electrons from the hydroxide (OH^-) ion:

$$H_3BO_{3(aq)} + OH^-_{(aq)} \rightleftharpoons B(OH)^-_{4(aq)} \qquad \text{eqn. 2}$$

For most applications the simple definition is sufficient, and we might think of bases (or alkalis) as those substances which yield OH^- in aqueous solution.

$$NaOH_{(aq)} \rightleftharpoons Na^+_{(aq)} + OH^-_{(aq)} \qquad \text{eqn. 3}$$

Acids and bases react to neutralize each other, producing a dissolved salt plus water.

$$HCl_{(aq)} + NaOH_{(aq)} \rightleftharpoons Cl^-_{(aq)} + Na^+_{(aq)} + H_2O_{(l)} \qquad \text{eqn. 4}$$

Two classes of acids and bases are recognized—strong and weak. Hydrochloric acid (HCl) and sodium hydroxide (NaOH) (eqns 1 and 4) are treated as if they dissociate completely in solution to form ions, so they are termed 'strong'. Weak acids and bases dissociate only partly.

$$HCOOH_{(aq)} \rightleftharpoons H^+_{(aq)} + HCOO^-_{(aq)} \qquad \text{eqn. 5}$$
(formic acid)

$$NH_4OH_{(aq)} \rightleftharpoons NH^+_{4(aq)} + OH^-_{(aq)} \qquad \text{eqn. 6}$$
$$\left(\begin{array}{c}\text{ammonium}\\\text{hydroxide}\end{array}\right)$$

Dissociation is an equilibrium process and is conveniently described in terms of equilibrium constants for the acid (K_a) and alkaline (K_b) dissociation:

$$K_a = \frac{aH^+ \cdot aHCOO^-}{aHCOOH} = 1.77 \times 10^{-4} \text{ mol l}^{-1} \qquad \text{eqn. 7}$$

$$K_b = \frac{aNH_4^+ \cdot aOH^-}{aNH_4OH} = 1.80 \times 10^{-5} \text{ mol l}^{-1} \qquad \text{eqn. 8}$$

* The acidity of the oceans is usually defined by its pH, which is discussed in Box 3.5.

---

is released in a single event than for many years previously. The eruptions occur in very specific locations where there are active volcanoes. In addition to massive eruptions that push great quantities of material into the upper parts of the atmosphere, we must not neglect smaller fumarolic emissions, from volcanic cracks and fissures, which gently release gases to the lower atmosphere over very long periods of time. The balance between these two volcanic sources is not accurately known, although for SO_2 it is probably about $50:50$.

Radioactive elements in rocks (see Section 2.8), most importantly potassium (K) and heavy elements such as radium (Ra), uranium (U) and thorium (Th), can release gases. Argon (Ar) arises from potassium decay and radon (Rn, a radioactive gas that has a half-life of 3.8 days) from radium decay. The uranium–thorium decay series results in the production of α particles, which are helium nuclei. Once these nuclei capture electrons, helium has effectively been added to the atmosphere.

Helium has not accumulated in the atmosphere over time because it is light enough to escape into space. The concentration of helium has been thus maintained in steady state through a balance of radioactive emanation from the crust and loss from the top of the atmosphere.

Table 3.2 Sources for particulate material in the atmosphere. From Brimblecombe (1986).

Source	Global flux (Tg yr^{-1})
Forest fires	35
Dust	750
Sea salt	1500
Volcanic dust	50
Meteoritic dust	1

3.4.2 Biological sources

Unlike the geological sources, biology does not appear to be a large direct source of particles to the atmosphere, unless we consider forest fires to be a biological source. Table 3.2 shows that forest fires are quite an important source of carbon (C), i.e. soot particles.

The living forest also plays an important role in exchanging gases with the atmosphere. The major gases O_2 and CO_2 are, of course, involved in respiration and photosynthesis. However, forests also emit enormous quantities of trace organic compounds. Terpenes, (a class of lipids) such as pinene and limonene, give forests their wonderful odour. Forests are also important sources of organic acids, aldehydes (see Table 2.1) and other organic compounds (see Section 2.7).

Although forests are obvious as sources of gas, it is the microorganisms that are especially important in generating atmospheric trace gases. Methane, which we have already discussed, is generated by reactions in anaerobic systems. Damp soils, as found in marshes or rice paddies, are important micro-biologically dominated environments, as are the digestive tracts of ruminants such as cattle.

The soils of the Earth are rich in nitrogen compounds, giving rise to a whole range of active nitrogen chemistry that generates many nitrogenous trace gases. We can consider urea (NH_2CONH_2), present in animal urine, as a typical bio-logically generated nitrogen compound in soil. Hydrolysis converts NH_2CONH_2 to ammonia (NH_3) and CO_2 according to the equation:

$$NH_2CONH_{2(aq)} + H_2O_{(l)} \rightarrow 2NH_{3(g)} + CO_{2(g)} \qquad \text{eqn. 3.7}$$

If the soil where this hydrolysis occurs is alkaline (Box 3.3), gaseous NH_3 can be released, although in acidic conditions it will react to form the non-volatile ammonium ion (NH_4^+):

$$NH_{3(g)} + H^+_{(aq)} \rightarrow NH^+_{4(aq)} \qquad \text{eqn. 3.8}$$

Plants can absorb soil NH_3 or NH_4^+ directly and some microorganisms, such as *Nitrosomonas*, oxidize NH_3, using it as an energy source for respiration, in the same way that other cells use reduced carbon compounds. One possible reaction would be:

$$2NH_{3(g)} + 2O_{2(g)} \rightarrow N_2O_{(g)} + 3H_2O_{(g)} \qquad \text{eqn. 3.9}$$

Here we can see a biological source for nitrous oxide (N_2O), an important and rather stable trace gas in the troposphere. In nature there are many other

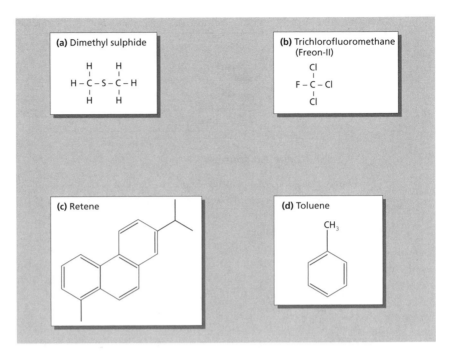

Fig. 3.4 Representations of some of the organic molecules discussed in Chapter 3. (a) Dimethyl sulphide. (b) Trichlorofluoromethane (Freon-11), one of the important CFCs. (c) Retene, a tricyclic compound derived from higher plant resins. (d) Toluene, a methylated aromatic compound.

reactions of nitrogen compounds in soils that produce the gases: NH_3, N_2, N_2O and nitric oxide (NO).

Microorganisms in the oceans also prove to be an enormous source of atmospheric trace gases. Seawater is rich in dissolved sulphate and chloride (and to a lesser extent salts of the other halogens: fluorine (F), bromine (Br) and iodine (I)). Marine microorganisms metabolize these elements, for reasons that are not properly understood, to generate sulphur (S)- and halogen-containing trace gases. However, the nitrate concentration of surface seawater is so low that the oceans are effectively a nitrogen desert. This means that seawater is not such a large source of nitrogen-containing trace gases.

Organosulphides produced by marine microorganisms make a particularly significant contribution to the atmospheric sulphur burden. The most characteristic compound is dimethyl sulphide (DMS; $(CH_3)_2S$; Fig. 3.4a). This volatile compound is produced by marine phytoplankton, such as *Phaeocystis pouchetii*, in the upper layers of the ocean by the hydrolysis of beta-dimethylsulphoniopropionate (DMSP; $(CH_3)_2S^+CH_2CH_2COO^-$) to DMS and acrylic acid ($CH_2CHCOOH$):

$$(CH_3)_2S^+CH_2CH_2COO^-_{(aq)} \rightarrow (CH_3)_2S_{(g)} + CH_2CHCOOH_{(aq)} \quad \text{eqn. 3.10}$$

Another important sulphur compound released from the oceans is carbonyl sul-

Box 3.4 Gas solubility

The solubility of gases in liquids is often treated as an equilibrium process. Take the dissolution of carbonyl sulphide (OCS) as an example:

$$OCS_{(g)} \approx OCS_{(aq)} \qquad \text{eqn. 1}$$

where $OCS_{(g)}$ and $OCS_{(aq)}$ represent the concentration of the substance carbonyl sulphide in the gas and liquid phase. This equilibrium relationship is often called Henry's law, after the English physical chemist who worked c. 1800. The Henry's law constant (K_H) describes the equilibrium. Using pressure (p) to describe the concentration (c) of $OCS_{(g)}$ in the gas phase, we have:

$$K_H = \frac{cOCS_{(aq)}}{pOCS_{(g)}} \qquad \text{eqn. 2}$$

If we take the atmosphere as the unit of pressure and $mol\,l^{-1}$ as the unit of concentration, the Henry's law constant will have the units $mol\,l^{-1}\,atm^{-1}$. The larger the values of this constant, the more soluble the gas. Table 1 shows that a gas like hydrogen peroxide is very soluble, oxygen very much less so.

Many quite important gases have only limited solubility, but often they can react in water, which enhances their solubility. Take the simple dissolution of formaldehyde $(HCHO)_2$ which readily hydrolyses to methylene glycol $(H_2C(OH)_2)$:

$$HCHO_{(g)} \approx HCHO_{(aq)} \qquad \text{eqn. 3}$$

$$HCHO_{(aq)} + H_2O_{(l)} \approx H_2C(OH)_{2(aq)} \qquad \text{eqn. 4}$$

The second equilibrium lies so far to the right that solubility is enhanced by a factor of about 2000 (Box 3.2).

Table 1 Some Henry's law constants at 15°C

Gas	K_H ($mol\,l^{-1}\,atm^{-1}$)
Hydrogen peroxide	2×10^5
Ammonia	90
Formaldehyde	1.7
Dimethyl sulphide	0.14
Carbonyl sulphide	0.035
Ozone	0.02
Oxygen	0.0015
Carbon monoxide	0.001

phide (OCS). This can be produced by reaction between carbon disulphide (CS_2) and water:

$$CS_{2(g)} + H_2O_{(g)} \rightarrow OCS_{(g)} + H_2S_{(g)} \qquad \text{eqn. 3.11}$$

and, although the flux to the atmosphere is smaller than that of DMS, its stability means that it will accumulate to higher concentrations. These sulphur gases have low solubility in water (Box 3.4), making them able easily to escape from the oceans into the atmosphere.

Halogenated organic compounds are well known in the atmosphere. Although these have an obvious human source, being present in cleaning fluids, fire extinguishers and aerosol propellants, they also have a wide range of biological sources. Methyl chloride (CH_3Cl) is the most abundant halocarbon in the atmosphere and arises primarily from poorly understood marine sources, although terrestrial microbiological processes and biomass burning also contribute. Bromine- and iodine-containing organic compounds are also released from the oceans and the distribution of this marine iodine over land-masses represents an important source of this essential trace element for mammals. As one might predict, the iodine-deficiency disease, goitre, has been common in regions remote from the oceans.

3.5 Reactivity of trace substances in the atmosphere

Gases with short residence times in the atmosphere are clearly those that can be removed easily. Some of these gases are removed by being absorbed by plants or solids or into water. However, chemical reactions are the usual reason for a gas having a short residence time.

What makes gases react in the atmosphere? It turns out that most of the trace gases listed in Table 3.3 are not very reactive with the major components of air. In fact, the most important reactive entity in the atmosphere is a fragment of a water molecule, the hydroxyl (OH) radical. This radical (a reactive molecular fragment) is formed by the photochemically initiated reaction sequence, started by the photon of light, hv:

$$O_{3(g)} + hv \rightarrow O_{2(g)} + O_{(g)} \qquad \text{eqn. 3.12}$$

$$O_{(g)} + H_2O_{(g)} \rightarrow 2OH_{(g)} \qquad \text{eqn. 3.13}$$

The OH radical can react with many compounds in the atmosphere and thus it has a short residence time. The rates are faster than with abundant gases such as O_2.

The reaction between nitrogen dioxide (NO_2) and the OH radical leads to the formation of HNO_3, an important contributor to acid rain.

$$NO_{2(g)} + OH_{(g)} \rightarrow HNO_{3(g)} \qquad \text{eqn. 3.14}$$

By contrast, kinetic measurements in the laboratory (which aim at determining the speed of reaction) show that gases that have slow rates of reaction with the OH radical have a long residence time in the atmosphere. Table 3.3 shows that

Table 3.3 Naturally occurring trace gases of the atmosphere. From Brimblecombe (1986).

	Residence time	Concentration (ppb)
Carbon dioxide	4 years	360 000
Carbon monoxide	0.1 year	100
Methane	3.6 years	1600
Formic acid	10 days	1
Nitrous oxide	20–30 years	300
Nitric oxide	4 days	0.1
Nitrogen dioxide	4 days	0.3
Ammonia	2 days	1
Sulphur dioxide	3–7 days	0.01–0.1
Hydrogen sulphide	1 day	0.05
Carbon disulphide	40 days	0.02
Carbonyl sulphide	1 year	0.5
Dimethyl sulphide	1 day	0.001
Methyl chloride	30 days	0.7
Methyl iodide	5 days	0.002
Hydrogen chloride	4 days	0.001

OCS, N_2O and even CH_4 have long residence times. The CFCs (chlorofluoro-carbons, Fig. 3.4b: refrigerants and aerosol propellants) also have very limited reactivity with OH. Gases like these build up in the atmosphere and eventually leak across the tropopause into the stratosphere. Here a very different chemistry takes place, no longer dominated by OH but by reactions which involve atomic oxygen (i.e. O). Gases that react with atomic oxygen in the stratosphere can interfere with the production of O_3:

$$O_{(g)} + O_{2(g)} \rightarrow O_{3(g)}$$ eqn. 3.15

and can be responsible for the depletion of the stratospheric O_3 layer. This means that CFCs are prime candidates for causing damage to stratospheric O_3 (Section 3.10).

We should note that nitrogen compounds are also damaging to O_3 if they can be transported to the stratosphere, because they are involved in similar reaction sequences. We have already seen that tropospheric NO_2 is unlikely to be transferred into the stratosphere (eqn. 3.14). It was, however, nitrogen compounds from the exhausts of commercial supersonic aircraft flying at high altitude that were the earliest suggested contaminants of concern. In this case the gases did not have to be unreactive and slowly transfer to the stratosphere, but were directly injected from aircraft engines. A large stratospheric transport fleet never came about, so attention has now turned to N_2O, a much more inert oxide of nitrogen produced at ground level and quite capable of getting into the stratosphere. This gas is produced both from biological activities in fertile soils (see Section 3.4.2, 5.5.1) and by a range of combustion processes—most interestingly, automobile engines with catalytic converters.

Finally, we should note that some reactions lead to the formation of particles in the atmosphere. Most particles are effectively removed by rainfall and thus have residence times close to the 4–10 days of atmospheric water. By contrast, very small particles in the 0.1–1 µm size range are not very effectively removed by rain droplets and have rather longer residence times.

3.6 The urban atmosphere

In the section above we began to look at human influence on the atmosphere. The changes wrought by humans are important, though often subtle on the global scale. It is in the urban atmosphere where human influence shows its clearest impact, so it is necessary to treat the chemistry of the urban atmosphere as a special case.

In urban environments there are pollutant compounds emitted to the atmosphere directly and these are called primary pollutants. Smoke is the archetypical example of a primary pollutant. However, many compounds undergo reactions in the atmosphere, as we have seen in the section above. The products of such reactions are called secondary pollutants. Thus, many primary pollutants can react to produce secondary pollutants. It is the distinction between primary and secondary pollution that now governs our understanding of the difference between two quite distinct types of air pollution that affect major cities of the world.

3.6.1 London smog—primary pollution

Urban pollution is largely the product of combustion processes. In ancient times cities such as Imperial Rome experienced pollution problems due to wood smoke. However, it was the transition to fossil fuel burning that caused the rapid development of air pollution problems. The inhabitants of London have burnt coal since the 13th century. Concern and attempts to regulate coal burning began almost immediately, as there was a perceptible and rather strange smell associated with it. Medieval Londoners thought this smell might be associated with disease.

Fuels usually consist of hydrocarbons, except in particularly exotic applications such as rocketry, where nitrogen, aluminium (Al) and even beryllium (Be) are sometimes used. Normal fuel combustion is an oxidation reaction (see Box 4.3) and can be described:

$$\text{`4CH'} + 5O_{2(g)} \rightarrow \quad 4CO_{2(g)} + 2H_2O_{(g)} \qquad \text{eqn. 3.16}$$
$$\text{fuel} + \text{oxygen} \rightarrow \text{carbon dioxide} + \text{water}$$

This would not seem an especially dangerous activity as neither CO_2 nor water is particularly toxic. However, let us consider a situation where there is not enough O_2 during combustion, i.e. as might occur inside an engine or boiler. The equation might now be written:

$$\text{`4CH'} + 3O_{2(g)} \rightarrow \quad 4CO_{(s)} + 2H_2O_{(g)} \qquad \text{eqn. 3.17}$$
$$\text{coal} + \text{oxygen} \rightarrow \text{carbon monoxide} + \text{water}$$

Here we have produced carbon monoxide (CO), a poisonous gas. With even less oxygen we can get carbon (i.e. smoke):

$$\text{`4CH'} + O_{2(g)} \rightarrow 4C_{(s)} + 2H_2O_{(g)} \qquad \text{eqn. 3.18}$$
$$\text{coal} + \text{oxygen} \rightarrow \text{`smoke'} + \text{water}$$

At low temperatures, in situations where there is relatively little O_2, pyrolysis reactions (i.e. reactions where decomposition takes place as a result of heat) may cause a rearrangement of atoms that can lead to the formation of polycyclic aromatic hydrocarbons (see Section 2.7) during combustion. The most notorious of these is benzo[*a*]pyrene (B[*a*]P; see Fig. 2.4), a cancer-inducing compound.

Thus, although the combustion of fuels would initially seem a harmless activity, it can produce a range of pollutant carbon compounds. When the earliest steam engines were being designed, engineers saw that an excess of oxygen would help convert all the carbon to CO_2. To overcome this they adopted a philosophy of 'burning your own smoke', even though this required considerable skill to implement and was consequently of only limited success.

In addition to these problems, contaminants within the fuel can also cause air pollution. The most common and worrisome impurity in fossil fuels is sulphur (S), partly present as the mineral pyrite, FeS_2. There may be as much as 6% sulphur in some coals and this is converted to SO_2 on combustion:

Table 3.4 Sulphur content of fuels.

Fuel	S (% by weight)
Coal	7.0–0.2
Fuel oils	4.0–0.5
Coke	2.5–1.5
Diesel fuel	0.9–0.3
Petrol	0.1
Kerosene	0.1
Wood	Very small
Natural gas	Very small

$$4FeS_{2(s)} + 11O_{2(g)} \rightarrow 8SO_{2(g)} + 2Fe_2O_{3(s)} \qquad \text{eqn. 3.19}$$

There are other impurities in fuels too, but sulphur has always been seen as most characteristic of the air pollution problems of cities.

If we look at the composition of various fuels (Table 3.4), we see that they contain quite variable amounts of sulphur. The highest amounts of sulphur are found in coals and in fuel oils. These are the fuels used in stationary sources such as boilers, furnaces (and traditionally steam engines), domestic chimneys, steam turbines and power stations. Thus, the main source of sulphur pollution, and indeed smoke, in the urban atmosphere is the stationary source. Smoke too is mainly associated with stationary sources. Steam trains and boats caused the occasional problem, but it was the stationary source that was most significant.

For many people, SO_2 and smoke came to epitomize the traditional air pollution problems of cities. Smoke and SO_2 are obviously primary pollutants because they are formed directly at a clearly evident pollutant source and enter the atmosphere in that form.

Classical air pollution incidents in London occurred under damp and foggy conditions in the winter. Fuel use was at its highest and the air near-stagnant. The presence of smoke and fog together led to the invention of the word smog (sm[oke and f]og), now often used to describe air pollution in general (Fig. 3.5). Sulphur dioxide is fairly soluble so could dissolve into the water that condensed around smoke particles.

$$SO_{2(g)} + H_2O_{(l)} \leftrightharpoons H^+_{(aq)} + HSO^-_{3(aq)} \qquad \text{eqn. 3.20}$$

Traces of metal contaminants (iron (Fe) or manganese (Mn)) catalysed the conversion of dissolved SO_2 to H_2SO_4 (see Box 4.4 for a definition of catalyst).

$$2HSO^-_{3(aq)} + O_{2(aq)} \rightarrow 2H^+_{(aq)} + 2SO^{2-}_{4(aq)} \qquad \text{eqn. 3.21}$$

Sulphuric acid has a great affinity for water so the droplet absorbed more water. Gradually the droplets grew and the fog thickened, attaining very low pH values (see Box 3.5).

Terrible fogs plagued London at the turn of the last century when Sherlock Holmes and Jack the Ripper paced the streets of the metropolis. The incidence of bronchial disease invariably rose at times of prolonged winter fog—little wonder, considering that the fog droplets contained H_2SO_4. Medical registrars in

Fig. 3.5 The London smog of 1952. Photograph courtesy of Popperfoto Northampton, UK.

Victorian England realized that the fogs were affecting health, but they, along with others, were not able to legislate smoke out of existence. Even where there was a will, and indeed there were enthusiasts in both Europe and North America who strove for change, the technology was far too naïve to achieve really noticeable improvements. The improvements that did come about were often due to changes in fuel, in location of industry or in climate.

3.6.2 Los Angeles smog—secondary pollution

The air pollutants that we have been discussing so far have come from stationary sources. Traditionally, industrial and domestic activities in large cities burnt coal. The transition to petroleum-derived fuels this century has seen the emergence of an entirely new kind of air pollution. This newer form of pollution is the result of the greater volatility of liquid fuels. The motor vehicle is such an important consumer of liquid fuels that it has become a major source of contemporary air pollution. However, the pollutants really responsible for causing the problems are not themselves emitted by motor vehicles. Rather, they form in the atmosphere. These secondary pollutants are formed from the reactions of primary pollutants, such as NO and unburnt fuel, which come directly from the automobiles. Chemical reactions that produce the secondary pollutants proceed most effectively in sunlight, so the resulting air pollution is called photochemical smog.

Box 3.5 The pH scale

The acidity of aqueous solutions is frequently described in terms of the pH scale. Acids (Box 3.3) give rise to hydrogen ions (H^+) in solution and the pH value of such a solution is defined:

$$pH = -\log_{10}\left(aH^+_{(aq)}\right) \qquad \text{eqn. 1}$$

We can write a similar relationship identifying pOH:

$$pOH = -\log_{10}\left(aOH^-_{(aq)}\right) \qquad \text{eqn. 2}$$

However, pH is related to pOH through the equilibrium describing the dissociation of water:

$$H_2O \rightleftharpoons H^+ + OH^-, \text{ i.e. } K_w = 10^{-14} = aH^+ . aOH^-$$
$$\text{eqn. 3}$$

such that $pH = 14 - pOH$.

It is important to notice that this is a logarithmic scale, so it is not appropriate to average pH values of solutions (although one can average H^+ concentrations).

On the pH scale, 7 is regarded as neutral. This is the point where $aH^+ = aOH^-$. There are a number of other important values on the scale (conventionally made to stretch from 0 to 14) that are relevant to the environment.

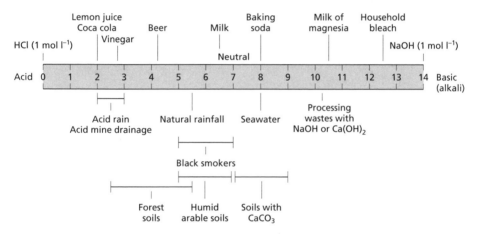

Fig. 1 pH scale showing values for familiar commodities (above the scale) and various environmental fluids discussed in this book (below the scale). Soil pH is measured on pure water (pH 7) equilibrated with the soil solids. Note that naturally alkaline fluids are rare. Industrial processing that involves strong bases like NaOH (e.g. bauxite processing) or $Ca(OH)_2$ (lime production) can contaminate river waters to around pH 10.

Photochemical smog was first noticed in Los Angeles during the Second World War. Initially it was assumed to be similar to the air pollution that had been experienced elsewhere, but conventional smoke abatement techniques failed to lead to any improvement. In the 1950s it became clear that this pollution was different, and the experts were baffled. A. Haagen-Smit, a biochemist studying vegetation damage in the Los Angeles basin, realized that the smog was caused by reactions of automobile exhaust vapours in sunlight.

Although air pollution and smoke have traditionally been closely linked, there were always those who thought there was more to air pollution than just smoke. We can now see how impurities in fuel give rise to further pollutants. In addition, the fact that we burn fuels, not in O_2, but in air has important consequences. We have learnt that air is a mixture of O_2 and N_2. At high temperature, in a flame, molecules in air may fragment, and even the relatively inert N_2 molecule can undergo reaction:

$$O_{(g)} + N_{2(g)} \rightarrow NO_{(g)} + N_{(g)} \hspace{3cm} \text{eqn. 3.22}$$

$$N_{(g)} + O_{2(g)} \rightarrow NO_{(g)} + O_{(g)} \hspace{3cm} \text{eqn. 3.23}$$

Equation 3.23 produces an oxygen atom, which can re-enter equation 3.22. Once an oxygen atom is formed in a flame, it will be regenerated and contribute to a whole chain of reactions that produce NO. If we add these two reactions we get:

$$N_{2(g)} + O_{2(g)} \rightarrow 2NO_{(g)} \hspace{3cm} \text{eqn. 3.24}$$

The equations show how nitrogen oxides are generated in flames. They arise because we burn fuels in air rather than just in O_2. In addition, some fuels contain nitrogen compounds as impurities, so the combustion products of these impurities are a further source of nitrogen oxides (i.e. NO_x, the sum of NO and NO_2).

Oxidation of nitric oxide in smog gives nitrogen dioxide (Box 3.6), which is a brown gas. This colour means that it absorbs light and is photochemically active and undergoes dissociation:

$$NO_{2(g)} + hv \rightarrow O_{(g)} + NO_{(g)} \hspace{3cm} \text{eqn. 3.25}$$

Equation 3.25 thus reforms the nitric oxide, but also gives an isolated and reactive oxygen atom, which can react to form O_3:

$$O_{(g)} + O_{2(g)} \rightarrow O_{3(g)} \hspace{3cm} \text{eqn. 3.26}$$

Ozone is the single pollutant that most clearly characterizes photochemical smog. However, O_3, which we regard as such a problem, is not emitted by automobiles (or any major polluter). It is a secondary pollutant.

The volatile organic compounds released through the use of petroleum fuels serve to aid the conversion of NO to NO_2. The reactions are quite complicated, but we can simplify them by using a very simple organic molecule such as CH_4, to represent the petroleum vapour from vehicles:

$$CH_{4(g)} + 2O_{2(g)} + 2NO_{(g)} \xrightarrow{hv} H_2O_{(g)} + HCHO_{(g)} + 2NO_{2(g)} \hspace{0.5cm} \text{eqn. 3.27}$$

We can see two things taking place in this reaction. Firstly, the automobile hydrocarbon is oxidized to an aldehyde (i.e. a molecule with a CHO functional group, see Table 2.1). In the reaction above it is formaldehyde (HCHO). Aldehydes are eye irritants and, at high concentrations, also carcinogens. This equation simply shows the net reactions in photochemical smog. In Box 3.6 the process is given in more detail. In particular, it emphasizes the role of the ubiquitous OH radical in promoting chemical reactions in the atmosphere.

The smog found in the Los Angeles basin (Plate 3.1, facing p. 138) is very different from that we have previously described as typical of coal-burning cities.

Box 3.6 Reactions in photochemical smog

Reactions involving nitrogen oxides (NO and NO_2) and ozone (O_3) lie at the heart of photochemical smog.

$$NO_{2(g)} + hv\ (<310\,nm) \rightarrow O_{(g)} + NO_{(g)} \qquad \text{eqn. 1}$$

$$O_{(g)} + O_{2(g)} + M_{(g)} \rightarrow O_{3(g)} + M_{(g)} \qquad \text{eqn. 2}$$

$$O_{3(g)} + NO_{(g)} \rightarrow O_{2(g)} + NO_{2(g)} \qquad \text{eqn. 3}$$

where M represents a 'third body' (Section 3.10.1)

It is conventional to imagine these processes that destroy and produce nitrogen dioxide (NO_2) as in a kind of equilibrium, which is represented by a notional equilibrium constant relating the partial pressures of the two nitrogen oxides and O_3:

$$K = \frac{pNO \cdot pO_3}{pNO_2} \qquad \text{eqn. 4}$$

If we were to increase NO_2 concentrations (in a way that did not use O_3), then the equilibrium could be maintained by increasing O_3 concentrations. This happens in the photochemical smog through the mediation of hydroxyl (OH) radicals in the oxidation of hydrocarbons. Here we will use methane (CH_4) as a simple example of the process:

$$OH_{(g)} + CH_{4(g)} \rightarrow H_2O_{(g)} + CH_{3(g)} \qquad \text{eqn. 5}$$

$$CH_{3(g)} + O_{2(g)} \rightarrow CH_3O_{2(g)} \qquad \text{eqn. 6}$$

$$CH_3O_{2(g)} + NO_{(g)} \rightarrow CH_3O_{(g)} + NO_{2(g)} \qquad \text{eqn. 7}$$

$$CH_3O_{(g)} + O_{2(g)} \rightarrow HCHO_{(g)} + HO_{2(g)} \qquad \text{eqn. 8}$$

$$HO_{2(g)} + NO_{(g)} \rightarrow NO_{2(g)} + OH_{(g)} \qquad \text{eqn. 9}$$

These reactions represent the conversion of nitric oxide (NO) to NO_2 and a simple alkane (see Section 2.7) such as CH_4 to an aldehyde (see Table 2.1), here formaldehyde (HCHO). Note that the OH radical is regenerated, so can be thought of as a kind of catalyst. Although the reaction will happen in photochemical smog, the attack of the OH radical is much faster on larger and more complex organic molecules. Aldehydes such as acetaldehyde (CH_3CHO) may also undergo attack by OH radicals:

$$CH_3CHO_{(g)} + OH_{(g)} \rightarrow CH_3CO_{(g)} + H_2O_{(g)} \qquad \text{eqn. 10}$$

$$CH_3CO_{(g)} + O_{2(g)} \rightarrow CH_3COO_{2(g)} \qquad \text{eqn. 11}$$

$$CH_3COO_{2(g)} + NO_{(g)} \rightarrow NO_{2(g)} + CH_3CO_{2(g)} \qquad \text{eqn. 12}$$

$$CH_3CO_{2(g)} \rightarrow CH_{3(g)} + CO_{2(g)} \qquad \text{eqn. 13}$$

The methyl radical (CH_3) in equation 13 may re-enter at equation 6. An important branch to this set of reactions is:

$$CH_3COO_{2(g)} + NO_{2(g)} \rightarrow CH_3COO_2NO_{2(g)} \qquad \text{eqn. 14}$$
$$\text{(PAN)}$$

leading to the formation of the eye irritant peroxyacetylnitrate (PAN).

There is no fog when Los Angeles smog forms, and visibility does not decline to just a few metres, as was typical of London fogs. Of course, the Los Angeles smog forms best on sunny days. London fogs are blown away by wind, but the gentle sea breezes in the Los Angeles basin can hold the pollution in against the mountains and prevent it from escaping out to sea. The pollution cannot rise in the atmosphere because it is trapped by an inversion layer: the air at ground level is cooler than that aloft, so that a cap of warm air prevents the cooler air from rising and dispersing the pollutants. A fuller list of the differences between Los Angeles- and London-type smogs is given in Table 3.5.

Table 3.5 Comparison of Los Angeles and London smog. From Raiswell *et al.* (1980).

Characteristic	Los Angeles	London
Air temperature	24 to 32°C	−1 to 4°C
Relative humidity	<70%	85% (+ fog)
Type of temperature inversion	Subsidence, at 1000 m	Radiation (near ground) at a few hundred metres
Wind speed	<3 m s^{-1}	Calm
Visibility	<0.8–1.6 km	<30 m
Months of most frequent occurrence	Aug. to Sept.	Dec. to Jan.
Major fuels	Petroleum	Coal and petroleum products
Principal constituents	O_3, NO, NO_2, CO, organic matter	Particulate matter, CO, S compounds
Type of chemical reaction	Oxidative	Reductive
Time of maximum occurrence	Midday	Early morning
Principal health effects	Temporary eye irritation (PAN)	Bronchial irritation, coughing (SO_2/smoke)
Materials damaged	Rubber cracked (O_3)	Iron, concrete

3.6.3 21st-century particulate pollution

It is valid to ask what might be different about pollution in the early 21st century. One of the most notable issues in the last decade or so has been a rise in concern about fine particles (or aerosols) in the atmosphere. Some of this concern has come about because fine particles are now more noticeable because we have lessened the emission of many pollutant gases and smoke into the atmosphere. In some cases the concentrations of these particles have increased in urban air. There has also been a growing awareness that fine particles have a significant impact on health.

Fine particles are those that are respirable. Traditionally this would have been particles less than 10 μm in diameter that can make their way into the respiratory system. These particles, often referred to as PM-10 (PM is short for particulate matter), are usually accompanied by even finer particles about 2.5 μm called PM-2.5. These finer particles can go deep into the lung and become deposited in the alveoli, the terminal sacs of the airways, where gases are exchanged with the blood. Once in the alveoli various biochemical processes seek to combat the invading particles which ultimately place the individual under increased stress and at risk from a range of health effects.

These fine particles come from a range of sources including some that come from combustion processes. In the late 20th century the increasing importance of the diesel engine in vehicles added to the fine particle concentrations of European cities. The diesel engine can emit very small particles, perhaps only 0.1 µm across, but these readily coagulate into somewhat larger particles. These particles can also become coated with a range of organic compounds, which have the potential to be carcinogenic contributing to long-term health impacts.

Reactions in the atmosphere lead to the formation of secondary particles. The best known of these are sulphate particles from the oxidation of SO_2 (eqns. 3.20 & 3.21). These particles are usually acid, although partial neutralization to ammonium bisulphate (NH_4HSO_4) is also possible. These particles have the potential to have additional irritant effects on the respiratory system. In recent years there has been a rising interest in the organic fraction of secondary aerosols with an awareness of the complexity of its chemistry. When volatile organic substances (see Box 4.14) react in the atmosphere they are typically converted to aldehydes, ketones (see Table 2.1) and organic acids. These more oxidized organic compounds are usually less volatile and can become associated with particles. Oxalic acid, a dicarboxylic acid ($HOOC.COOH$), is a highly oxidized small organic compound and is typical of the oxidation products found in the modern urban atmosphere. It is a relatively strong acid and not at all volatile, so readily incorporated into fine particles. This acid is also able to form complexes (see Box 6.4) with metals such as iron in the aerosol particles. Concern about the health impacts of small primary and secondary particles has driven much research into aerosols in urban air.

The 1990s was also a period when there was an increased awareness of the transport of pollutants from large-scale forest fires into areas with large populations. This was most notably reported in terms of smoke from tropical fires in South East Asia, although there were also worries about carbon monoxide from fires spreading into cities of the USA. In China, Korea and Japan there have been observations of increased haze in the air as particulates drift eastward from central China. Some of the particulate material is from wind-blown dust, but this is mixed with agricultural and industrial pollutants and even the soot from cooking stoves. Although there has been much discussion of the health effects of the smoke from such sources, studies have typically had to rely on information about urban aerosols, which are likely to be rather different. Biomass burning yields many millions of tonnes of soot, which has a graphitic structure and characteristic organic compounds such as abietic acid and retene (Fig. 3.4c) derived from plant resins. Potassium and zinc are also likely to be found in the particles from forest fires.

3.7 Air pollution and health

We saw in Section 3.6.1 that the acid-laden smoke particles in the London atmosphere caused great harm to human health in the past. Pollutants in the atmosphere still cause concern because of their effect on human health, although today we need to consider a wider range of potentially harmful trace substances. The photochemical smog encountered ever more widely in modern cities gives urban atmospheres that are unlike the smoky air of cities in the past. Petrol as a fuel, unlike coal, produces little smoke.

The two gases that particularly characterize photochemical smog, O_3 and nitrogen oxides, caused particular concern because of their potential to induce respiratory problems. Ozone impairs lung function, while nitrogen oxides at high concentrations are particularly likely to affect asthmatics. Oxygen-containing compounds, such as aldehydes, cause eye, nose and throat irritation, as well as headaches, during periods of smog. Eye irritation is a frequent complaint in Los Angeles and other photochemically polluted cities. This eye irritation is particularly associated with a group of nitrogen-containing organic compounds. They are produced in reactions of nitrogen oxides with various organic compounds in the smog (Box 3.6). The best known of these nitrogen-containing eye irritants is peroxyacetylnitrate, often called PAN.

Photochemical smog is not the only pollution problem created by vehicles. Automobiles are also associated with other pollutants such as lead (Pb) and benzene (C_6H_6). The success of lead tetralkyl compounds as antiknock agents for improving the performance of automotive engines has meant that, in countries with high car use, very large quantities of lead have been mobilized. This lead has been widely dispersed, but particularly large quantities have been deposited in cities and near heavily used roads. Lead is a toxin and has been linked with several environmental health problems. Perhaps the most worrying evidence has come from studies (although difficult to reproduce) which suggest a decline in intelligence among children exposed to quite low concentrations of lead.

Unleaded petrol was introduced in the USA in the 1970s so that catalytic converters could be used on cars. Since then, unleaded petrol has become used more widely. There is evidence that blood lead concentrations have dropped in parallel with the declining automotive source of lead. Nevertheless, the decrease in atmospheric lead may not yet be enough to reduce possible subtle health effects in children to a satisfactory level. This is because children have a high intake of food relative to their body weight. Thus children are more likely than adults to consume a significant amount of their intake of lead with food and water. Although some of the lead in foodstuffs may have come from the atmosphere, lead in foods may also result from processing.

Benzene (see Fig. 2.4) is another pollutant component of automotive fuels. It occurs naturally in crude oil and is a useful component because it can prevent pre-ignition in unleaded petrol (the production process is usually adjusted so that the benzene concentration is about 5%). There is evidence that in some locations, where there has been a switch to fuels with high concentrations of aromatic hydrocarbons, there has been a sharp increase in photochemical smog. This is due to the high reactivity of these hydrocarbons in the urban atmosphere. This problem should draw our attention to the way in which the solution of one obvious environmental problem (lead from petrol) may introduce a second rather more subtle problem (i.e. increased photochemical smog from reactive aromatic compounds).

Benzene is also a potent carcinogen. It appears that more than 10% of the benzene used by society (33 M tonne yr^{-1}) is ultimately lost to the atmosphere. High concentrations of benzene can be found in the air of cities and these concentrations may increase the number of cancers. Exposure is complicated by the importance of other sources of benzene to humans, for example tobacco smoke. Toluene ($C_6H_5CH_3$; Fig. 3.4d) is another aromatic compound present in large concentrations in petrol. Toluene is less likely to be a carcinogen than benzene but it has some undesirable effects. Perhaps most importantly it reacts to form a PAN-type compound, peroxybenzoyl nitrate, which is a potent eye irritant.

As emphasized in the previous section, particles have increasingly come to be seen as an important influence on the environmental health of modern populations. The fine PM-2.5 is able to penetrate through the respiratory system all the way to the alveoli. Normally particles are removed from the respiratory system in the mucus which is driven upwards by fine hairs or cilia. There are no cilia in the air sacs, so roving amoeba-like cells (macrophages) engulf the particles. They can migrate upwards to the ciliated parts of the respiratory system or through the alveolar walls. The activities of the macrophages, although important, release inflammatory compounds. The inflammatory effects can easily be transmitted to the blood, such that pulmonary inflammation can readily become associated with cardiovascular problems. This explains the enhanced death rate often observed when human populations are exposed to particle-laden air. In addition to these immediate effects the particles are rich in polycyclic aromatic hydrocarbons (PAHs), which are carcinogens and could explain some of the cancer incidence seen in urban populations.

3.8 Effects of air pollution

In the past, when smoke was the predominant air pollutant, its effects were easy to see. Even today, black incrustations on older buildings in many large cities are still evident. In addition, clothes were soiled, curtains and hangings were blackened and plant growth was affected. City gardeners carefully chose only the most resistant plants. Early last century, the trees around industrial centres became so blackened that light-coloured moths were no longer camouflaged. Melanic (dark) forms became more common because predators could see them less easily. Plants are also very sensitive to SO_2 and one of the first effects seems to be the inhibition of photosynthesis.

The traditional smog generated by coal burning contained SO_2 and its oxidation product, H_2SO_4, in addition to smoke. Sulphuric acid is a powerful corrosive agent and rusts iron bars and weathers building stones. Architects sometimes complained of layers of sulphate damage 10 cm thick on calcareous stone through the reaction:

$$H_2SO_{4(aq)} + CaCO_{3(s)} + H_2O_{(l)} \rightarrow CO_{2(g)} + CaSO_{4(s)}.2H_2O_{(l)} \qquad \text{eqn. 3.28}$$

Sulphuric acid converts limestone ($CaCO_3$) into gypsum ($CaSO_4.2H_2O$). The deterioration is severe because gypsum is soluble and dissolves in rain. Perhaps more importantly, gypsum occupies a larger volume than limestone, which adds mechanical stress so that the stone almost explodes from within.

The diesel engine is no longer confined to large vehicles in Europe, as passenger cars have taken advantage of potentially lower fuel costs. The fuel injection process of the diesel engine leads to the fuel dispersing as droplets within the engine. These may not always burn completely, so diesel engines can produce large quantities of smoke if not properly maintained. Diesel smoke now makes a significant contribution to the soiling quality of urban air.

In the modern urban atmosphere, O_3 may be the pollutant of particular concern for health. However, it is a reactive gas that will also attack the double bonds of organic molecules (see Section 2.7) very readily. Rubber is a polymeric material with many double bonds, so it is degraded and cracked by O_3. Tyres and windscreen wiper blades are especially vulnerable to oxidants, although newer synthetic rubbers have double bonds protected by other chemical groups, which can make them more resistant to damage by O_3.

Many pigments and dyes are also attacked by O_3. The usual result of this is that the dye fades. This means that it is important for art galleries in polluted cities to filter their air, especially where they house collections of paintings using traditional colouring materials, which are especially sensitive. Nitrogen oxides, associated with photochemical smogs, can also damage pigments. It is possible that nitrogen oxides may also increase the rate of damage to building stone, but it is not really clear how this takes place. Some have argued that NO_2 increases the efficiency of production of H_2SO_4 on stone surfaces in those cities that have moderate SO_2 concentrations.

$$SO_{2(g)} + NO_{2(g)} + H_2O_{(l)} \rightarrow NO_{(g)} + H_2SO_{4(aq)} \qquad \text{eqn. 3.29}$$

Others have suggested that the nitrogen compounds in polluted atmospheres enable microorganisms to grow more effectively on stone surfaces and enhance the biologically mediated damage. There is also the possibility that gas-phase reactions produce HNO_3 (eqn. 3.14) and that this deposits directly on to calcareous stone. Diesel soot that increasingly disfigures buildings may also carry organic nutrients to the surface that could enhance biological damage.

Finally, we should remember that it is not just materials that are damaged by photochemical smog, since plants are especially sensitive to the modern atmospheric pollutants. Recollect that it was this sensitivity that led Haagen-Smit to recognize the novelty of the Los Angeles smog. Ozone damages plants by changing the 'leakiness' of cells to important ions such as potassium. Early symptoms of such injury appear as water-soaked areas on the leaves.

Urban air pollution remains an issue of much public concern. While it is true that in many cities the traditional problems of smoke and SO_2 from stationary sources are a thing of the past, new problems have emerged. In particular, the automobile and heavy use of volatile fuels have made photochemical smog a widespread occurrence. This has meant that there has been a parallel rise in legislation to lower the emission of these organic compounds to the atmosphere.

3.9 Removal processes

So far, we have examined the sources of trace gases and pollutants in the atmosphere and the way in which they are chemically transformed. Now we need to look at the removal process to complete the source–reservoir–sink model of trace gases that we have adopted.

Our discussions have emphasized the importance of the OH radical as a key entity in initiating reactions in the atmosphere. Attack often occurs through hydrogen abstraction, and subsequent reactions with oxygen and nitrogen oxides (as illustrated in Box 3.6). This serves to remind us that the basic transformation that takes place in the atmosphere is oxidation (see also Box 4.3). This is hardly unexpected in an atmosphere dominated by oxygen, so we can argue that reactions within the atmosphere generally oxidize trace gases.

Oxidation of non-metallic elements yields acidic compounds, and it is this that explains the great ease with which acidification occurs in the atmosphere. Carbon compounds can be oxidized to organic compounds, such as formic acid (HCOOH) or acetic acid (CH_3COOH) or, more completely, to carbonic acid (H_2CO_3, i.e. dissolved CO_2). Sulphur compounds can form H_2SO_4 and, in the case of some organosulphur compounds, methane sulphonic acid (CH_3SO_3H). Nitrogen compounds can ultimately be oxidized to HNO_3. The solubility of many of these compounds in water makes rainfall an effective mechanism for their removal from the atmosphere. The process is known as 'wet removal'.

It is important to note that, even in the absence of SO_2, atmospheric droplets will be acidic through the dissolution of CO_2 (Box 3.7). This has implications for the geochemistry of weathering (see Section 4.4). The SO_2, however, does make a substantial contribution to the acidity of droplets in the atmosphere. It can, so to speak, acidify rain (Box 3.7). However, let us consider the possibility of subsequent reactions that can cause even more severe acidification:

$$H_2O_{2(aq)} + HSO_{3(aq)}^- \rightarrow SO_{4(aq)}^{2-} + H_{(aq)}^+ + H_2O_{(l)} \qquad \text{eqn. 3.30}$$

$$O_{3(aq)} + HSO_{3(aq)}^- \rightarrow SO_{4(aq)}^{2-} + H_{(aq)}^+ + O_{2(aq)} \qquad \text{eqn. 3.31}$$

Hydrogen peroxide (H_2O_2) and O_3 are the natural strong oxidants present in rainwater. These oxidants can potentially oxidize nearly all the SO_2 in a parcel of air. Box 3.8 shows that under such conditions rainfall may well have pH values lower than 3. This illustrates the high acid concentrations possible in the atmosphere as trace pollutants are transferred from the gas phase to droplets. Liquid water in the atmosphere has a volume about a million times smaller than the gas phase; thus a substantial increase in concentration results from dissolution.

After the water falls to the Earth, further concentration enhancement can take place if it freezes as snow. When snow melts the dissolved ions are lost preferentially, as they tend to accumulate on the outside of ice grains which make up snowpacks. This means that at the earliest stages of melting it is the dissolved H_2SO_4 that comes out. Concentration factors of as much as 20-fold are possible. This has serious consequences for aquatic organisms, and especially their young, in the spring as the first snows thaw. It is not just acid rain, but acid rain amplified.

It is also possible for gaseous or particulate pollutants to be removed directly from the atmosphere to the surface of the Earth under a process known as dry deposition. This removal process may take place over land or the sea, but it is

Box 3.7 Acidification of rain droplets

In Box 3.4 we saw the way reactions affect the solubility of gases. It is possible for some gases to undergo more complex hydration reactions in water, which influence its pH (see Box 3.5). The best known of these is the dissolution of carbon dioxide (CO_2), which gives natural rainwater its characteristic pH.

$$CO_{2(g)} + H_2O_{(l)} \rightleftharpoons H_2CO_{3(aq)} \qquad \text{eqn. 1}$$

$$H_2CO_{3(aq)} \rightleftharpoons H^+_{(aq)} + HCO^-_{3(aq)} \qquad \text{eqn. 2}$$

$$HCO^-_{3(aq)} \rightleftharpoons H^+_{(aq)} + CO^{2-}_{3(aq)} \qquad \text{eqn. 3}$$

Equation 3 is not important in the atmosphere, so the pH of a droplet of water in equilibrium with atmospheric CO_2 can be determined by combining the first two equilibrium constant equations that govern the dissolution (i.e. Henry's law, as discussed in Box 3.4) and dissociation. If carbonic acid (H_2CO_3) is the only source of protons, then aH^+ must necessarily equal $aHCO^-_3$. Thus the equilibrium equation for equation 2 can be written:

$$K' = \frac{aH^+ . aHCO^-_3}{aH_2CO_3} = \frac{(aH^+)^2}{aH_2CO_3} \qquad \text{eqn. 4}$$

The Henry's law constant defined by equation 1 is:

$$K_H = \frac{aH_2CO_3}{pCO_2} \qquad \text{eqn. 5}$$

which defines aH_2CO_3 as $K_H . pCO_2$, which can now be substituted in equation 4:

$$K' = \frac{(aH^+)^2}{K_H . pCO_2} \qquad \text{eqn. 6}$$

Rearranging gives:

$$aH^+ = (K_H K' pCO_2)^{1/2} \qquad \text{eqn. 7}$$

Substituting the appropriate values of the equilibrium constants (Table 1) and using a CO_2 partial pressure (pCO_2) of 360 ppm, i.e. 3.6×10^{-4} atm, will yield a hydrogen ion (H^+) activity of 2.4×10^{-6} mol l^{-1} or a pH of 5.6.

Sulphur dioxide (SO_2) is at much lower activity in the atmosphere, but it has a greater solubility and dissociation constant. We can set equations analogous to those for CO_2:

$$aSO_{2(g)} + aH_2O \rightleftharpoons aH_2SO_{3(aq)} \qquad \text{eqn. 8}$$

$$aH_2SO_{3(aq)} \rightleftharpoons aH^+_{(aq)} + aHSO_{3(aq)} \qquad \text{eqn. 9}$$

and once again rearranging gives:

$$aH^+ = (K_H K' pSO_2)^{1/2} \qquad \text{eqn. 10}$$

If a small amount of SO_2 is present in the air at an activity of 5×10^{-9} atm (not unreasonable over continental land masses), we can calculate a pH value of 4.85. So even low activity of SO_2 can have a profound effect on droplet pH.

Table 1 Henry's law constants and first dissociation constants for atmospheric gases that undergo hydrolysis (25°C).

Gas	K_H (mol l^{-1} atm^{-1})	K' (mol l^{-1})
Sulphur dioxide	2.0	2.0×10^{-2}
Carbon dioxide	0.04	4.0×10^{-7}

still termed 'dry deposition'. It is really a bit of a misnomer because the surfaces available for dry deposition are often most effective when wet.

3.10 Chemistry of the stratosphere

The stratosphere has been emphasized as a reservoir of smaller size than the troposphere and hence more easily affected by small amounts of trace gases. These can be injected by high flying aircraft or powerful volcanic explosions. However,

Box 3.8 Removal of sulphur dioxide from an air parcel

A parcel of air over a rural area of an industrial continent would typically be expected to contain sulphur dioxide (SO_2) at a concentration of 5×10^{-9} atm. This means that a cubic metre of air contains 5×10^{-9} m^3 of SO_2. We can convert this to moles quite easily because a mole of gas occupies 0.0245 m^3 at 15°C and atmospheric pressure. Thus our cubic metre of air contains $5 \times 10^{-9}/0.0245 = 2.04 \times 10^{-7}$ mol of SO_2. In a rain-laden cloud we can expect one cubic metre to contain about 1 g of liquid water, i.e. 0.001 dm^3.

If the SO_2 were all removed into the droplet and oxidized to sulphuric acid

(H_2SO_4), we would expect the 2.04×10^{-7} mol to dissolve in 0.001 dm^3 of liquid water, giving a liquid-phase activity of 2.04×10^{-4} mol l^{-1}. The H_2SO_4 formed is a strong acid (Box 3.3), so dissociates with the production of two protons under atmospheric conditions:

$$H_2SO_4 \rightarrow 2H^+ + SO_4^{2-} \qquad \text{eqn. 1}$$

Thus the proton activity will be 4.08×10^{-4} mol l^{-1}, or the pH 3.4. Evaporation of water from the droplet and removal of further SO_2 as the droplet falls through air below the cloud can lead to even further reduction in pH.

---

there is another source that is related to the low reactivity of some gases. Gases in the atmosphere tend to react with the OH radical. Gases that do not react with OH in the troposphere can survive long enough to be transferred into the stratosphere. This includes OCS, N_2O and to a lesser extent CH_4. Once in the stratosphere these gases become involved in reactions involving atomic oxygen (O). In addition to these natural trace gases there are a number of anthropogenic trace gases that are resistant to attack by OH. Among these the CFCs have become infamous because of their effects on stratospheric chemistry, particularly that of ozone (O_3). The discovery in 1984 that there was a hole in the ozone layer over Antarctica emphasized the threat imposed by these gases.

Although O_3 is a toxin in the troposphere (Section 3.6.2), it plays a vital role in shielding organisms on the Earth from damaging UV radiation. There are only very small amounts of O_3 in the upper atmosphere. If all the O_3 in the Earth's atmosphere, most of which is found in the stratosphere, were brought to ground level it would constitute a layer of pure O_3 only 3 mm thick. The tenuous nature of the O_3 layer means that for some decades scientists have been concerned that O_3 in the stratosphere could be damaged by the presence of CFCs. However, calculations of gas-phase chemistry suggested that changes in the atmosphere as a whole would be small. This explains why the detection of an O_3 hole over Antarctica in 1984 came as a surprise (Fig. 3.6). The rapid destruction of O_3 in the polar stratosphere in the 1970s and 1980s (Fig. 3.6) proved the chemistry of the O_3 layer to be much more complex than had previously been thought.

3.10.1 Stratospheric ozone formation and destruction

The formation of ozone is a photochemical process that uses the energy involved in light. The shorter the wavelength of light, the larger the amount of energy it

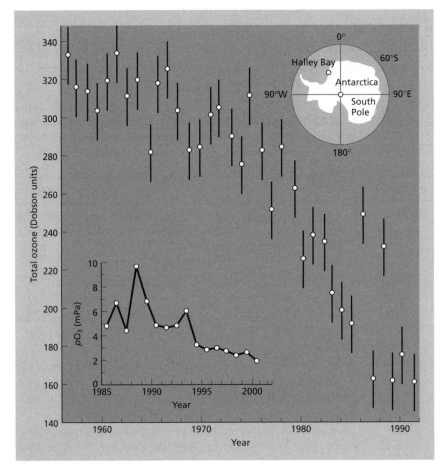

Fig. 3.6 Mean October levels of total ozone above Halley Bay (76°S), Antarctica, since 1957. The 1986 value is anomalous due to deformation of the ozone hole, which left Halley Bay temporarily outside the circumpolar vortex (a tight, self-contained wind system). Dobson units represent the thickness of the ozone layer at sealevel temperature and pressure (where 1 Dobson unit is equivalent to 0.01 mm). Data courtesy of the British Antarctic Survey. Inset shows seasonally averaged (Sep.–Nov.) ozone partial pressure at about 17 km at 70°S. Data courtesy of G. König-Langlo.

carries. It requires ultraviolet (UV) radiation of wavelength less than 242 nm to have sufficient energy to split the oxygen molecule (O_2) apart:

$$O_{2(g)} + h\upsilon \rightarrow O_{(g)} + O_{(g)} \qquad\qquad \text{eqn. 3.32}$$

The UV photon here is symbolized by $h\upsilon$. Once oxygen atoms (O) have been formed, they can react with O_2.

$$O_{2(g)} + O_{(g)} \rightarrow O_{3(g)} \qquad\qquad \text{eqn. 3.33}$$

The production of O_3 by this photochemical process can be balanced against the reactions that destroy O_3. The most important is photolysis:

$$O_{3(g)} + h\upsilon \rightarrow O_{2(g)} + O_{(g)} \qquad\qquad \text{eqn. 3.34}$$

$$O_{3(g)} + O_{(g)} \rightarrow 2O_{2(g)} \qquad \text{eqn. 3.35}$$

together with an additional reaction describing the destruction process for oxygen atoms:

$$O_{(g)} + O_{(g)} + M_{(g)} \rightarrow O_{2(g)} + M_{(g)} \qquad \text{eqn. 3.36}$$

Note the presence of the 'third body' M, which carries away excess energy during the reaction. The third body would typically be O_2 or a nitrogen molecule (N_2). Without this 'third body' the O_2 that formed might split apart again. Calculations that balance the production and destruction of O_3 considering only reactions that involve the element oxygen (i.e. oxygen-only paths) give a fair description of the O_3 observed in the stratosphere. The results of these calculations produce the correct shape for the vertical profile of O_3 in the atmosphere and the peak O_3 concentration occurs at the correct altitude, but the predicted concentrations are too high. This is because there are other pathways that destroy O_3. Some involve hydrogen-containing species:

$$OH_{(g)} + O_{3(g)} \rightarrow O_{2(g)} + HO_{2(g)} \qquad \text{eqn. 3.37}$$

$$HO_{2(g)} + O_{(g)} \rightarrow OH_{(g)} + O_{2(g)} \qquad \text{eqn. 3.38}$$

which sum:

$$O_{3(g)} + O_{(g)} \rightarrow 2O_{2(g)} \qquad \text{eqn. 3.39}$$

Similar reactions can be written for nitrogen-containing species, for example nitric oxide (NO), which arises from supersonic aircraft, or nitrous oxide (N_2O), which crosses the tropopause into the stratosphere:

$$NO_{(g)} + O_{3(g)} \rightarrow O_{2(g)} + NO_{2(g)} \qquad \text{eqn. 3.40}$$

$$NO_{2(g)} + O_{(g)} \rightarrow NO_{(g)} + O_{2(g)} \qquad \text{eqn. 3.41}$$

and $N_2O_{(g)}$ can enter reaction 3.40 via the initial step:

$$N_2O_{(g)} + O_{(g)} \rightarrow 2NO_{(g)} \qquad \text{eqn. 3.42}$$

Reactions involving these species sum in such a way as to destroy O_3 and atomic oxygen while restoring the OH or NO molecules. They can thus be regarded as catalysts for O_3 destruction. In this case the catalysts are chemical species that facilitate a reaction, but undergo no net consumption or production in the reaction (see also Box 4.4). The important point of these catalytic reaction chains in the chemistry of stratospheric O_3 is that a single pollutant molecule can be responsible for the destruction of a large number of O_3 molecules.

It is now very well established that the most important of these catalytic reaction chains affecting polar ozone loss are the ones based around chlorine-containing species, as detailed below.

3.10.2 Ozone destruction by halogenated species

Natural chlorine in the stratosphere is mainly transferred there as methyl chloride (CH_3Cl), which probably comes from marine and terrestrial biological sources (Section 3.4.2). This natural source, however, accounts for only 25% of the chlorine which is transported across the tropopause. By the early 1970s the CFCs used as aerosol propellants and refrigerants had become widely distributed through the troposphere. There appeared to be no obvious mechanism for the destruction of these highly stable compounds in the lower atmosphere. However, the knowledge that CFCs were being transported to the stratosphere raised concern over their effect on the O_3 layer. These compounds, for example $CFCl_3$ (Freon-11; Fig. 3.4b) and CF_2Cl_2 (Freon-12), absorb UV radiation in the 190–220 μm range, which results in the photodissociation reactions:

$$CFCl_{3(g)} + hv \rightarrow CFCl_{2(g)} + Cl_{(g)} \qquad \text{eqn. 3.43}$$

$$CF_2Cl_{2(g)} + hv \rightarrow CF_2Cl_{(g)} + Cl_{(g)} \qquad \text{eqn. 3.44}$$

These reactions produce the free chlorine atoms that react with O_3 in the catalytic manner, i.e.:

$$O_{3(g)} + Cl_{(g)} \rightarrow O_{2(g)} + ClO_{(g)} \qquad \text{eqn. 3.45}$$

$$ClO_{(g)} + O_{(g)} \rightarrow O_{2(g)} + Cl_{(g)} \qquad \text{eqn. 3.46}$$

which sum:

$$O_{3(g)} + O_{(g)} \rightarrow 2O_{2(g)} \qquad \text{eqn. 3.47}$$

However, ClO produced on reaction with O_3 may not always react with atomic oxygen (eqn. 3.46), but can interact with nitrogen compounds instead:

$$ClO_{(g)} + NO_{2(g)} + M \rightarrow ClONO_{2(g)} + M \qquad \text{eqn. 3.48}$$

This reaction is of considerable importance because it effectively removes the nitrogen and chlorine species involved in the cycles that destroy O_3. If, however, there are solid surfaces present, the chlorine sequestered in this reaction can be released:

$$ClONO_{2(g)} + HCl_{(s)} \rightarrow Cl_{2(g)} + HNO_{3(s)} \qquad \text{eqn. 3.49}$$

$$Cl_{2(g)} + hv \rightarrow 2Cl_{(g)} \qquad \text{eqn. 3.50}$$

$$2Cl_{(g)} + 2O_{3(g)} \rightarrow 2ClO_{(g)} + 2O_{2(g)} \qquad \text{eqn. 3.51}$$

$$2ClO_{(g)} + M \rightarrow Cl_2O_{2(g)} + M \qquad \text{eqn. 3.52}$$

$$Cl_2O_{2(g)} + hv \rightarrow ClO_{2(g)} + Cl_{(g)} \qquad \text{eqn. 3.53}$$

$$ClO_{2(g)} + M \rightarrow Cl_{(g)} + O_{2(g)} + M \qquad \text{eqn. 3.54}$$

Equations 3.51–3.54 then sum to:

$$2O_{3(g)} \rightarrow 3O_{2(g)} \qquad \text{eqn. 3.55}$$

The reaction sequence (eqns. 3.49–3.54) is particularly fast at low temperature. Moreover, the square dependence on chlorine concentration implicit within equation 3.50 makes the reaction very sensitive to chlorine concentration. It is these low-temperature processes on particle surfaces that offer the best explanation for the dramatic decrease in O_3 observed over the Antarctic continent (Fig. 3.6). Future modelling of O_3 depletion will have to allow increasingly for the heterogeneous aspect of its chemistry. It may well be that, in addition to solid surfaces for reactions, liquid droplets also provide an important medium for reaction.

In the 1990s it became clear that bromine-containing compounds (halons) also played an important role in stratospheric chemistry. Bromoform ($CHBr_3$), a gas naturally released from the oceans, played some role in this but there were also significant emissions from human activities. For example, halons are used in some fire extinguishers, being non-toxic and leaving no residue after evaporation. This class of compounds is typified by the simple halon 1211 (CF_2ClBr). These materials can supply both bromine and chlorine to the stratosphere, so they have also been regulated under protocols to reduce human impact on the ozone layer.

3.10.3 Saving the ozone layer

The clear links between CFCs, depletion in stratospheric O_3, increased UV radiation reaching the Earth's surface and possible increased incidence of skin cancer in humans have not escaped the media, who have been able, at times during the 1970s and 1980s, to capture the imagination of the general public. It is probably correct that the CFC issue aroused immediate concern because its cause was apparently obvious—in the shape of the aerosol can! Although it is true that aerosol propellant was only a contributor to CFC build up in the atmosphere (refrigerant coolants and industrial uses being other important sources), there is little doubt that the aerosol can became a late 20th-century 'icon' for environmental activism. It is this public awareness that has made the CFC–stratospheric O_3 story such a good example of how environmental chemistry research can lead to major international legislation.

Against the odds, the anti-aerosol lobby took on the multimillion-dollar aerosol industry and achieved real success. By the late 1970s, CFCs were at least partially banned in deodorant and hair sprays in the USA; Canada imposed similar controls in the early 1980s. It was, however, the discovery of the Antarctic O_3 hole that provoked stronger action. In 1987 a meeting of the United Nations Environment Programme in Montreal resulted in 31 countries agreeing to the so-called 'Montreal Protocol', under which developed countries agreed to a 50% cut in CFCs. Following this agreement, further meetings in Helsinki (1989) and Copenhagen (1992) made the conditions of the Montreal Protocol more stringent, resulting in an agreement to ban production of CFCs in developed countries.

Response to the Montreal Protocol by industry was positive, with agreements to phase out CFC production, resulting in a search for viable safe alternatives; decline in some atmospheric CFCs is now evident (Fig. 3.7). In developed countries, hydrocarbons or alternative means of pressuring containers have largely replaced CFCs in aerosol cans, hydrochlorofluorocarbons (HCFCs)—which are 95% less damaging to O_3 than CFCs—are used in the production of polystyrene foams and as refrigerant coolants and a propane/butane mixture is being developed as an alternative refrigerant coolant. Even HCFCs are gradually being phased out and replaced by substances that are less likely to cause ozone deple-

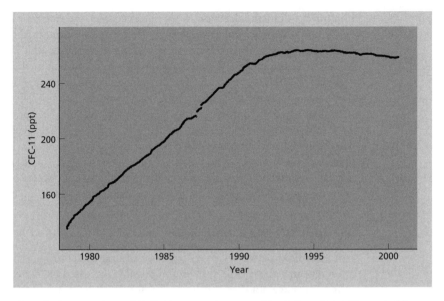

Fig. 3.7 Concentrations of CFC-11 measured at ground level, Cape Grim, Tasmania. Note that concentrations of CFC-11 have been falling in the 1990s following the rapid increase during the 1980s. Copyright CSIRO Australia, May 2002.

tion. Attention must now turn to assisting developing countries, which still use CFCs to switch to alternate compounds.

In the case of the halons, replacements are also being phased in. For example, halon 1301 (bromotrifluoromethane, CF_3Br), widely used as a fire-extinguishing agent to protect sensitive electronic equipment, is being replaced by HFC-227 (CF_3CHFCF_3), which contains no chlorine or bromine.

It is sobering to remember that despite the recent success in limiting production of CFCs, these stable substances have a long residence time in the atmosphere, between 40 and 150 years. This means their effects on the stratospheric O_3 will continue for some time after the bans on their production. Current estimates suggest that the policies in place should see a decline in the stratospheric bromine and chlorine concentrations over the next 50 years, paralleled by a rise in ozone concentrations.

3.11 Further reading

Brimblecombe, P. (1987) *The Big Smoke*. Methuen, London.

Brimblecombe, P. (1996) *Air Composition and Chemistry*, 2nd edn. Cambridge University Press, Cambridge.

Brimblecombe, P. & Maynard, R.L. (2001) *The Urban Atmosphere and Its Effects*. Imperial College Press, London.

Elsom, D.M. (1992) *Atmospheric Pollution A Global Problem*. Blackwell Scientific Publications, Oxford.

Jacobson M.Z. (2002) *Atmospheric Pollution*. Cambridge University Press, Cambridge.

3.12 Internet search keywords

chemical composition atmosphere
troposphere
partial pressure atmosphere
Dalton's Law
chemical equilibrium
equilibrium constant
residence time atmosphere
reactive trace gases
OH radical
acid base
gas solubility
Henry's Law
pH scale
atmospheric aerosols
urban air pollution
primary pollutant atmosphere
secondary pollutant atmosphere
London air pollution
Los Angeles air pollution
ozone photochemical smog

PM 10
PM 2.5
biomass burning soot
biomass burning Asia
air pollution health
ozone respiratory
nitrogen oxides respiratory
PAN atmospheric pollution
benzene atmospheric pollution
acidification rain
acidification CO_2
acidification SO_2
stratospheric chemistry
stratospheric ozone
stratospheric ozone depletion
CFCs ozone depletion
ozone hole
halogens ozone
Montreal Protocol

The Chemistry of Continental Solids

4

4.1 The terrestrial environment, crust and material cycling

Terrestrial environments consist of solid (rocks, sediments and soils), liquid (rivers, lakes and groundwater) and biological (plants and animals) components. The chemistry of terrestrial environments is dominated by reactions between the Earth's crust and fluids in the hydrosphere and atmosphere.

The terrestrial environment is built on continental crust, a huge reservoir of igneous and metamorphic rock (mass of continental crust = 23.6×10^{24} g). This rock, often called crystalline basement, forms most of the continental crust. About 80% of this basement is covered by sedimentary rocks contained in sedimentary basins, with average thicknesses of around 5 km. About 60% of these sedimentary rocks are mudrocks (clay minerals and quartz—SiO_2), with carbonates (limestones—calcium carbonate $(CaCO_3)$—and dolostones—$MgCa(CO_3)_2$) and sandstones (mainly quartz) accounting for most of the rest (Fig. 4.1).

Mud, silt and sandy sediments form mainly by weathering—the breakdown and alteration of solid rock. Usually, these sedimentary particles are transported by rivers to the oceans, where they sink onto the seabed. Here, physical and biological processes and chemical reactions (collectively known as diagenesis) convert sediment into sedimentary rock. Eventually these rocks become land again, usually during mountain building (orogenesis).

The geological record shows that this material-transport mechanism has operated for at least 3.8 billion years. New sediments are derived either from older sedimentary rocks or from newly generated or ancient igneous and metamorphic rock. The average chemical composition of suspended sediment in rivers, sedimentary mudrock and the upper continental crust is quite similar (Table 4.1).

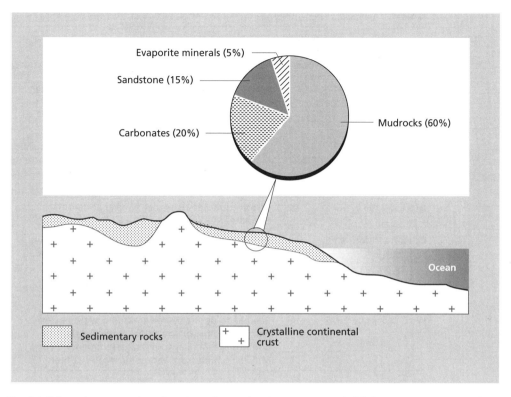

Fig. 4.1 Schematic cross-section of continental crust showing geometry and global average composition of sedimentary cover.

Table 4.1 Average chemical composition of upper continental crust, sedimentary mudrock and suspended load of rivers. Data from Wedepohl (1995) and Taylor and McLennan (1985).

	Average upper continental crust (wt%)*	Average sedimentary mudrock† (wt%)	Average suspended load (rivers)‡ (wt%)
SiO_2	65.0	62.8	61.0
TiO_2	0.6	1.0	1.1
Al_2O_3	14.7	18.9	21.7
FeO	4.9	6.5	7.6
MgO	2.4	2.2	2.1
CaO	4.1	1.3	2.3
Na_2O	3.5	1.2	0.9
K_2O	3.1	3.7	2.7
Σ	98.3	99.9	99.4

*A silicate analysis is usually given in units of weight% of an oxide (grams of oxide per 100 g of sample). As most rocks consist mainly of oxygen-bearing minerals, this convention removes the need to report oxygen separately. The valency of each element governs the amount of oxygen combined with it. A good analysis should sum (Σ) to 100 wt%.

† This analysis represents terrigenous mudrock (i.e. does not include carbonate and evaporite components), a reasonable representation of material weathered from the upper continental crust.

‡ Average of Amazon, Congo, Ganges, Garronne and Mekong data.

Table 4.2 Agents of material transport to the oceans. After Garrels *et al.* (1975).

Agent	Percentage of total transport	Remarks
Rivers	89	Present dissolved load 17%, suspended load 72%. Present suspended load higher than geological past due to human activities (e.g. deforestation) and presence of soft glacial sediment cover
Glacier ice	7	Ground rock debris plus material up to boulder size. Mainly from Antarctica and Greenland. Distributed in seas by icebergs. Composition similar to average sediments
Groundwater	2	Dissolved materials similar to river composition. Estimate poorly constrained
Coastal erosion	1	Sediments eroded from cliffs, etc. by waves, tides, storms, etc. Composition similar to river suspended load
Volcanic	0.3 (?)	Dusts from explosive eruptions. Estimate poorly constrained
Wind-blown dust	0.2	Related to desert source areas and wind patterns, e.g. Sahara, major source for tropical Atlantic. Composition similar to average sedimentary rock. May have high (<30%) organic matter content

This suggests that rivers represent an important pathway of material transport (Table 4.2) and that sedimentary mudrocks record crustal composition during material cycling.

This chapter focuses on components of the solid terrestrial environment that are chemically reactive. Silicates that formed deep in the crust, at high temperature and pressure, are unstable when exposed to Earth surface environments during weathering. The minerals adjust to the new set of conditions to regain stability. This adjustment may be rapid (minutes) for a soluble mineral such as halite (sodium chloride, $NaCl$) dissolving in water, or extremely slow (thousands or millions of years) in the weathering of resistant minerals such as quartz. Although the emphasis is on understanding how minerals are built and how they weather, it is clear that water—a polar solvent (Box 4.1)—exerts a major influence on the chemical reactions, while organisms, particularly plants and bacteria, influence the types and rates of chemical reactions in soils.

Soils occupy an interface between the atmosphere and the lithosphere and the biogeochemical processes that occur there are tremendously important. Soils are a precious resource: indeed human existence—along with many other organisms—is dependent upon them. Soils provide habitat for organisms and allow the growth of vegetation, which in turn provides food and additional habitat for other organisms. Soils thus support many global food webs and their associated

Box 4.1 Properties of water and hydrogen bonds

The water molecule H_2O is triangular in shape, with each hydrogen (H) bonded to the oxygen (O) as shown in Fig. 1. The shape results from the geometry of electron orbits involved in the bonding. Oxygen has a much higher electronegativity (Box 4.2) than hydrogen and pulls the bonding electrons toward itself and away from the hydrogen atom. The oxygen thus carries a partial negative charge (usually expressed as $\delta-$), and each hydrogen a partial positive charge ($\delta+$), creating a dipole (i.e. electrical charges of equal magnitude and opposite sign a small distance apart). At any time a small proportion of water molecules dissociate completely to give H^+ and OH^- ions.

$$H_2O_{(l)} \rightleftharpoons H^+_{(aq)} + OH^-_{(aq)} \qquad \text{eqn. 1}$$

for which the equilibrium constant is:

$$K_w = \frac{aH^+.aOH^-}{aH_2O} = 10^{-14}\,mol^2\,l^{-2} \qquad \text{eqn. 2}$$

The activity of pure water is by convention unity (1), so equation 2 simplifies to:

$$K_w = aH^+.aOH^- = 10^{-14}\,mol^2\,l^{-2} \qquad \text{eqn. 3}$$

The polar nature of the water molecule allows the ions of individual water molecules to interact with their neighbours. The small hydrogen atom can approach and interact with the oxygen of a neighbouring molecule particularly effectively. The interaction between the hydrogen atom, with its partial positive charge, and oxygen atoms of neighbouring water molecules with partial negative charges, is particularly strong—by the standards of intermolecular interactions – though weaker than within covalent bonding. This type of interaction is called hydrogen bonding.

The molecules in liquid water are less randomly arranged than in most liquids because of hydrogen bonds. The polarity of the bonds makes water an effective solvent for ions; the water molecules are attracted to the ion by electrostatic force to form a cluster around it. Moreover, ionic-bonded compounds, with charge separation between component ions, are easily decomposed by the force of attraction of the water dipole. Hydrogen bonding gives water a relatively high viscosity and heat capacity in comparison with other solvents. Hydrogen bonds also allow water to exist as a liquid over a large temperature range. Since most biological transport systems are liquid, this latter property is fundamental to supporting life.

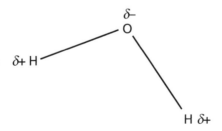

Fig. 1 The water molecule.

vegetation influences the global cycling of atmospheric gases and hence climate. It is clear that humans have a duty to manage soils judiciously. If soils are lost through erosion, bad agricultural practice or contamination they no longer fulfil their fundamental functions. While soils can be destroyed over a period of years or decades, their regeneration may require hundreds or even thousands of years.

4.2 The structure of silicate minerals

Most of the Earth's crust is composed of silicate minerals, for example feldspars, and quartz, which crystallized from magma or formed deep in the crust, at high temperature and pressure during metamorphism. Silicate minerals are compounds principally of silicon (Si) and oxygen (O), combined with other metals. The basic building block of silicates is the SiO_4 tetrahedron, in which silicon is situated at the centre of a tetrahedron of four oxygen ions (Fig. 4.2). This arrangement of ions is caused by the attraction—and strength of bonding—between positively charged and negatively charged ions (see Section 2.3), and the relative size of the ions—which determines how closely neighbouring ions can approach one another (Section 4.2.1).

4.2.1 Coordination of ions and the radius ratio rule

In crystals where bonding is largely ionic (see Section 2.3.2), the densest possible packing of equal-sized anions (represented by spheres) is achieved by stacks of regular planar layers, as shown in Fig. 4.3. Spheres in a single layer have hexagonal symmetry, i.e. they are in symmetrical contact with six spheres. The layers are stacked such that each sphere fits into the depression between three other spheres in the layer below.

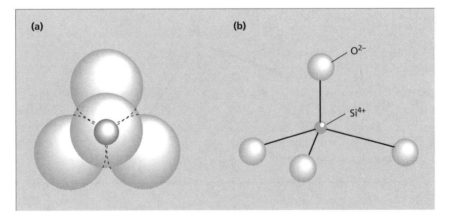

Fig. 4.2 Structure of SiO_4 tetrahedron. (a) Silicon and oxygen packing. The shaded silicon atom lies below the central oxygen atom, but above the three oxygens that lie in a single plane. (b) SiO_4 tetrahedron with bond length exaggerated.

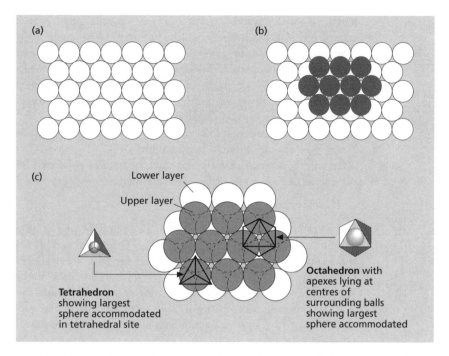

Fig. 4.3 (a) Spheres in planar layers showing hexagonal symmetry. (b) An upper layer of spheres (shaded) is stacked on the layer in (a), such that each upper sphere fits into the depression between three spheres in the lower layer. (c) Enlargement of (b), where heavy lines show coordination polyhedra, joining the centres of adjacent spheres, delineating two geometries, tetrahedra and octahedra. After McKie and McKie (1974) and Gill (1996), with kind permission of Kluwer Academic Publishers.

The gaps between neighbouring spheres have one of two possible three-dimensional geometries. The first geometry is delineated by the surfaces of four adjacent spheres. A three-dimensional shape constructed from the centre of each adjacent sphere (Fig. 4.3) has the form of a tetrahedron; consequently these gaps are called tetrahedral sites. The second type of gap is bounded by six adjacent spheres and a three-dimensional shape constructed from the centre of these spheres has the form of a regular octahedron. These are called octahedral sites. In ionic crystals, cations occupy some of these tetrahedral and octahedral sites. The type of site a cation occupies is determined by the radius ratio of the cation and anion, i.e.:

$$\text{Radius ratio} = r_{\text{cation}}/r_{\text{anion}} \qquad \text{eqn. 4.1}$$

where r = ionic radius.

To fit exactly into an octahedral site delineated by six spheres of radius r, a cation must have a radius of $0.414r$. With this radius ratio the cation touches all six of the surrounding anions in octahedral coordination. The short distance between ions means that the bond length is short and strong (optimum bond length). In real crystals, radius ratios are usually smaller or larger than this critical value of 0.414. If smaller, the optimum bond length is exceeded, and the structure collapses into a new stable configuration where the cation maintains optimum bond length with fewer, more closely packed anions. If the radius ratio is larger than 0.414, octahedral coordination is maintained, but the larger cation prevents the anions from achieving their closest possible packing. The upper limit for octahedral coordination is the next critical radius ratio of 0.732, at which point the cation is large enough to simultaneously touch eight equidistant anion neighbours, reachieving optimum bond length.

In silicate minerals the layered stack of spheres is formed by oxygen anions (O^{2-}) and the radius ratio rule can be defined as:

$$\text{Radius ratio} = r_{\text{cation}}/rO^{2-} \qquad \text{eqn. 4.2}$$

Radius ratio values relative to O^{2-} are given in Table 4.3. The table shows that silicon (Si) exists in four-fold (tetrahedral) coordination with oxygen (O), i.e. it will fit into a tetrahedral site. This explains the existence of the SiO_4 tetrahedron. Octahedral sites, being larger than tetrahedral sites, accommodate cations of larger radius. However, some cations, for example strontium (Sr^{2+}) and caesium (Cs^+) (radius ratio >0.732), are too big to fit into octahedral sites. They exist in eight-fold or 12-fold coordination and usually require minerals to have an open, often cubic, structure.

Table 4.3 Radius ratio values for cations relative to O^{2-}. From Raiswell *et al.* (1980).

Critical radius ratio	Predicted coordination	Ion	Radius ratio r_k/rO^{2-}	Commonly observed coordination numbers
	3	C^{4+}	0.16	3
	3	B^{3+}	0.16	3, 4
0.225				
	4	Be^{2+}	0.25	4
	4	Si^{4+}	0.30	4
	4	Al^{3+}	0.36	4, 6
0.414				
	6	Fe^{3+}	0.46	6
	6	Mg^{2+}	0.47	6
	6	Li^+	0.49	6
	6	Fe^{2+}	0.53	6
	6	Na^+	0.69	6, 8
	6	Ca^{2+}	0.71	6, 8
0.732				
	8	Sr^{2+}	0.80	8
	8	K^+	0.95	8–12
	8	Ba^{2+}	0.96	8–12
1.000				
	12	Cs^+	1.19	12

The radius ratio rule is only applicable to ionic compounds. In silicate minerals, however, it is the bonds between oxygen and silicon and between oxygen and aluminium (Al) which are structurally important. These bonds are almost equally ionic and covalent in character and the radius ratio rule predicts the coordination of these ions adequately.

4.2.2 The construction of silicate minerals

The SiO_4 tetrahedron has a net 4^- charge, since silicon has a valency of 4^+, and each oxygen is divalent (2^-). This means that the silicon ion (Si^{++}) can satisfy only half of the bonding capacity of its four oxygen neighbours. The remaining bonds are used in one of two ways as silicates crystallize from magma:

1 Some magmas are rich in elements which are attracted to the electronegative tetrahedral oxygen (Box 4.2). The bonds between these elements (e.g. magnesium (Mg)) and oxygen have ionic character (Box 4.2) and result in simple crystal structures, for example olivine (the magnesium-rich form is called forsterite) (Section 4.2.3). The cohesion of forsterite relies on the Mg^{2+}–SiO_4 ionic bond. Bonding *within* the SiO_4 tetrahedron has a more covalent character. During weathering, water, a polar solvent (Box 4.1), severs the weaker metal–SiO_4 tetrahedron ionic bond, rather than bonds within the tetrahedron itself. This releases metals and free SiO_4^+ as silicic acid (H_4SiO_4).

2 In some magmas, electropositive elements (opposite behaviour to electronegative elements) like magnesium are scarce. In these magmas each oxygen ion is likely to bond to two silicon ions, forming bonds of covalent character. The formation of extended networks of silicon–oxygen is called polymerization, and is used to classify structural organization in silicate minerals (Section 4.2.3).

4.2.3 Structural organization in silicate minerals

Silicate minerals are classified by the degree to which silicon–oxygen bonded networks (polymers) form (Fig. 4.4). The degree of polymerization is measured by the number of non-bridging oxygens (i.e. those bonded to just one Si^{++}).

Monomer silicates

These are built of isolated SiO_4 tetrahedra, bonded to metal cations as in olivine (Fig. 4.4b) and garnet. The basic unit of the polymer, the SiO_4 tetrahedra, is uncombined or single (mono), giving rise to the term monomer. These minerals have four non-bridging oxygens and are also known as orthosilicates.

Chain silicates

If each SiO_4 tetrahedron shares two of its oxygens, chains of linked tetrahedra form (Fig. 4.4c). Chain silicates have two non-bridging oxygens and an overall Si:O ratio of 1:3, giving the general formula SiO_3. The pyroxene group of minerals provides the most important chain silicates—for example, enstatite

Box 4.2 Electronegativity

Electronegativity is a measure of the tendency of an atom to attract an additional electron. It is used as an index of the covalent (see Section 2.3.1) or ionic (see Section 2.3.2) nature of bonding between two atoms. Atoms with identical electronegativity, or molecules such as nitrogen (N_2) consisting of two identical atoms, share their bonding electrons equally and so form pure covalent bonds. When component atoms in a compound are dissimilar, the bonds may become progressively polar. For example, in hydrogen chloride (HCl) the chlorine (Cl) atoms have a strong affinity for electrons, which are to a small extent attracted away from the hydrogen (H) towards chlorine. The bonding

electrons are still shared, but not equally as in N_2.

H   : Cl   N   :   N   (: = bonding electrons)
(polar bond)   (covalent bond)

Hence the chlorine atom carries a slight negative charge and the hydrogen atom a slight positive charge. Extreme polarization means that the bond becomes ionic in character. A bond is considered ionic if it has more than 50% ionic character.

Elements that donate electrons (e.g. magnesium, calcium, sodium and potassium) rather than attract them are called electropositive.

Table 1 Partial list of electronegativities and percentage ionic character of bonds with oxygen.

Ion	Electro-negativity	% Ionic character	Ion	Electro-negativity	% Ionic character	Ion	Electro-negativity	% Ionic character
Cs^+	0.7	89	Zn^{2+}	1.7	63	P^{5+}	2.1	35
K^+	0.8	87	Sn^{2+}	1.8	73	Au^{2+}	2.4	62
Na^+	0.9	83	Pb^{2+}	1.8	72	Se^{2-}	2.4	—
Ba^+	0.9	84	Fe^{2+}	1.8	69	C^{4+}	2.5	23
Li^+	1.0	82	Si^{4+}	1.8	48	S^{2-}	2.5	—
Ca^{2+}	1.0	79	Fe^{3+}	1.9	54	I^-	2.5	—
Mg^{2+}	1.2	71	Ag^+	1.9	71	N^{5+}	3.0	9
Be^{2+}	1.5	63	Cu^+	1.9	71	Cl^-	3.0	—
Al^{3+}	1.5	60	B^{3+}	2.0	43	O^{2-}	3.5	—
Mn^{2+}	1.5	72	Cu^{2+}	2.0	57	F^-	4.0	—

Measurements of % ionic character are not applicable for anions since their bonds with oxygen are predominantly covalent.

($MgSiO_3$). As in monomer silicates, bonding within chains is stronger than bonding between chains, which is between metal ions and non-bridging oxygens.

Double-chain silicates

In this structure the single chains are cross-linked, such that alternate tetrahedra share an oxygen with the neighbouring chain (Fig. 4.4d). Consequently, this structure has 1.5 non-bridging oxygens, since, for every four tetrahedra, two share two oxygens and the other two share three oxygens. The overall Si:O

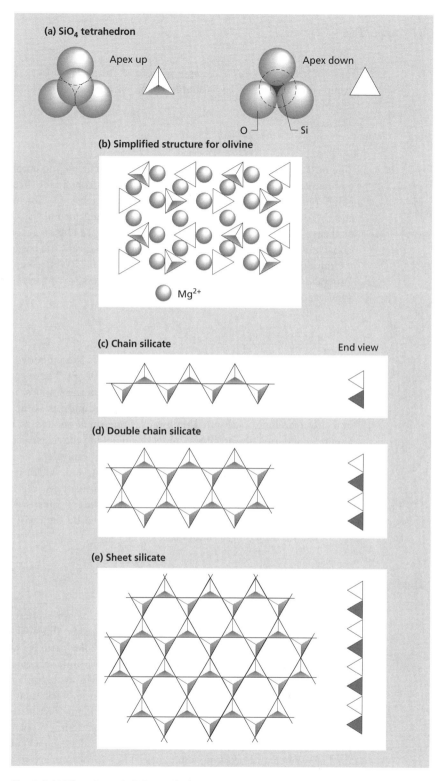

Fig. 4.4 (a) The schematic SiO_4 tetrahedron in Fig. 4.2a can be represented as a tetrahedron, each tip representing the position of oxygen anions. The sketches represent monomer structure in (b) and progressive polymerization of adjacent tetrahedra to form (c) chains, (d) cross-linked double chains and (e) sheets. The cross-linked structures form hexagonal rings which can accommodate anions such as OH^-. After Gill (1996), with kind permission of Kluwer Academic Publishers.

ratio is therefore $4:11$, giving a general formula Si_4O_{11}. The amphibole group of minerals has double-chain structure—for example, tremolite ($Ca_2Mg_5Si_8O_{22}(OH)_2$).

Sheet silicates

The next step in polymerization is to cross-link chains into a continuous, semi-covalently bonded sheet, such that every tetrahedron shares three oxygens with neighbouring tetrahedra (Fig. 4.4e). This structure has one non-bridging oxygen and the overall $Si:O$ ratio is $4:10$, giving a general formula Si_4O_{10}. The hexagonal rings formed by the cross-linkage of chains are able to accommodate additional anions, usually hydroxide (OH^-). This structure is the basic framework for the mica group—for example, muscovite ($Mg_3(Si_4O_{10})(OH)_4$)—and all of the clay minerals. These minerals are thus stacks of sheets, giving rise to their 'platy' appearance.

Framework silicates

In this class of silicates every tetrahedral oxygen is shared between two tetrahedra, forming a three-dimensional semicovalent network. There are no non-bridging oxygens, the overall $Si:O$ ratio being $1:2$, as in the simplest mineral formula of the class, quartz (SiO_2). Substitution of aluminium into some of the tetrahedral sites (the ionic radius of aluminium is just small enough to fit) gives rise to a huge variety of aluminosilicate minerals, including the feldspar group, the most abundant mineral group in the crust. Substituting tetravalent silicon for trivalent aluminium causes a charge imbalance in the structure, which is neutralized by the incorporation of other divalent or monovalent cations. For example, in the feldspar orthoclase ($KAlSi_3O_8$), one in four tetrahedral sites is occupied by aluminium in place of silicon. The charge is balanced by the incorporation of one K^+ for each tetrahedral aluminium.

4.3 Weathering processes

The surface of the continental crust is exposed to the atmosphere, making it vulnerable to physical, biological and chemical processes. Physical weathering is a mechanical process which fragments rock into smaller particles without substantial change in chemical composition. When the confining pressure of the crust is removed by uplift and erosion, internal stresses within the underlying rocks are removed, allowing expansion cracks to open. These cracks may then be prised apart by thermal expansion (caused by diurnal fluctuations in temperature), by the expansion of water upon freezing and by the action of plant roots. Other physical processes, for example glacial activity, landslides and sandblasting, further weaken and break up solid rock. These processes are important because they vastly increase the surface area of rock material exposed to the agents of chemical weathering, i.e. air and water.

Chemical weathering is caused by water—particularly acidic water—and gases, for example oxygen, which attack minerals. However, many of these reactions are catalysed by bacteria (biological weathering). Chemoautotrophic bacteria, for example *Thiobacillus*, derive their energy directly from the weathering of minerals (see Section 5.4.2). Some ions and compounds of the original mineral are removed in solution, percolating through the mineral residue to feed groundwater and rivers. Fine-grained solids may be washed from the weathering site, leaving a chemically modified residue, the basis of soils. We can view weathering processes—physical, biological and chemical weathering usually occur together —as the adjustment of rocks and minerals formed at high temperatures and pressures to Earth surface conditions of low temperature and pressure. The mineralogical changes occur to regain stability in a new environment.

4.4 Mechanisms of chemical weathering

Different mechanisms of chemical weathering are recognized and various combinations of these occur together during the breakdown of most rocks and minerals.

4.4.1 Dissolution

The simplest weathering reaction is the dissolution of soluble minerals. The water molecule (Box 4.1) is effective in severing ionic bonds (see Section 2.3.2), such as those that hold sodium (Na^+) and chlorine (Cl^-) ions together in halite (rock salt). We can express the dissolution of halite in a simple way, i.e.:

$$NaCl_{(s)} \overset{H_2O}{\rightleftharpoons} Na^+_{(aq)} + Cl^-_{(aq)} \qquad \text{eqn. 4.3}$$
(halite)

This reaction shows the dissociation (breaking of an entity into parts) of halite into free ions, forming an electrolyte solution. The reaction does not contain hydrogen ions (H^+), showing that the process is independent of pH.

4.4.2 Oxidation

Free oxygen is important in the decomposition of reduced materials (Box 4.3), typically iron, sulphur and organic matter. For example, the oxidation of reduced iron (Fe^{2+}) and sulphur (S) in the common sulphide, pyrite (FeS_2), results in the formation of sulphuric acid (H_2SO_4), a strong acid:

$$2FeS_{2(s)} + 7\tfrac{1}{2}O_{2(g)} + 7H_2O_{(l)} \rightarrow 2Fe(OH)_{3(s)} + 4H_2SO_{4(aq)} \qquad \text{eqn. 4.4}$$

Where mining exposes large amounts of pyrite to weathering, the generation of this acid creates significant environmental problems (see Section 5.4.2).

Reduced iron-bearing silicate minerals, for example some olivines, pyroxenes and amphiboles, may also undergo oxidation, as depicted for the iron-rich olivine, fayalite:

Box 4.3 Oxidation and reduction (redox)

Oxidation and reduction (redox) reactions are driven by electron transfers (see Section 2.2). Thus the oxidation of iron by oxygen:

$$4Fe_{(metal)} + 3O_{2(g)} \rightarrow 2Fe_2O_{3(s)} \qquad \text{eqn. 1}$$

can be considered to consist of two half-reactions:

$$4Fe - 12e^- \rightarrow 4Fe^{3+} \qquad \text{eqn. 2}$$

$$3O_2 + 12e^- \rightarrow 6O^{2-} \qquad \text{eqn. 3}$$

where e^- represents one electron.

Oxidation involves the loss of electrons and reduction involves the gain of electrons

In equations 1–3, oxygen—the oxidizing agent (also called an electron acceptor)—is reduced because it gains electrons.

Equation 1 shows that elements like oxygen can exist under set conditions of temperature and pressure in more than one state, i.e. as oxygen gas and as an oxide. The oxidation state of an element in a compound is assigned using the following rules:
1 The oxidation number of all elements is 0.
2 The oxidation number of a monatomic ion is equal to the charge on that ion, for example: $Na^+ = Na(+1)$, $Al^{3+} = Al(+3)$, $Cl^- = Cl(-1)$.
3 Oxygen has an oxidation number of –2 in all compounds except O_2, peroxides and superoxides.
4 Hydrogen has an oxidation number of +1 in all compounds.
5 The sum of the oxidation numbers of the elements in a compound or ion equals the charge on that species.
6 The oxidation number of the elements in a covalent compound can be deduced by considering the shared electrons to belong exclusively to the more electronegative atom (see Box 3.2). Where both atoms have the same electronegativity the electrons are considered to be shared equally. Thus the oxidation numbers of carbon and chloride in CCl_4 are +4 and –1 respectively and the oxidation number of chlorine in Cl_2 is 0.

Oxidation states are important when predicting the behaviour of elements or compounds. For example, chromium is quite insoluble and non-toxic as chromium (III), while as chromium (VI) it forms the soluble complex anion CrO_4^{2-}, which is toxic. As with most simple rules, those for oxidation state assignment apply to most but not all compounds.

Since redox half-reactions involve electron transfer, they can be measured electrochemically as electrode potentials, which are a measure of energy transfer (Box 4.8). The reaction:

$$\tfrac{1}{2}H_2 - e^- \rightarrow H^+ \qquad \text{eqn. 4}$$

is assigned an electrode potential ($E°$) of zero (at standard temperature and pressure) by international agreement. All other electrode potentials are measured relative to this value and are readily available as tabulated values in geochemical texts and on the Internet.

A positive $E°$ shows that the reaction proceeds spontaneously (e.g. the reduction of fluorine gas (oxidation state 0) to fluoride (F^-, oxidation state –1). A negative $E°$ shows that the reaction is spontaneous in the reverse direction (e.g. the oxidation of Li to Li^+).

To calculate the overall $E°$ for a reaction the relevant half-reactions are combined (regardless of the stoichiometry of the reactions). For example, the reaction of Sn^{2+} solution with Fe^{3+} solution involves two half-reactions:

$$Fe^{3+} + e^- \rightarrow Fe^{2+} \quad E° = 0.77\,V \text{ (tabulated value)} \qquad \text{eqn. 5}$$

$$Sn^{4+} + 2e^- \rightarrow Sn^{2+} \quad E° = 0.15\,V \text{ (tabulated value)} \qquad \text{eqn. 6}$$

These combine to give a positive $E°$, which shows that the forward reaction (eqn. 7) is favoured:

$$2Fe^{3+} + Sn^{2+} \rightarrow 2Fe^{2+} + Sn^{4+} \qquad \text{eqn. 7}$$

$E°$ for this reaction $= 0.77 - 0.15 = 0.62\,V$.

The ability of any natural environment to bring about oxidation or reduction processes is measured by a quantity called its *redox potential* or Eh (see Box 5.4).

$$Fe_2SiO_{4(s)} + \tfrac{1}{2}O_{2(g)} + 5H_2O_{(l)} \rightarrow 2Fe(OH)_{3(s)} + H_4SiO_{4(aq)} \qquad \text{eqn. 4.5}$$
(fayalite)
(Fe(II))   (Fe(III))

The products are silicic acid (H_4SiO_4; see below) and colloidal hydrated iron oxide ($Fe(OH)_3$), a weak alkali, which dehydrates to yield a variety of iron oxides, for example Fe_2O_3 (haematite—dull red colour) and $FeOOH$ (goethite and lepidocrocite—yellow or orange-brown colour). The common occurrence of these iron oxides reflects their insolubility under oxidizing Earth surface conditions.

The water in equation 4.5 acts as an intermediary to speed up oxidation, as shown by the everyday experience of metallic iron oxidation (rusting). In this role water acts as a kind of catalyst (Box 4.4). The oxidation potential of equation 4.5 depends on the partial pressure of gaseous oxygen and the acidity of the solution. At pH 7, water exposed to air has an Eh of 810 mV (Box 4.3), an oxidation potential well above that necessary to oxidize ferrous iron.

Oxidation of organic matter

The oxidation of reduced organic matter in soils is catalysed (Box. 4.4) by heterotrophic microorganisms. Bacterially mediated oxidation of organic matter to CO_2 is important because it generates acidity. In biologically active (biotic) soils, CO_2 may be concentrated by 10–100 times the amount expected from equilibrium with atmospheric CO_2, yielding carbonic acid (H_2CO_3) and H^+ via dissociation; to simplify the equations organic matter is represented by the generalized formula for carbohydrate, CH_2O (see Section 2.7.2):

$$CH_2O_{(s)} + O_{2(g)} \longrightarrow CO_{2(g)} + H_2O_{(l)} \qquad \text{eqn. 4.6}$$

$$CO_{2(g)} + H_2O_{(l)} \rightleftharpoons H_2CO_{3(aq)} \qquad \text{eqn. 4.7}$$

$$H_2CO_{3(aq)} \rightleftharpoons H^+_{(aq)} + HCO^-_{3(aq)} \qquad \text{eqn. 4.8}$$

These reactions may lower soil water pH from 5.6 (its equilibrium value with atmospheric CO_2; see Box 3.5) to 4–5. This is a simplification since soil organic matter (humus) is not often completely degraded to CO_2. The partial breakdown products, however, possess carboxyl (COOH) or phenolic functional groups (see Section 2.7.1), which dissociate (Box 4.5) to yield H^+ ions that lower the pH even more:

$$RCOOH_{(s)} \rightarrow RCOO^-_{(aq)} + H^+_{(aq)} \qquad \text{eqn. 4.9}$$

$$+ \; H^+ \qquad \text{eqn. 4.10}$$

Phenol   Phenate

Box 4.4 Metastability, reaction kinetics, activation energy and catalysts

Some reduced compounds appear to be stable at Earth surface temperatures despite the presence of atmospheric oxygen. Graphite, for example, is a reduced form of carbon which we might expect to react with oxygen, i.e.:

$$C_{graphite} + O_{2(g)} \rightarrow CO_{2(g)} \qquad \text{eqn. 1}$$

Although the reaction of oxygen with graphite is energetically favoured, graphite exists because the reaction is kinetically very slow. Many natural materials are out of equilibrium with their ambient environment and are reacting imperceptibly slowly. These materials are metastable. Metastability can be demonstrated using a graph of energy in a chemical system in which substances A and B react to give C and D (Fig. 1). In order for reaction to take place, A and B must come into close association and this usually requires an input of energy (activation energy). Under cold (low-energy) conditions a small number of A and B will occasionally have the energy to overcome the activation energy, but this

will be rare and the reaction will proceed slowly. If the energy of the reactants is increased (e.g. by heating), then the reaction will be able to proceed more quickly because more A and B will have the required activation energy.

An alternative way to increase the rate of a reaction is to lower the activation energy. This is done by a catalyst—a substance that alters the rate of a chemical reaction without itself undergoing any overall chemical change. In our hypothetical reaction, the catalyst acts as an intermediate compound that allows A and B to come together more readily. In the environment biological enzyme systems—especially those of microorganisms—catalyse reactions that would not otherwise proceed spontaneously because of kinetic inhibition. Water, strictly speaking, is not a chemical catalyst; however its solvent properties provide a kind of catalytic role as an intermediary, allowing closer interaction of ions and substances than is possible under dry conditions.

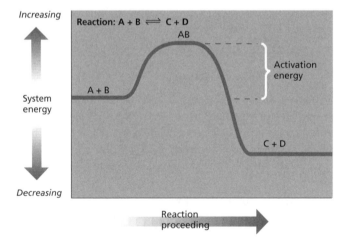

Fig. 1 Schematic representation of energy in a chemical system.

Box 4.5 Dissociation

Dissociation and dissociation constants (K_d and K_a)

Many chemical components can be described as 'ionizable' meaning they can dissociate to form charged species (ions). In order to describe the degree to which compounds are susceptible to ionization, dissociation constants (K_d) are derived at equilibrium. These dissociation constants are a ratio of dissociated species to undissociated species. Thus, the larger the value of K_d the greater the proportion of dissociated species at equilibrium. Where protons are generated the term K_d can be replaced with the K_a (the acid dissociation constant, see Box 3.3). Thus, K_a provides an indication of a given acid's strength. In general:

$$K_a = \frac{cA^- \cdot cH^+}{cHA}$$ eqn. 1

where cHA is the concentration of the undissociated species, cA^- is the concentration of the dissociated base and cH^+ is the concentration of dissociated protons.

For example, phenol dissociates to form phenate ions and a proton:

| Phenol | Phenate | Proton |

eqn. 2

For this equilibrium the dissociation constant is derived using the expression:

$$K_d = K_a = \frac{cPhenate \cdot cH^+}{cPhenol}$$ eqn.3

In the case of phenol the value of $K_a = 1.1 \times 10^{-10}$. This very small value shows that phenol is a weak acid as there is a far greater proportion of undissociated phenol than there is of phenate ions and protons. It is more common to convert K_a values to a negative log scale (pK_a), similar to that for pH (see Box 3.5), i.e.:

$$pK_a = -\log_{10} K_a$$ eqn. 4

Thus, for phenol:

$$pK_a = -\log_{10} 1.1 \times 10^{-10} = 9.96$$ eqn. 5

Dissociation of a hydroxyl functional group is dependent upon the nature of the rest of the molecule to which it is attached.

Figure 1 indicates how pK_a values vary for phenols containing different additional functional groups (see Section 2.7.1). In the case of o-nitrophenol the presence of the nitro functional group ($-NO_2$) increases the pK_a value, making it more acidic than phenol. In the case of o-cresol the presence of the methyl functional group ($-CH_3$) decreases the pK_a value, making it less acidic than phenol.

These differences are due to the electron-withdrawing or electron-donating nature of the functional groups. The electron-withdrawing/donating nature of a functional group is governed by the electronegativity (Box 4.2) of the atoms comprising the function group, or more specifically by the polarity of the bonds between these atoms. The different pK_a values in Fig. 1 are related to the stability of the phenate ion that forms as a product of dissociation (Fig. 2). In o-cresol the negative charge can not be delocalized (see Section 2.7) onto the methyl functional group (as it is electron-donating). Thus, delocalization of the negative charge is only possible over the aromatic ring. In o-nitrophenol the negative charge of the phenate ion can be drawn over a greater number of atoms as a result of the nitro functional group being electron-withdrawing. Thus, the negative charge can be delocalized over both the aromatic ring and the nitro functional group. It is this greater delocalization of the negative charge in o-nitrophenol that increases the stability of the phenate ion. As a consequence of this greater stability the dissociation of the OH-group on o-nitrophenol is energetically more favourable then the dissociation of the OH-group on o-cresol, making o-nitrophenol more acidic than o-cresol.

(continued on p. 82)

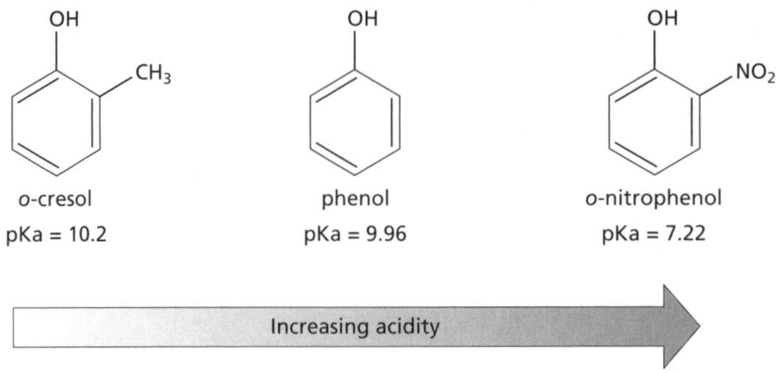

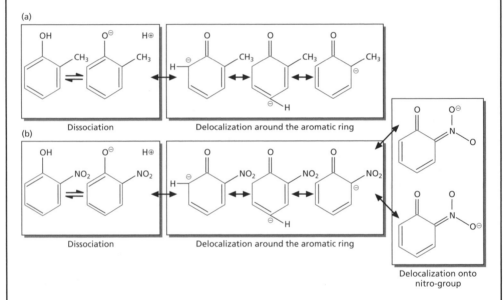

o-cresol
pKa = 10.2

phenol
pKa = 9.96

o-nitrophenol
pKa = 7.22

Increasing acidity

Fig. 1 Variation in pK_a values for phenols containing different additional functional groups.

(a)

Dissociation

Delocalization around the aromatic ring

(b)

Dissociation

Delocalization around the aromatic ring

Delocalization onto
nitro-group

Fig. 2 Delocalization of negative charge in the phenate ion where (a) methyl and (b) nitro functional groups are present.

Dissociation and pH

As dissociation is an equilibrium process it is logical that where parameters influence equilibrium the degree of dissociation will also alter. Thus, pH has a huge bearing on the extent to which ionizable species dissociate since pH is a measure of free proton concentrations (or more correctly the free proton activity) (see Section 2.6).

Re-arranging equation 1:

$$cH^+ = K_a \cdot \frac{cHA}{cA^-} \qquad \text{eqn. 6}$$

Taking the log of this equation yields:

$$\log_{10} cH^+ = \log_{10} K_a + \log_{10} \frac{cHA}{cA^-} \qquad \text{eqn. 7}$$

multiplying this equation by –1 yields:

$$-\log_{10} cH^+ = -\log_{10} K_a - \log_{10} \frac{cHA}{cA^-} \qquad \text{eqn. 8}$$

and thus:

$$pH = pK_a + \log_{10} \frac{cA^-}{cHA} \qquad \text{eqn. 9}$$

(continued)

This is the *Henderson-Hasselbalch* equation; it indicates the relationship between pH and pK_a. Notice that where the concentration of undissociated acid cHA and the concentration of its dissociated base cA^- are equal then pH $= pK_a$. Thus, the pK_a value is the pH at which there is an equal proportion of dissociated and undissociated acid.

The relationship between extent of dissociation and pH is important because pH can vary sizeably in natural environments. For example, variability in soil pH is marked, suggesting that the behaviour of ionizable species within soils will be different also.

where R denotes aliphatic or aromatic hydrocarbon groups (see Section 2.7). The acidity generated by organic matter decomposition is used to break down most silicate minerals by the process of acid hydrolysis.

4.4.3 Acid hydrolysis

Continental water contains dissolved species that render it acidic. The acidity comes from a variety of sources: from the dissociation of atmospheric CO_2 in rainwater—and particularly from dissociation of soil-zone CO_2 (Section 4.4.2)—to form H_2CO_3, and natural and anthropogenic sulphur dioxide (SO_2) to form H_2SO_3 and H_2SO_4 (see Boxes 3.7 & 3.8). Reaction between a mineral and acidic weathering agents is usually called acid hydrolysis. The weathering of $CaCO_3$ demonstrates the chemical principle involved:

$$CaCO_{3(s)} + H_2CO_{3(aq)} \rightleftharpoons Ca^{2+}_{(aq)} + 2HCO^-_{3(aq)} \qquad \text{eqn. 4.11}$$

The ionic Ca–CO_3 bond in the calcite crystal is severed and the released CO_3^{2-} anion attracts enough H^+ away from the H_2CO_3 to form the stable bicarbonate ion HCO_3^-. Note that the second HCO_3^- formed in equation 4.11 is left over when H^+ is removed from H_2CO_3. Bicarbonate is a very weak acid, since it dissociates very slightly into H^+ and CO_3^{2-}, but it is not quite dissociated enough to react with carbonate. Overall, the reaction neutralizes the acid contained in water. The reaction is dependent on the amount of CO_2 available: adding CO_2 causes the formation of more H_2CO_3 (eqn. 4.7), which dissolves more $CaCO_3$ (forward reaction in eqn. 4.11); conversely, lowering the amount of CO_2 encourages the reverse reaction and precipitation of $CaCO_3$. Stalactites and stalagmites forming in caves are an example of $CaCO_3$ precipitation induced by the degassing of CO_2 from groundwater. This response to varying CO_2 is a clear example of Le Chatelier's Principle (see Box 3.2).

Acid hydrolysis of a simple silicate, for example the magnesium-rich olivine, forsterite, is summarized by:

$$Mg_2SiO_{4(s)} + 4H_2CO_{3(aq)} \rightarrow 2Mg^{2+}_{(aq)} + 4HCO^-_{3(aq)} + H_4SiO_{4(aq)} \qquad \text{eqn. 4.12}$$

Note that the dissociation of H_2CO_3 forms the ionized HCO_3^-, which is a slightly stronger acid than the neutral molecule (H_4SiO_4) released by the destruction of the silicate.

The combined effects of dissolving CO_2 into soilwater (eqn. 4.7), the subsequent dissociation of H_2CO_3 (eqn. 4.8) and the production of HCO_3^- by acid hydrolysis weathering reactions (eqns. 4.11 & 4.12) mean that surface waters have near-neutral pH, with HCO_3^- as the major anion.

4.4.4 Weathering of complex silicate minerals

So far we have considered the weathering of monomer silicates (e.g. olivine), which dissolve completely (congruent solution). This has simplified the chemical reactions. The presence of altered mineral residues during weathering, however, suggests that incomplete dissolution is more usual. Upper crustal rocks have an average composition similar to the rock granodiorite (Table 4.4). This rock is composed of the framework silicates, plagioclase feldspar, potassium feldspar and quartz (Table 4.4), plagioclase feldspar being most abundant. Thus, a simplified weathering reaction for plagioclase feldspar might best represent average chemical weathering. We can illustrate this using the calcium (Ca)-rich plagioclase feldspar, anorthite.

$$CaAl_2Si_2O_{8(s)} + 2H_2CO_{3(aq)} + H_2O_{(l)} \rightarrow$$
$$Ca^{2+}_{(aq)} + 2HCO_{3(aq)} + Al_2Si_2O_5(OH)_{4(s)} \qquad \text{eqn. 4.13}$$

The formula of the solid product ($Al_2Si_2O_5(OH)_4$) is that for kaolinite, an important member of the serpentine–kaolin group of clay minerals (Section 4.5.1). This reaction demonstrates incongruent dissolution of feldspar, i.e. dissolution with *in situ* reprecipitation of some compounds from the weathered mineral.

Schematic representation of the anorthite chemical weathering reaction (Fig. 4.5) shows the edge of the anorthite crystal in contact with an H_2CO_3 weathering solution. Natural crystal surfaces have areas of excess electrical charge at their corners and edges, usually because of imperfections (rows of atoms slightly out

Table 4.4 Percentage mineral composition of the upper continental crust. After Nesbit and Young (1984).

	Average upper continental crust	Average exposed continental crust surface
Plagioclase feldspar	39.9	34.9
Potassium feldspar	12.9	11.3
Quartz	23.2	20.3
Volcanic glass	—	12.5
Amphibole	2.1	1.8
Biotite mica	8.7	7.6
Muscovite mica	5.0	4.4
Chlorite	2.2	1.9
Pyroxene	1.4	1.2
Olivine	0.2	0.2
Oxides	1.6	1.4
Others	3.0	2.6

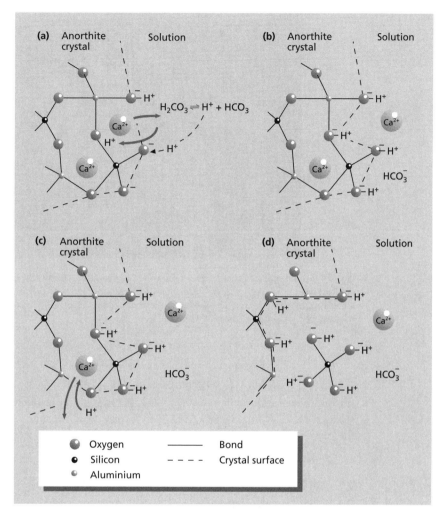

Fig. 4.5 Weathering reactions at the surface of a feldspar (after Raiswell *et al*. 1980). (a) Broken bonds become protonated by H^+ dissociated from carbonic acid and ionic-bonded Ca^{2+} is released to solution. (b) Protonated lattice. (c) Further severing of ionic bonds causes complete protonation of the edge tetrahedron. (d) Edge tetrahedron is completely removed to solution as H_4SiO_4.

of place) or damage (broken bonds). Areas of excess negative charge are preferentially attacked by soil acids, resulting in the formation of etch pits on the mineral surface (Fig. 4.6). Hydrogen ions dissociated from H_2CO_3 hydrate the silicate surface. The ionic bonds between Ca^{2+} and SiO_4 tetrahedra are easily severed, releasing Ca^{2+} into solution. The result is a metal-deficient hydrated silicate and a calcium bicarbonate ($Ca^{2+} + 2HCO_3^-$) solution. Continued reaction may break the more covalent bonds within the tetrahedral framework. The

Fig. 4.6 Scanning electron micrograph showing square-shaped etch pits developed on dislocations in a feldspar from a southwestern England granite. Note that in places the pits are coalescing, causing complete dissolution of the feldspar. Scale bar = 10 μm. Photograph courtesy of ECC International, St Austell, UK.

tetrahedral framework is particularly weak where aluminium has substituted for silicon, since the aluminium–oxygen bond has more ionic character. The product released to solution is H_4SiO_4 (Fig. 4.5). Equation 4.14 expresses quantitatively the reaction for the sodium (Na)-rich feldspar, albite.

$$2NaAlSi_3O_{8(s)} + 9H_2O_{(l)} + 2H_2CO_{3(aq)} \rightarrow$$
$$Al_2Si_2O_5(OH)_{4(s)} + 2Na^+_{(aq)} + 2HCO^-_{3(aq)} + 4H_4SiO_{4(aq)} \qquad \text{eqn. 4.14}$$

We might conclude that the dominant weathering mechanism of the upper crust is acid hydrolysis, resulting in a partially degraded and hydrated residue, with silicic acid and metal bicarbonate dissolved in water. We should, however, remember that much of the Earth's continental area is mantled by younger sedimentary rocks, including relatively soluble ones, for example limestone. Limestones are common in young mountain belts, such as the European Alps and the Himalayas, where rates of physical weathering are high. It is therefore probable that average weathering reactions are really biased toward the weathering of the sediment cover, rather than average continental crust. Studies of Alpine weathering seem to confirm this. Many Alpine streamwaters are low in dissolved sodium and H_4SiO_4 but high in calcium and HCO^-_3. These results suggest that the dissolution of limestone—not feldspar—is the locally important weathering reaction.

4.5 Clay minerals

The reactions to illustrate weathering of complex silicates during acid hydrolysis (eqns. 4.13 & 4.14) predict that clay minerals will be an important solid product and this is confirmed by looking at soils. Clay minerals are important constituents in most soils. These sheet silicates that are less than $2\,\mu m$ (Section 4.2.3) are constructed of layers of atoms in tetrahedral and octahedral coordination, known as tetrahedral and octahedral sheets.

The tetrahedral sheets are layers of SiO_4 tetrahedra which share three oxygens with neighbouring tetrahedra. These basal oxygens form a hexagonal pattern (Section 4.2.3). The fourth tetrahedral (apical) oxygen of each tetrahedron is arranged perpendicular to the basal sheet (Fig. 4.7a). The sheet carries a net negative charge.

The octahedral sheet is composed of cations, usually aluminium, iron or magnesium, arranged equidistant from six oxygen (or OH) anions (Fig. 4.7b). Aluminium is the common cation and the ideal octahedral sheet has the composition of the aluminium hydroxide mineral, gibbsite ($Al(OH)_3$). Where octahedral sites are filled by trivalent aluminium, only two of every three sites are occupied to

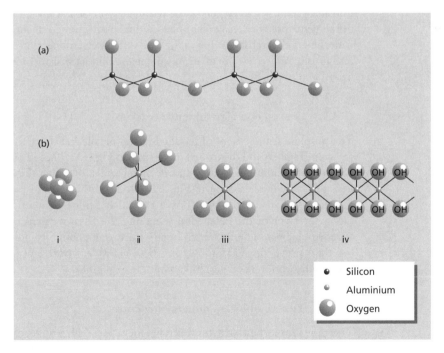

Fig. 4.7 (a) A sheet of SiO_4 tetrahedra linked via basal oxygens, with apical oxygens pointing upward. (b) Octahedra and the octahedral sheet: (i) the atoms packed together; (ii) the octahedron expanded; (iii) conventional representation of an octahedron; (iv) conventional representation of an octahedral sheet, showing aluminium equidistant between six hydroxyls — forming the mineral gibbsite.

Table 4.5 Simplified classification of clay minerals. After Martin *et al.* (1991) with kind permission from the Clay Minerals Society.

Layer type	Group	Common minerals	Octahedral character	Interlayer material
1:1	Serpentine–kaolin	Kaolinite	Dioctahedral	None
2:1	Smectite	Montmorillonite	Dioctahedral	Hydrated exchangeable cations
	True (flexible) mica	Biotite	Trioctahedral	Non-hydrated monovalent cations
		Muscovite, illite	Dioctahedral	
	Chlorite	Chamosite	Trioctahedral	Hydroxide sheet

maintain electrical neutrality and the sheet is classified as dioctahedral (Table 4.5). Where divalent cations fill octahedral sites, all available sites are filled and the sheet is classified as trioctahedral (Table 4.5).

Combining these sheets gives the basic clay mineral structure. The combination allows the apical oxygen of the tetrahedral sheet and the OH groups lodged in the centre of the hexagonal holes of the basal tetrahedral sheet to be shared with the octahedral sheet (Fig. 4.8). The various clay mineral groups (Table 4.5) result from different styles of arrangement and mutual sharing of ions in the tetrahedral and octahedral sheets.

4.5.1 One to one clay mineral structure

The simplest arrangement of tetrahedral and octahedral sheets is a 1:1 layering, shown in Fig. 4.8 and developed more fully in Fig. 4.9. These 1:1 minerals comprise the serpentine-kaolin group of clay minerals, of which the mineral kaolinite is probably the best known (Fig. 4.10). In kaolinite, the 1:1 packages are held together by hydrogen bonds (Box 4.1), which bridge between OH groups of the upper layer of the octahedral sheet and the basal oxygens of the overlying tetrahedral sheet. The hydrogen bonds are strong enough to hold the 1:1 units together, preventing cations getting between layers (interlayer sites). Isomorphous substitution (Box 4.6) in kaolinite is negligible.

4.5.2 Two to one clay mineral structure

The other important structural arrangement is a 2:1 structure, comprising an octahedral layer, sandwiched between two tetrahedral sheets with apical oxygens pointing inward on each side of the octahedral sheet (Fig. 4.11). The mutual sharing of two layers of apical oxygens in the octahedral sheets implies a higher oxygen:OH ratio in the structure of the 2:1 vs. the 1:1 octahedral sheets. All of the other clay mineral groups share this structure, the most important being the

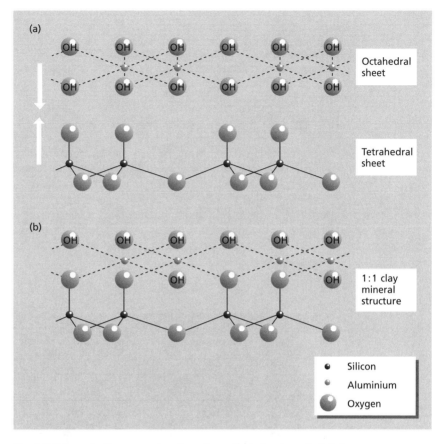

Fig. 4.8 Schematic diagram to show how the octahedral and tetrahedral sheets, seen as separate entities in (a), can be merged to form 1:1 clay mineral structure shown in (b).

mica group, which includes the common mica minerals and illite, the smectite group and the chlorite group.

Illite is a term used to describe clay-sized mica-type minerals and is not a specific mineral name; however, in general, illite composition is similar to muscovite mica (Fig. 4.11). In the muscovite structure, one of every four tetrahedral silicons is replaced by aluminium. The regular replacement of tetravalent silicon by trivalent aluminium means that the tetrahedral sheet in muscovite carries a strong net negative charge. Ideally, illite has dioctahedral structure, but some of the octahedral aluminium is substituted by Fe^{2+} and Mg^{2+} (Box 4.6), resulting in a net negative charge for the octahedral sheet. In total, the illite 2:1 unit has a strong net negative charge, known as the layer charge. This negative charge is neutralized by large cations, usually K^+, which sit between the 2:1 units and bond ionically with basal oxygens of the opposing tetrahedral sheets in six-fold coordination. The radius ratio rule predicts that K^+ should exist in eight-fold or

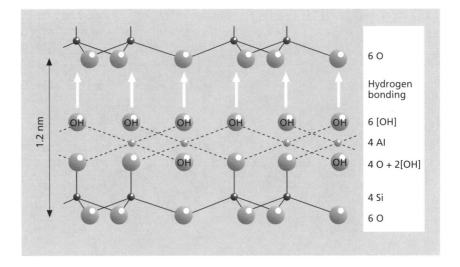

Fig. 4.9 The structure of a 1:1 clay mineral (kaolinite).

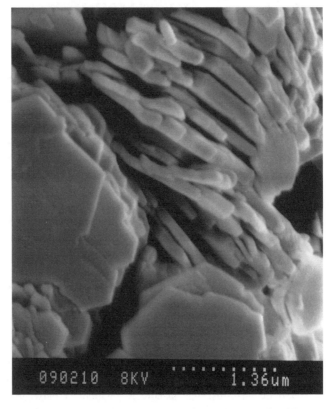

Fig. 4.10 Scanning electron microscope image of kaolinite showing plate-like crystals arranged in stacks. Scale bar = 1.36 μm. Photograph courtesy of S. Bennett.

Box 4.6 Isomorphous substitution

Isomorphism describes substances which have very similar structure. The carbonate mineral system is a good example, where some minerals differ only on the basis of the cation, for example, $CaCO_3$ (calcite), $MgCO_3$ (magnesite), $FeCO_3$ (siderite). The basic similarity of structure allows interchangeability of cations between end-member minerals. For example, most natural calcite has a measurable amount of both Mg^{2+} and Fe^{2+} substituted for some of the Ca^{2+}. The amount of isomorphous substitution is shown by the following notation, $(Ca_{0.85}Mg_{0.1}Fe_{0.05})CO_3$. In other words, 85% of the Ca^{2+} sites are occupied by Ca^{2+}, 10% of the Ca^{2+} sites are occupied by Mg^{2+}, and 5% of the Ca^{2+} sites are occupied by Fe^{2+}. The complex chemistry of freshwater and

seawater means that natural minerals incorporate many trace elements and rarely conform to their ideal formulae.

The radius ratio rule (Section 4.2.1) predicts that divalent Ca^{2+}, Mg^{2+} and Fe^{2+} will have six-fold coordination because of their similar ionic radii (0.106 nm Ca^{2+}, 0.078 nm Mg^{2+}, 0.082 nm Fe^{2+}). They are therefore interchangeable without upsetting either the physical packing or the electrical stability of an ionic compound.

In compounds where bonding has covalent character, isomorphous substitution is prevented. This is because the need for electron sharing in the bond modifies structures away from the simple packing geometries predicted by the radius ratio rule.

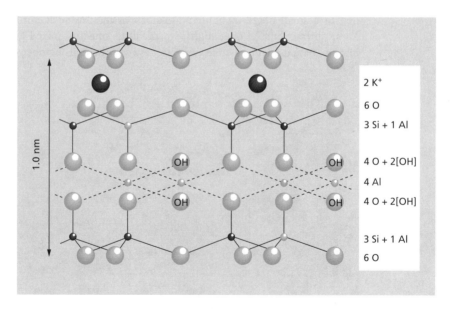

1.0 nm

2 K$^+$

6 O

3 Si + 1 Al

4 O + 2[OH]

4 Al

4 O + 2[OH]

3 Si + 1 Al

6 O

Fig. 4.11 The structure of muscovite mica.

12-fold coordination with oxygen (Section 4.2.1), but this does not occur, due to slight distortion in the illite structure.

It is important to note that bonding between 2:1 illite units cannot be fulfilled by hydrogen bonds associated with OH groups (as in kaolinite), since each

2:1 unit presents only basal tetrahedral oxygens on its outer surfaces. Moreover, the ionic bonding between K^+ in the interlayer site and the tetrahedral oxygens is a relatively strong bond, making illitic clays stable minerals. This accounts for their abundance as weathering products, particularly in temperate and colder climates.

The smectite group of clay minerals are structurally similar to illites (Fig. 4.12). In octahedral sites, substitution of Al^{3+} by Mg^{2+} or Fe^{3+} is common and some substitution of Si^{4+} by Al^{3+} in the tetrahedral sites also occurs, resulting in a net negative layer charge. This charge is, however, only about one-third the strength of the illite layer charge. Consequently, smectite is not able to bond interlayer cations effectively and the 2:1 units are not tightly bonded together. This allows water and other polar solvents to penetrate interlayer sites, causing the mineral to swell. Cations, principally H^+, Na^+, Ca^{2+} and Mg^{2+}, also enter the interlayer site with water and neutralize the negative charge. Bonding between the 2:1 units is effected by the hydrated cation interlayer, by a combination of hydrogen bonds and van der Waals' forces (Box 4.7). This weak bonding holds cations loosely in the interlayer sites, making them prone to replacement by other cations. As a consequence, smectites have high cation exchange capacity (Section 4.8).

The similar structure of illite and smectite allows mixing or interstratification of 2:1 units to form mixed-layer clays. Most illites and smectites are interstratified to a small degree, but they are not classified as such until detectable by X-ray diffraction. As one might expect, illite–smectite mixed-layer clays have intermediate cation exchange capacity between the end-member compositions.

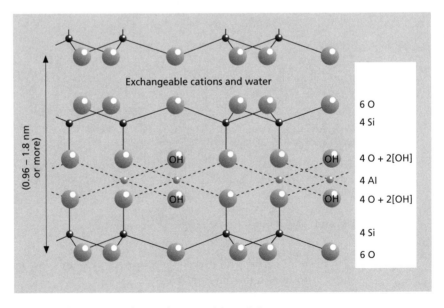

Fig. 4.12 The structure of a 2:1 clay mineral (smectite).

Box 4.7 Van der Waals' forces

Non-polar molecules have no permanent dipole and cannot form normal bonds. The non-polar noble gases, however, condense to liquid and ultimately form solids if cooled sufficiently. This suggests that some form of intermolecular force holds the molecules together in the liquid and solid state. The amount of energy (Box 4.8) required to melt solid xenon is $14.9\,kJ\,mol^{-1}$, demonstrating that cohesive forces operate between the molecules.

Weak, short-range forces of attraction, independent of normal bonding forces, are known as van der Waals' forces, after the 19th-century Dutch physicist. These forces arise because, at any particular moment, the electron cloud around a molecule is not perfectly symmetrical. In other words, there are more electrons (thus net negative charge) on one side of a molecule than on the other, generating an instantaneous electrical dipole.

This dipole induces dipoles in neighbouring molecules, the negative pole of the original molecule attracting the positive pole of the neighbour. In this way, weak induced dipole-induced dipole attractions exist between molecules.

Induced dipoles continually arise and disappear as a result of electron movement, but the force between neighbouring dipoles is always attractive. Thus, although the average dipole on each molecule measured over time is zero, the resultant forces between molecules at any instant are not zero.

As the size of molecules increase, so do the number of constituent electrons. As a result, larger molecules have stronger induced dipole-induced dipole attractions. It must be stressed, however, that van der Waals' forces are much weaker than both covalent and ionic bonds.

4.6 Formation of soils

So far we have discussed the *mechanisms* and *solid products* of chemical weathering without precise consideration of the environment in which these reactions occur. While chemical attack of exposed bedrock surfaces can happen, most weathering reactions occur in (or under) soils. We have already noted that the oxidation of soil organic matter causes acidity of natural waters (Section 4.4.2), promoting chemical weathering. This acknowledges the important role of soils in environmental chemistry. So what exactly are soils? A glance at a dictionary suggests that soils constitute the upper layer of the Earth's continental crust in which plants grow, usually consisting of disintegrated rock with admixture of organic remains.

Soil formation is influenced by geological (G), environmental (E) and biological (B) factors such that the product, soil (S), is a function of all of these factors with respect to time (t), i.e.:

$$S = f(G, E, B)dt \qquad\qquad \text{eqn. 4.15}$$

In fact a number of key factors can be identified in soil formation, including parent material (p), climate (cl), relief (r), vegetation (v) and the influence of organisms (o). Thus, equation 4.15 above can be more precisely written:

$$S = f(p, cl, r, v, o)dt \qquad\qquad \text{eqn. 4.16}$$

The key factors in equation 4.16 are summarized in Fig. 4.13 where the soil is depicted as a simple box. Various inputs and outputs govern the box's content while processes within the box generate the outputs. It is important to note that of all of these factors only time is an independent variable, the others being inextricably linked. For clarity, these factors are discussed individually below, although in nature they operate together.

4.6.1 Parent (bedrock) material (p)

Parent material is the material from which soils are derived. The main constituent of most soil (excluding peat soils) is inorganic mineral material. Crustal rocks are the main source of these mineral components, and the rate of rock weathering is strongly dependent on the solubility and stability of the constituent minerals. The mineral fraction in soils is dominated by silicate minerals, and their susceptibility to weathering follows a sequence which is roughly the reverse of the original crystallization order or 'Bowen's reaction series' (Fig. 4.14). High-temperature silicates such as olivine and calcium feldspar are furthest from stability at Earth surface temperatures (and pressures) and are easily weathered, whereas lower temperature minerals, such as quartz, are quite resistant.

There is experimental evidence that dissolution rates of specific monomer silicates (e.g. Ca_2SiO_4, Mg_2SiO_4, etc.) are proportional to the rate of reaction between the divalent cation and soil water molecules during hydration (see Section 5.2). The rate of reaction between water molecules and alkaline earth ions (see Fig. 2.2) is related to ionic size ($Ca(H_2O)_6^{2+} > Mg(H_2O)_6^{2+} > Be(H_2O)_6^{2+}$). This is mirrored by experimental dissolution rates, where $Ca_2SiO_4 > Mg_2SiO_4 > Be_2SiO_4$, and is controlled by the relative strength of the cation–oxygen bond.

From a global perspective, the weathering of average upper-crustal granodiorite will produce two types of solid product. Quartz being quite resistant to weathering (Fig. 4.14) will be released into the soil as a major component of the sand (60–2000 μm) and silt (2–60 μm) fractions. Although not strictly true, we will assume that quartz is chemically inert and takes no further part in chemical reactions. Feldspars, however, are weatherable (Fig. 4.15) and break down to form clay minerals (Section 4.5) as part of the soil clay (<2 μm) fraction. The relative proportions of sand, silt and clay size classes are important because they influence the water-holding capacity of soils. Sandy soils tend to be free draining and dry, whereas clay soils are usually poorly drained and wetter.

In general, the mineralogical and elemental composition of a soil will reflect that of the parent rock. For example, a soil forming on limestone ($CaCO_3$) will have a high calcium (Ca^{2+}) content, like the limestone itself. It is, however, important to note that both solids and solutes are transported to some degree during rock weathering. As a result, soil compositions may not directly match those of the rocks beneath them.

4.6.2 Climate (cl)

Wind, rain, temperature and evaporation all play a part in soil formation. Temperature is a major factor in determining the rate of chemical weathering, as heat

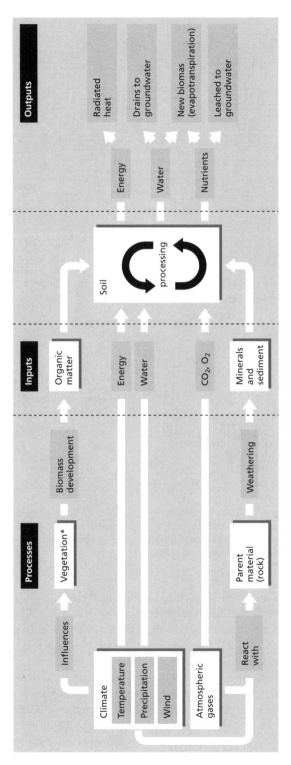

Fig. 4.13 Flow diagram summarizing soil-forming processes, inputs and outputs. * Note that vegetation is influenced by parent material as well as climate.

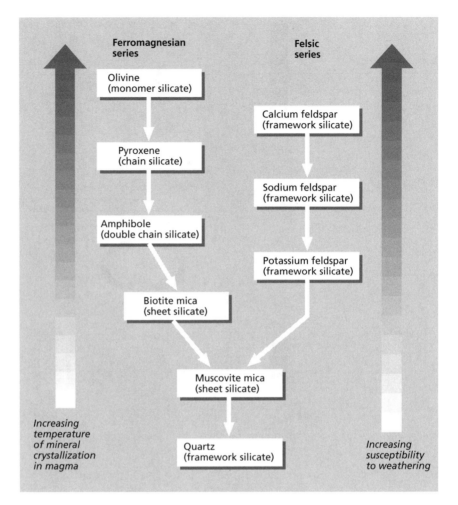

Fig. 4.14 Common silicate minerals ranked in Bowen's reaction series. Note that minerals formed under high temperatures with more ionic bonding are more susceptible to weathering. Ferromagnesian—minerals which contain essential iron and magnesium. Felsic—a rock containing feldspars and quartz.

speeds up chemical reactions by supplying energy (Box 4.8). For most reactions a 10°C rise in temperature causes at least a doubling of reaction rate. This suggests that weathering in the tropics, where mean annual temperatures are around 20°C, will be about double the rate of weathering in temperate regions where mean annual temperatures are around 12°C.

The effect of temperature is linked to the availability of water. The dry air of hot, arid environments is an ineffective weathering agent. Vegetation and hence soil organic matter are sparse, and this reduces the concentration of organic acids. Moreover, close contact between rock particles and acid is prevented by the lack of water. Short-lived rainfall events may move surface salts into the soil, but the

Fig. 4.15 Highly weathered granite, showing residual materials—mainly kalolinized feldspar and quartz. The granite has broken down principally due to the weathering of feldspars. Note the 'corestone' of less weathered granite by the figure. Two Bridges Quarry, Dartmoor, UK. Photograph courtesy of J. Andrews.

general dominance of evaporation over rainfall means that soluble salts tend to precipitate on the land surface, forming crusts of gypsum, calcium carbonate and other salts.

In humid, tropical climates, weathering is rapid, partly because the high temperatures speed up reactions, but mainly because the consistent supply of heavy rainfall allows rapid flushing and removal of all but the most insoluble compounds, for example oxides of aluminium and iron (Section 4.7). Flushing constantly removes (leaches) soluble components and is particularly important in the undersaturated zone of soils (Box 4.9).

The presence of water is critically important to most of the physical, chemical and biological processes that occur within soils and is expressed as a soil 'water balance'. Water balance expresses the difference between water use and water need, i.e. the water stored in soil that affects soil (micro)biology and plant growth (Section 4.6.4).

4.6.3 Relief (r)

Relief describes the form and gradient of slopes and is thus a component of topography (Section 4.7). Slope steepness governs the extent of water infiltration into soils. Where slopes are steep water infiltration is generally lower, although overland-flow and through-flow of water is greater. Thus, the potential for mass flow

Box 4.8 Chemical energy

The study of energy change is called thermodynamics. For example, the combustion of graphite carbon yields energy:

$$C_{(graphite)} + O_{2(g)} \rightarrow CO_{2(g)} \qquad \text{eqn. 1}$$

The total energy released, or the energy change in going from reactants to products, is termed the change in Gibbs free energy (ΔG), which is measured in kilojoules per mole (kJ mol^{-1}). If energy is released, i.e. the products have lower free energy than the reactants, ΔG is considered negative. $\Delta G°$ for the burning of graphite under standard temperature (25°C) and pressure (1 atm), indicated by the superscript °, is −394.4 kJ mol^{-1}. Tables of $\Delta G°$ for various reactions are widely available and values of $\Delta G°$ for different reactions can be calculated by simple arithmetic combination of tabulated values. Any reaction with a negative ΔG value will in theory proceed spontaneously—the chemical equivalent of water flowing downhill—releasing energy. The reverse reaction requires an input of energy, i.e.:

$$CO_{2(g)} \rightarrow C_{(graphite)} + O_{2(g)} \quad \Delta G° + 394.4 \text{ kJ mol}^{-1}$$
$$\text{eqn. 2}$$

Since an energetically favoured reaction proceeds from reactants to products, there is a relationship between ΔG and the equilibrium constant (K) for a reaction.

$$\Delta G = -RT \ln K \qquad \text{eqn. 3}$$

where T is the absolute temperature (measured in Kelvin (K)) and R is the universal gas constant (8.314 J mol^{-1} K^{-1}), relating pressure, volume and temperature for an ideal gas (see Box 3.1).

Converting equation 3 to decimal logarithms gives:

$$\Delta G = -RT 2.303 \log_{10} K \qquad \text{eqn. 4}$$

which at 25°C (298 K) yields:

$$\Delta G° = -5.707 \log_{10} K \qquad \text{eqn. 5}$$

or:

$$\log_{10} K = \frac{-\Delta G°}{5.707} \qquad \text{eqn. 6}$$

The total energy released in a chemical reaction has two components, enthalpy and entropy. Change in enthalpy (ΔH, measured in J mol^{-1}) is a direct measure of the energy emitted or absorbed by a reaction. Change in entropy (ΔS, measured in J mol^{-1} K^{-1}) is a measure of the degree of disorder. Most reactions proceed to increase disorder, for example by splitting a compound into constituent ions or atoms. Enthalpy and entropy are related:

$$\Delta G = \Delta H - T\Delta S \qquad \text{eqn. 7}$$

In most reactions the enthalpy term dominates, but in some reactions the entropy term is important. For example, the dissolution of the soluble fertilizer potassium nitrate (KNO_3) occurs spontaneously. However, ΔH for the reaction

$$KNO_{3(s)} \rightarrow K^{+}_{(aq)} + NO^{-}_{3(aq)} \qquad \text{eqn. 8}$$

is +35 kJ and the solution absorbs heat (gets colder) as KNO_3 dissolves. Despite the positive enthalpy, the large increase in disorder (entropy) in moving from a crystalline solid to ions in a solution, gives an overall favourable energy balance or negative ΔG for the reaction.

Electrode potentials ($E°$, Box 4.3) are a measure of energy transfer and so can be related to G:

$$G° = -nFE° \qquad \text{eqn. 9}$$

where n is the number of electrons transferred and F is the universal Faraday constant (the quantity of electricity equivalent to one mole of electrons = 6.02×10^{23} e$^-$).

of material is greater on steeper slopes. Conversely, the potential for dissolution and transport of dissolved material is lower on steep slopes because the contact time between soil water and mineral solids is lower. The form of a slope—whether it is linear, concave or convex—also influences water movement, and potentially

---

Box 4.9 Mineral reaction kinetics and solution saturation

The discussion in Section 4.6.2 suggests that weathering rates of minerals are proportional to the water flow rate. This is only true if the waters are close to saturation (Box 4.12) with respect to the weathering mineral. If water flow is continuous and sufficiently high, a limit is reached beyond which further flushing is no longer a rate-controlling factor.

With insoluble minerals (solubility $<10^{-4}\,mol\,l^{-1}$ (Box 4.12)), including all silicates and carbonates, ion detachment from mineral surfaces is very slow, such that ions never build up in solution close to the crystal surface. The weathering rate of these minerals thus depends mainly on the rate of ion detachment from the crystal surface, rather than the efficiency of flushing (water flow rate). Only in very soluble minerals (solubility $> 2 \times 10^{-3}\,mol\,l^{-1}$), for example evaporite minerals, can ions detach rapidly from the mineral surface to form a microenvironment close to the crystal surface which is saturated with respect to the dissolving mineral. The rate of dissolution is then controlled by the efficiency of dispersal of these ions and water flushing effects then become important.

---

microclimate too; these factors in turn affect the potential for erosion and dissolution.

4.6.4 Vegetation (v)

The development of vegetation is clearly influenced by parent material, relief and climate. However, the development of vegetation also feeds back positively to modulate these factors. For example, soils will tend to form where plants help stabilize the substrate, preventing erosion by surface water or wind. At the same time plant roots may help break up parent material, while evapotranspiration will influence microclimate. Vegetation is the main contribution of organic matter to soil and although most soils contain less than 5% organic matter (by weight), it is an *extremely* important component. The type of vegetation developed controls the nature of the soil organic matter (SOM), comprising a complex mixture of various biopolymers (Box 4.10) including cellulose and lignin (Table 4.6). SOM, often called humus, is composed of the elements carbon, hydrogen, oxygen, phosphorus, nitrogen and sulphur bonded together to form huge macromolecules. The presence of SOM influences soil structure and thereby its water-holding capacity, keeping mineral surfaces and soilwater in close contact.

4.6.5 Influence of organisms (o)

Soil dwelling macro- and micro-fauna process soil organic matter by feeding on it, deriving energy from the oxidation of the reduced substrate (Box 4.3). Phytophagous organisms consume living plant material, while sacrophagous organisms consume dead plant material. Electrons removed in oxidation processes pass down an electron transport chain inside cells. Ultimately, the energy from these

Box 4.10 Biopolymers

Cellulose

Cellulose is the most common plant polymer forming the structural fibres of many plants. Cellulose is constructed solely of glucose monomers (Fig. 1a), i.e. isolated or single units with consistent structure, in this case based on a carbon ring. The glucose monomers are linked by an ether bond (C–O–C; Fig. 1a). Thus, cellulose has a simple straight chain structure. Degradation of cellulose occurs stepwise. Initially the large polymer chain is cleaved into smaller units

containing, for example, two or three glucose monomers (cellobiose and cellotriose, respectively Fig. 1b) by the action of cellulase enzymes. It is this depolymerization reaction that is the rate-limiting step in cellulose biodegradation. The smaller units are then further degraded to produce glucose monomers that are then completely mineralized (converted to their inorganic constituents, e.g. CO_2 and H_2O) under aerobic conditions (Fig. 1c) or fermented to ethanol and CO_2 under anaerobic conditions.

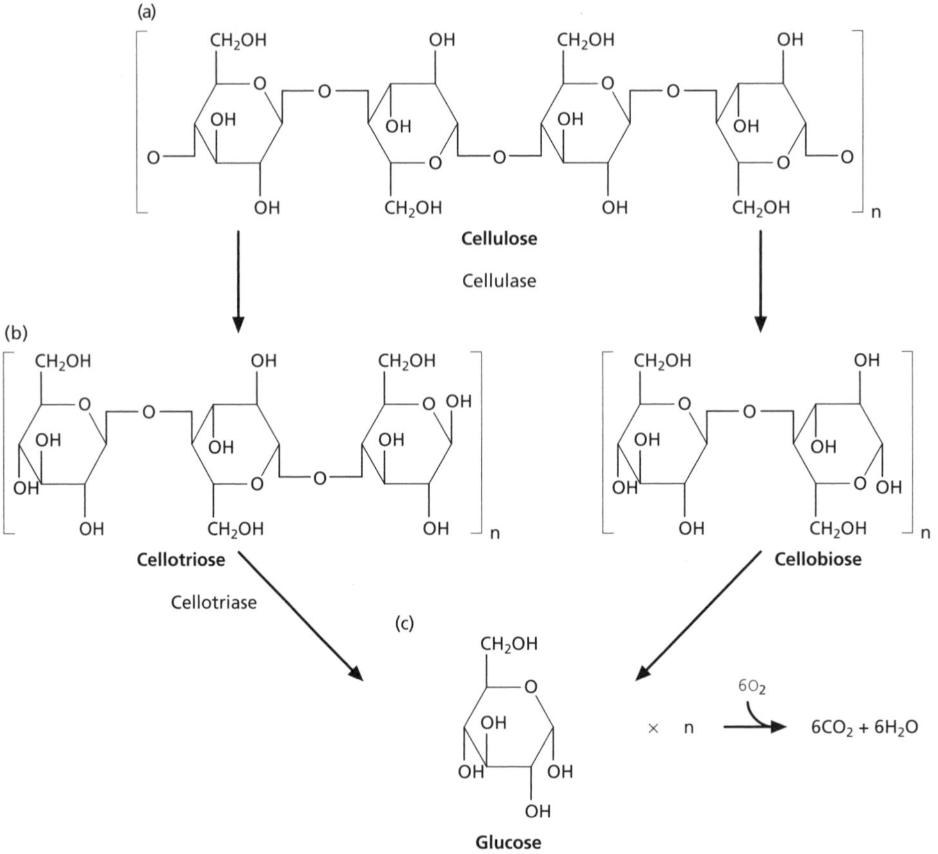

Fig. 1 Steps in the aerobic degradation of cellulose.

(continued)

Hemicelluloses

Hemicelluloses are the second most common class of plant polymer. Hemicellluloses are more complex than cellulose, being polymers of hexose sugars (six-atom ring, e.g. glucose, galactose, mannose), and pentose sugars (five-atom ring, e.g. ribose, xylose and arabinose) as well as uronic acids including glucuronic and galacturonic acid. An example of the degradation of a hemicellulose is that of pectin (Fig. 2), which is a polymer of galacturonic acid. Pectin is important in plant structure, being involved in plant cell wall formation. Decomposition of pectin occurs in three stages. Initially, pectin esterase enzymes attack esters ($COOCH_3$) on side-chains (Fig. 2a) resulting in carboxylic acid ($COOH$) where esters were originally (Fig. 2b). The second stage of degradation is depolymerization to form glucuronic acid monomers (Fig. 2c). These monomers are then mineralized to CO_2 and H_2O.

Lignin

Lignin is the third most common plant polymer after cellulose and hemicelluloses. Lignin gives wood its toughness and structural rigidity. Lignin is formed from the metabolic processing of glucose (non-

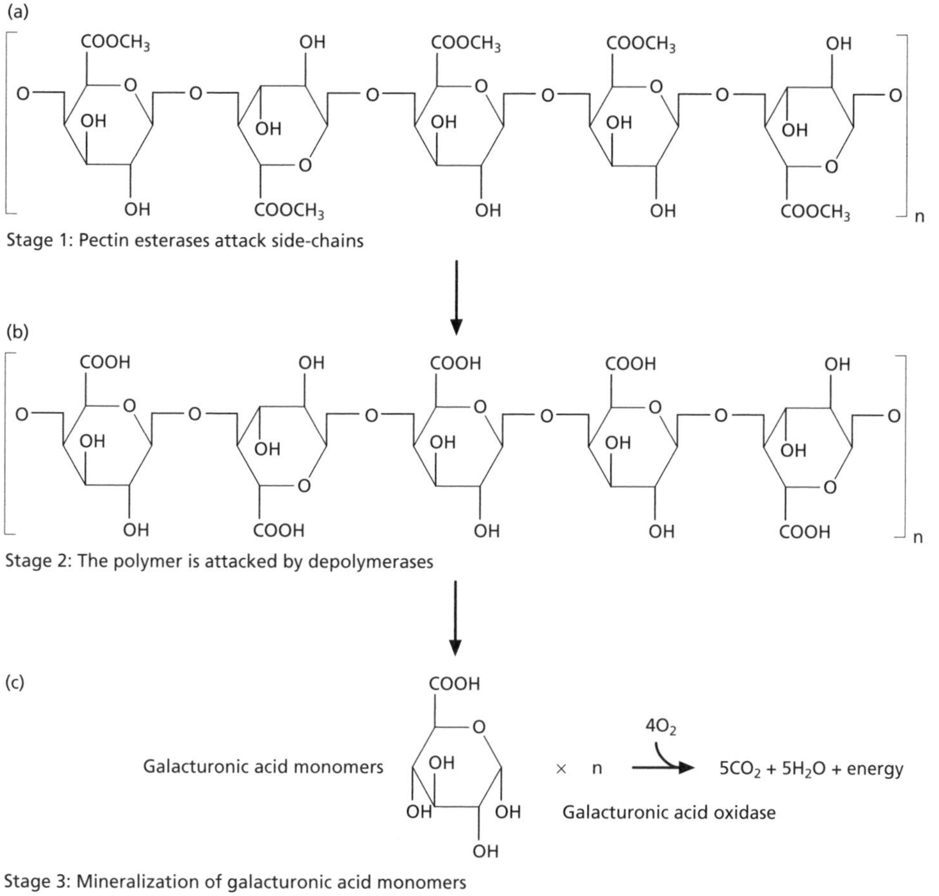

(a)

Stage 1: Pectin esterases attack side-chains

(b)

Stage 2: The polymer is attacked by depolymerases

(c)

Galacturonic acid monomers

$\times n \xrightarrow{4O_2} 5CO_2 + 5H_2O + \text{energy}$

Galacturonic acid oxidase

Stage 3: Mineralization of galacturonic acid monomers

Fig. 2 Steps in the aerobic degradation of the hemicellulose pectin.

(continued on p. 102)

aromatic) that is converted into three basic aromatic (see Section 2.7) monomers, coumaryl alcohol, coniferyl alcohol and sinapyl alcohol. These monomers react together and with their precursors to produce lignin. Thus, lignin is a tremendously complex polymer with a random structure (Fig. 3). Lignin has a condensed structure that is highly aromatic, making it the most resistant component of plant tissues to degradation. Only a few soil organisms are capable of degrading lignin. These organisms belong to a group of fungi known as white rot fungi or basidiomycetes. They produce powerful non-specific extracellular enzymes called peroxidases which, with the aid of H_2O_2, O_2^- and $^1O_2$, cause the depolymerization of lignin. Carbon–carbon bonds and ether (C–O–C) bonds are cleaved initially resulting in the formation of monomeric phenols, aromatic acids and alcohols; these are, in turn, mineralized to CO_2 and H_2O.

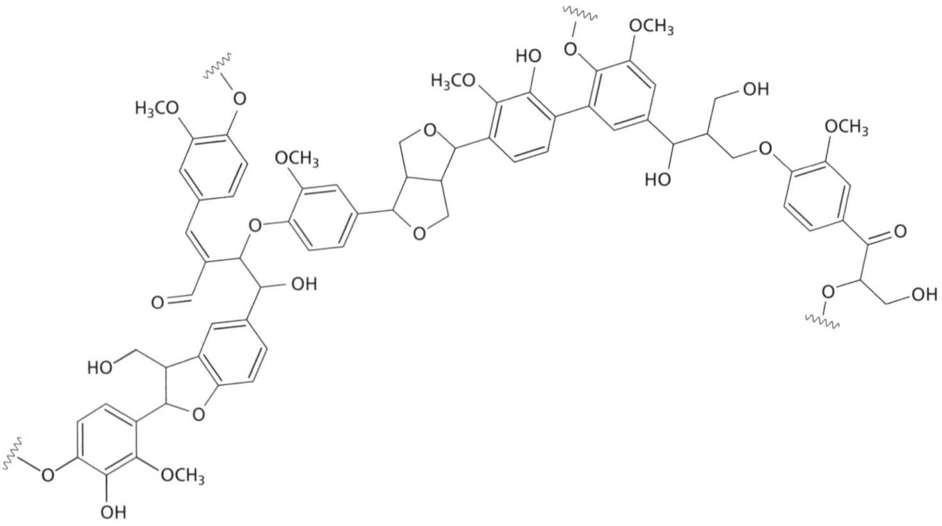

Fig. 3 Partial structure of lignin. After Killops and Killops (1993).

Table 4.6 Relative proportions of biopolymers in plant-derived soil organic matter.

Plant residues	Percentage in soils
Cellulose	50
Hemicelluloses	20
Lignin	15
Protein	5
Carbohydrates and amino acids	5
Pectin	1
Waxes and pigments	1

electrons is harnessed in the production of adenosine triphosphate (ATP), the cell's energy-storing compound. Terminal electron acceptors (e.g. oxygen) are the final place the electrons arrive in the electron transport chain. Although oxygen is not the only electron acceptor, it is the most thermodynamically favoured

Table 4.7 Order of bacterial reactions during microbial respiration of organic matter based on energy yield. Modified from Berner (1980), reprinted by permission of Princeton University Press.

Bacterial reaction	$\Delta G°$ (kJ mol^{-1} of CH$_2$O)
Aerobic respiration: important in all oxygenated Earth surface environments	
$CH_2O + O_2 \rightarrow CO_2 + H_2O$	−475
Denitrification: most important in terrestrial and marine environments impacted by anthropogenic inputs from fertilizers	
$5CH_2O + 4NO_3^- \rightarrow 2N_2 + 4HCO_3^- + CO_2 + 3H_2O$	−448
Manganese reduction: minor reaction important in some marine sediments	
$CH_2O + 3CO_2 + H_2O + 2MnO_2 \rightarrow 2Mn^{2+} + 4HCO_3$	−349
Iron reduction: can be significant in some soils and marine sediments with high iron contents from contamination or weathering flux (e.g. Amazon Delta)	
$CH_2O + 7CO_2 + 4Fe(OH)_3 \rightarrow 4Fe^{2+} + 8HCO_3^- + 3H_2O$	−114
Sulphate reduction: major process in anaerobic marine sediments, especially on continental shelves	
$2CH_2O + SO_4^{2-} \rightarrow H_2S + 2HCO_3^-$	−77
Methanogenesis: important process in freshwater wetlands, waterlogged soils and in deeply buried low-sulphate marine sediments	
$2CH_2O \rightarrow CH_4 + CO_2$	−58

Note: Free energy value for organic matter (CH$_2$O) is that of sucrose.

(Table 4.7), and is therefore utilized first by aerobic organisms. Once oxygen has been used up other electron acceptors are used in order of energy efficiency (Table 4.7), by nitrate, manganese, iron and sulphate reducers. This order of electron acceptor usage is common to most biotic Earth surface environments (see Sections 5.5 & 6.2.4), although the relative importance of each electron acceptor depends on its concentration in specific environments. Sulphate reduction, for example, is not a common process in most soils because sulphate is in low concentrations in most continental waters except in some coastal areas (see Section 5.5). Methanogenesis can, however, be an important microbial reaction in waterlogged soils, especially paddy fields and marshes.

Soil microorganisms (fungi, bacteria and actinomycetes) play a major role in the degradation of organic matter, ultimately releasing nutrient elements — about 98% nitrogen, 5–60% phosphorus and 10–80% sulphur to the soil nutrient pool — along with micronutrients such as boron and molybdenum, into the soil for reuse by plants and animals. The role of biotic soils as sources of N and CH$_4$

to the atmosphere, and N and P to the hydrosphere is discussed in Sections 3.4.2 and 5.5.1 respectively.

The biosphere clearly influences rates of weathering reactions but there is debate as to how much. Some estimates suggest the presence of biotic soils enhance weathering rates by 100–1000 times over abiotic weathering rates. There is also interest in whether microbial participation in weathering is simply a co-incidental byproduct of metabolism. Is it possible that the microbes get something for their trouble? If soil bacteria do gain something by colonizing specific mineral substrates then the likely controls are nutrient element availability, absence of toxic elements or capacity to help buffer microenvironmental pH, all parameters that contribute to the success or failure of a microbial population. Recent work shows that in some weathering systems specific silicate minerals are heavily col-onized and broken down by bacteria, whereas other minerals are left untouched. The colonized minerals contain the nutrient elements phosphorus (P) and iron (Fe) (see Section 5.5.1), while the uncolonized minerals are typically aluminium-rich but lacking nutrient potential. If bacteria are widely proven to select bene-ficial minerals to weather, we will need to rethink the traditional view that mineral weathering is mainly controlled by relative instability, as shown for example in Fig. 4.14.

The debate surrounding weathering rates is important, since the consumption of CO_2 by soil weathering reactions (Section 4.4.3) lowers atmospheric partial pressure of CO_2 (pCO_2). Some researchers argue that, prior to the evolution of vascular land plants some 400 million years ago, weathering rates may have been much lower, giving rise to a higher atmospheric pCO_2 and enhanced greenhouse warming (see Section 7.2.4). Others, however, believe that thin soils, stabilized by primitive lichens and algae, covered the land surface billions of years before the evolution of vascular plants. These primitive biotic soils may have been quite effective in enhancing weathering rates, acting in a 'Gaian' way (see Section 1.3.3) by consuming atmospheric CO_2 and lowering global temperatures. This cooling effect may have helped improve the habitability of the early Earth for other organisms.

4.7 Wider controls on soil and clay mineral formation

In an average upper-crustal granodiorite, it is mainly feldspars that weather to form clay minerals (eqns. 4.13 & 4.14). Since feldspars are framework silicates, the formation of clay minerals (sheet silicates) must involve an intermediate step. This step is not at all well understood although it has been proposed that fulvic acids, from the decay of organic matter in soil, may react with aluminium to form a soluble aluminium–fulvic acid complex, with aluminium in six-fold coordina-tion. This gibbsitic unit may then have SiO_4 tetrahedra adsorbed on to it to form clay mineral structures.

A minority of unweathered rock-forming silicates, for example the micas, are already sheet silicates. It is not difficult to envisage that alteration could transform these to clay minerals. Alteration is most likely in the interlayer areas, especially at damaged crystal edges. The clay mineral formed will depend on the composition of both the original mineral and the ions substituted during alteration. For example, the replacement of K^+ in muscovite by Mg^{2+} would lead to the formation of magnesium smectite.

Topography influences most of the soil-forming factors (Section 4.6). Where all the soils in an area of varied topography have formed on the same parent material, the soils differ principally due to changes in relief and drainage. These soils are described as part of a catena ('chain') and are intrinsically linked to the landscape of a region or area. Two key processes control the development of a soil catena: (i) erosion and subsequent transport and deposition of eroded material; and (ii) leaching and subsequent transport and deposition of dissolved materials. The latter process is dependent on soil chemistry and in the tropics where rainfall is high, a chemical catena develops.

Wet tropical environments promote dissolution and transport of a number of soil constituents. Chemical species have been categorized into four groups by their relative mobility during rock weathering (Table 4.8). At high altitude in the tropics, high rainfall causes mobilization of Group I soluble ions and oxyanions from the soil. These easily dissolved species (see Section 5.2) are carried in solution down slope to lower altitude where they accumulate. Thus, the tropical catena is characterized by distinct soil end-members (Fig. 4.16). The high-altitude end-member is iron- and silica-rich Groups III and IV and base cation-deficient (Box 4.11), and called an oxisol or ferralsol depending on the classification system used (Plate 4.1, facing p. 138). By contrast, the low altitude end-member is base cation-rich and also clay-rich and known as a vertisol (swelling clay soil).

The high-altitude oxisols (ferralsols) are of low fertility on account of their low cation exchange capacity (CEC) (Section 4.8). Furthermore, the high iron concentrations can prove toxic to some plants and animals while the formation

Table 4.8 Mobility of different chemical species in relation to rock weathering. Modified from Polynov (1937).

	Species	Mobility*	Comments†
Group I	Cl^-	100	Soluble anions easily leached by water
	SO_4^{2-}	57	
Group II	Ca^{2+}	3.00	Relatively soluble cations, easily leached by water
	Na^+	2.40	
	Mg^{2+}	1.30	
	K^+	1.25	
Group III	SiO_2	0.20	Relatively insoluble element, typically present as quartz grains
Group IV	Fe_2O_3	0.04	Highly insoluble elements present as Fe and Al oxides
	Al_2O_3	0.02	

* Most mobile = 100.
† The reasons for differing solubility are discussed in Section 5.2 and depicted in Fig. 5.2.

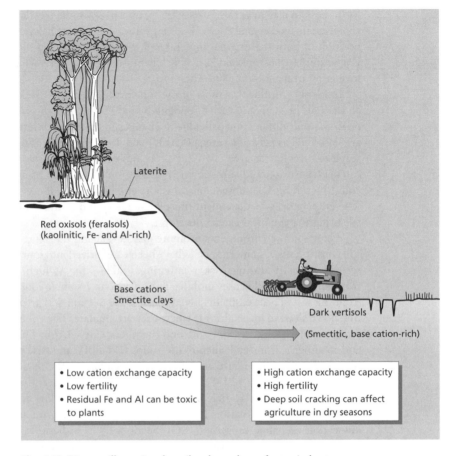

Fig. 4.16 Diagram illustrating the soil end-members of a tropical catena.

Box 4.11 Base cations

The term base cation, or non-acid cation, is often used in soil chemistry and refers to cations of the alkali metals and alkali earth metals (see Section 2.2), most importantly Ca^{2+}, Na^+, K^+ and Mg^{2+}. The weathering of these metals from crustal minerals is a slow, but very important process that helps neutralize acidity. For example, weathering of the (Na)-rich feldspar, albite, proceeds as in equation 1.

The acidity (H^+) contained in H_2CO_3 is neutralized and dissolved cations (Na^+), bicarbonate (HCO_3^-) and silicic acid (H_4SiO_4) are released. These 'released' base cations thus become available in soilwater to take part in exchange reactions (Section 4.8).

$$2NaAlSi_3O_{8(s)} + 9H_2O_{(l)} + 2H_2CO_{3(aq)} \rightarrow Al_2Si_2O_5(OH)_{4(s)} + 2Na^+_{(aq)} + 2HCO^-_{3(aq)} + 4H_4SiO_{4(aq)}$$

eqn. 1

of impenetrable siliceous-iron layers (laterite) inhibit plant growth. By contrast, the low-altitude vertisols have high fertility on account of their high CEC. These clay-rich soils do not suffer from metal-associated toxicity, although deep cracking during dry periods can be a problem for agriculture.

The marked contrast in solubility (Box 4.12) between insoluble oxides or oxyhydroxides of aluminium and iron and other more soluble soil cations and H_4SiO_4

Box 4.12 Solubility product, mineral solubility and saturation index

The dynamic equilibrium between a mineral and its saturated solution (i.e. the point at which no more mineral will dissolve), for example:

$$CaCO_{3(calcite)} \rightleftharpoons Ca^{2+}_{(aq)} + CO^{2-}_{3(aq)} \qquad \text{eqn. 1}$$

is quantified by the equilibrium constant (K), in this case:

$$K = \frac{aCa^{2+} \cdot aCO_3^{2-}}{aCaCO_3} \qquad \text{eqn. 2}$$

Since the $CaCO_3$ is a solid crystal of calcite, it is difficult to express its presence in terms of activity (see Section 2.6). This is overcome by recognizing that reaction between a solid and its saturated solution is not affected by the amount of solid surface presented to the solution (as long as the mixture is well stirred). Thus the activity of the solid is effectively constant; it is assigned a value of 1 or unity (see eqn. 3), and makes no contribution to the value of K in equation 3.

The equilibrium constant for a reaction between a solid and its saturated solution is known as the solubility product and is usually given the notation K_{sp}. Solubility products have been calculated for many minerals, usually using pure water under standard conditions (1 atm pressure, 25°C temperature).

The solubility product for calcite (eqn. 1) is thus:

$$K_{sp} = \frac{aCa^{2+} \cdot aCO_3^{2-}}{1} = aCa^{2+} \cdot aCO_3^{2-} = 3.3 \times 10^{-9} \, mol^2 \, l^{-2}$$

$$\text{eqn. 3}$$

The solubility product can be used to calculate the solubility (mol l^{-1}) of a mineral in pure water. The case for calcite is simple since each mole of $CaCO_3$ that dissolves produces one mole of Ca^{2+} and one mole of CO_3^{2-}. Thus:

$$\text{Calcite solubility} = aCa^{2+} = aCO_3^{2-} \qquad \text{eqn. 4}$$

and therefore:

$$(\text{Calcite solubility})^2 = aCa^{2+} \cdot aCO_3^{2-} = K_{sp} = 3.3 \times 10^{-9} \, mol^2 \, l^{-2} \qquad \text{eqn. 5}$$

Thus:

$$\text{Calcite solubility} = \sqrt{3.3 \times 10^{-9}} = 5.7 \times 10^{-5} \, mol \, l^{-1} \qquad \text{eqn. 6}$$

The degree to which a mineral has dissolved in water can be calculated using the saturation index, i.e.:

$$\text{Degree of saturation} = \Omega = \frac{IAP}{K_{sp}} \qquad \text{eqn. 7}$$

IAP is the ion activity product, i.e. the numerical product of ion activity in the water. An Ω value of 1 indicates saturation, values greater than 1 indicate supersaturation and values less than 1 indicate undersaturation.

For example, groundwater in the Cretaceous chalk aquifer of Norfolk, UK, has a calcium ion (Ca^{2+}) activity of $1 \times 10^{-3} \, mol \, l^{-1}$ and a carbonate ion (CO_3^{2-}) activity of $3.5 \times 10^{-6} \, mol \, l^{-1}$. The saturation state of the water with respect to calcite is:

$$\Omega = \frac{aCa^{2+} \cdot aCO_3^{2-}}{K_{sp(calcite)}} = \frac{1 \times 10^{-3} \cdot 3.5 \times 10^{-6}}{3.3 \times 10^{-9}} = 1.06$$

$$\text{eqn. 8}$$

i.e. the water is slightly supersaturated with respect to calcite.

Table 4.9 Chemical index of alteration (CIA) values for various crustal materials. Data from Nesbitt and Young (1982), Maynard *et al.* (1991) and Taylor and McLennan (1985).

Material	CIA
Clay minerals	
Kaolinite	100
Chlorite	100
Illite	75–85
Smectite	75–85
Other silicate minerals	
Plagioclase feldspar	50
Potassium feldspar	50
Muscovite mica	75
Sediments	
River Garonne (southern France) suspended load	75*
Barents Sea (silt)	65*
Mississippi delta average sediment	64*
Amazon delta muds	70–75
Amazon weathered residual soil clay	85–100
Rocks	
Average continental crust (granodiorite)	50
Average shales	70–75
Basalt	30–40
Granite	45–50

Value calculated using total CaO rather than CaO (see text).

under normal soil pH ranges (Table 4.8) has been used to formulate a chemical index of alteration (CIA); using molecular proportions:

$$CIA = (Al_2O_3/Al_2O_3 + CaO^* + Na_2O_3 + K_2O) \times 1000 \qquad \text{eqn. 4.17}$$

where CaO* is CaO in silicate minerals (i.e. excludes Ca-bearing carbonates and phosphates). Thus, CIA values approaching 100 are typical of materials formed in heavily leached conditions where soluble calcium, sodium and potassium have been transported away from the weathering site. Kaolinite clays attain such values (Table 4.9), whereas illites and smectites have CIA values around 75–85 (Table 4.9). In comparison, unleached feldspars have CIA values around 50.

The CIA predicts that kaolinite will form under heavily leached conditions, and this is confirmed by observations in tropical weathering regimes. On stable well-drained land surfaces where weathering and leaching have been prolonged, the oxisols (ferralsols) develop kaolinitic and, in extreme cases, gibbsitic clay mineralogies (Fig. 4.17). Such sites are mantled by iron-rich (laterite) and aluminous (bauxite) surface deposits (Plate 4.1). These surface deposits can become thick enough to prevent further interaction between surface waters and bedrock, lowering the rate of subsequent bedrock weathering.

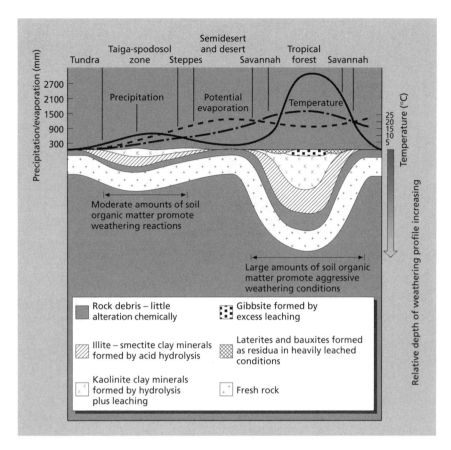

Fig. 4.17 Schematic relationship between weathering zones and latitude–vegetation–climatic zones. The influence of relief (mountainous areas), where soils are typically thin, is not included. After Strakhov (1967). With kind permission of Kluwer Academic Publishers.

By contrast, smectite clays develop in poorly drained sites. On the basaltic island of Hawaii, soil clay mineral type changes in the sequence smectite–kaolinite–gibbsite as rainfall amount increases (Fig. 4.18). A similar, generalized zonation has been proposed for clay mineral distribution with depth in soils, again based on the degree of leaching (Fig. 4.19).

Intense leaching favours kaolinite formation since cations and H_4SiO_4 are removed, lowering the silicon:aluminium ratio and favouring the 1:1 structural arrangement. Less intense leaching favours a higher silicon: aluminium ratio, allowing the formation of various 2:1 clay minerals, depending on the supply of cations. For example, the weathering of basalt provides abundant magnesium for the formation of magnesium smectite. In the most intense tropical weathering environments, all of the silica is removed, favouring the formation of gibbsite, which can be thought of as a 0:1 arrangement (i.e. only octahedral sheet present; Fig. 4.19).

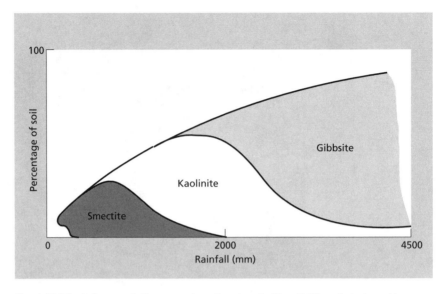

Fig. 4.18 The influence of climate on clay mineralogy in Hawaii. The relatively rapid water flow rates associated with high rainfall result in the preferential removal of cations and silica. After Sherman (1952).

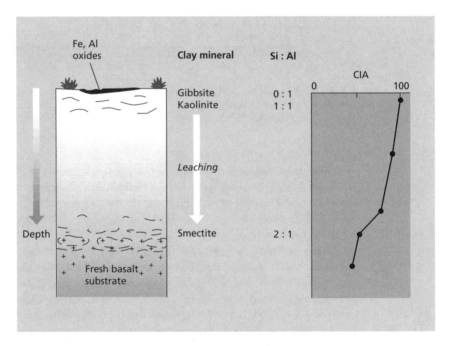

Fig. 4.19 Idealized vertical distribution of clay minerals formed under leaching conditions in soils developed on basalt. CIA values increase from 30–40 in fresh rock to near 100 in heavily leached surface soils.

4.8 Ion exchange and soil pH

Exchangeable ions are those that are held temporarily on materials by weak, electrostatic forces. If particles with one type of adsorbed ion are added to an electrolyte solution containing different ions, some of the particle-surface adsorbed ions are released into solution and replaced by those from the solution (Fig. 4.20).

We have seen that the interlayer sites of clay minerals, particularly smectites, hold ions weakly, giving these minerals a capacity for ion exchange. Clay mineral ion exchange can also be a surface phenomenon. Edge damage to minerals can break bonds to expose either uncoordinated oxygens (sites of net negative charge) or uncoordinated silicon or other metal ions (sites of net positive charge). These surface charges are balanced by electrostatic adsorption of cations and anions respectively.

In soils with neutral or alkaline pH (see Box 3.5), ion-exchange sites are typically occupied by exchangeable base cations, for example Ca^{2+}, Mg^{2+}, Na^+ and K^+ (Box 4.11). These exchangeable base cations are in equilibrium with the soil pore water and H^+ ions. However, changes in soil pH, for example to more acidic conditions (higher solution H^+ concentration), encourage exchange of base cations for H^+ ions, causing H^+ pore water concentrations to fall (pH increases) along with an increase in pore water base cation concentration. The presence of the base cations effectively buffers the soil water pH (see Section 5.3.1), minimizing change. Clearly, a lack of exchangeable base cations, for example in oxisols (Section 4.7), will often lead to low soilwater pH as there is little or no buffer capacity. Understanding the buffer capacity of a soil to pH fluctuations is important, as some fundamental chemical processes (e.g. dissociation, Box 4.5), chemical species, for example phosphate (see Section 5.2) and elements, for example aluminium (see Section 5.4), are very sensitive to pH change.

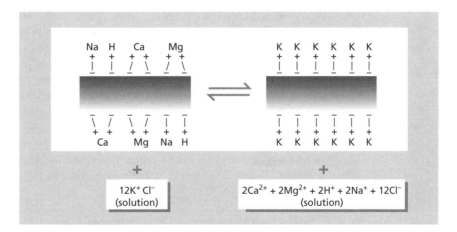

Fig. 4.20 Schematic diagram to show ion-exchange equilibria on the surface of a clay particle. Potassium ions in solution are exchanged for other cations, causing the exchange equilibrium to move from left to right.

Table 4.10 Representative cation exchange capacity, CEC (meq* $100\,g^{-1}$ dry wt), for various soil materials. Table adapted from *Pedology, Weathering and Geomorphological Research* by P. W. Birkeland, copyright 1974 by Oxford University Press, Inc. Used by permission of Oxford University Press, Inc.

Non-clay materials	CEC	Clay minerals	CEC	Cation exchange site
Quartz, feldspars	1–2			
Hydrous oxides of Al and Fe	4	Kaolinite	3–15	Edge effects
		Illite	10–40	Mainly edge effects, plus some interlayer
		Chlorite	10–10	
Organic matter	150–500	Smectite	80–150	Mainly interlayer plus some edge effects

*A milliequivalent (meq) is the charge carried by 1.008 milligrams (mg) of H^+ (i.e. the atomic mass of hydrogen), or the charge carried by any ion (measured in $mg\,l^{-1}$), divided by its relative atomic mass and multiplied by the numerical value of the charge. If, for example, divalent Ca^{2+} displaces H^+, it occupies the charged sites of $2H^+$ ions. Thus the amount of Ca^{2+} required to displace 1 meq of H^+ is 40 (atomic mass of Ca) divided by 2 (charge) = 20 mg, i.e. the mass of 1 meq of Ca^{2+}.

Under high soil pH conditions, damaged soil clays with exposed OH groups in octahedral layers may dissociate, forming a negative charge, which is neutralized by other cations in the soil water, for example:

$$\text{Clay–mineral–O}^- \text{–H}_{(s)} + \text{K}^+_{(aq)} \rightleftharpoons \text{clay–mineral–O}^- \text{–K}_{(s)} + \text{H}^+_{(aq)} \qquad \text{eqn. 4.18}$$

Micrometre-sized clay minerals have a large surface area : volume ratio and constitute a significant sink for some anions and cations in soil environments. Despite this, in smectite clays, surface ion exchange is much less important than interlayer site exchange (Table 4.10). The cation exchange capacity (CEC) of a soil provides a useful indication of the availability of essential trace metals (which are fundamental for plant health) and is often considered the best index of soil fertility. In soils it is usually impossible to calculate the amount of ion exchange caused by clay minerals alone, since other components, especially organic matter, have significant CECs (Table 4.10). Organic matter provides a significant proportion of exchange sites in many soils, created by the dissociation of acidic or alcoholic/phenolic functional groups at pH above 5. The high CEC of soil organic matter ($150–500\,meq/100\,g_{soil}$; Table 4.10) means that organic matter content, rather than clay mineral content, often provides the most significant contribution to a soil's CEC. Addition of organic matter to soil is thus an easy and effective means of increasing the CEC, and thereby the soil fertility.

4.9 Soil structure and classification

As a result of the various factors and processes outlined in Sections 4.6.1–4.6.5, over time soils develop stable and diagnostic features, many of which are recognizable in the field. These features, particularly specific layers called 'soil horizons', are the basis for soil classification. An idealized soil profile, i.e. a vertical section, is shown in Fig. 4.21. Soil horizons are described using an internationally agreed system of abbreviations that are shown on Fig. 4.21 and used in the descriptions below.

The classification of soils is potentially complex. Two systems of soil classification are common, that of the US Department of Agriculture (USDA) and that of UNESCO's Food and Agriculture Organisation (FAO). In this book we mostly use the USDA's first tier of classification known as soil orders (Fig. 4.22), which can be related to factors such as degree of weathering, type of parent material and climate. Where possible the USDA name is followed by the equivalent FAO name in brackets. In some cases, for example vertisol, the name is the same in both systems. Most of the order names have the common ending -*sol*, from the Latin *solum*, meaning soil. Although soils are classified using stable features developed over time, the soils formed under dynamic conditions. It is variations in soil-forming processes that give rise to the vast number of different soil types. Three contrasting soils with distinctive diagnostic horizons are shown in Plate 4.2 (facing p. 138). Clearly a range of dynamic influences have controlled the development of each of these soils, and these are discussed separately below, emphasizing the role of soil chemistry.

4.9.1 Soils with argillic horizons

The term 'argillic' indicates that a soil has a clay-rich horizon (Bt in Fig. 4.23). The downward percolation of water through the soil controls most of the significant processes in the development of the clay-rich (argillic) horizon (Bt). The movement of water causes; (i) the leaching of calcium ions (Ca^{2+}) from the A horizon; (ii) the washing of materials down profile (eluvation); and (iii) the deposition of these 'washed-down' materials at depth (illuvation) (Fig. 4.21). The leaching (decalcification) of Ca^{2+} that had bound together clay particles with excess negative charge (Fig. 4.23), causes destabilization of clay aggregates, allowing them to fall apart (deflocculate). The disaggregated clay particles are themselves then susceptible to translocation down profile (eluvation). The clay particles will re-flocculate lower in the profile, to form the argillic horizon, where sufficient divalent cations are present to re-bind them (Fig. 4.23). In the soil shown in Plate 4.2 the Ca^{2+} is supplied from weathering a $CaCO_3$-rich 'C' horizon below the Bt horizon. Note that the chain of events leading to the development of the argillic horizon is analogous to those that translocate clays over larger scales in the formation of vertisols (Section 4.7). Argillic horizons can form in a number of soil types, for example ultisols (acrisols), mollisols (chernozems and kastanozems) and alfisols (luvisols).

4.9.2 Spodosols (podzols)

Spodosols (podzols) contain separate horizons from which material has been both removed and deposited (Fig. 4.24). However, spodosols form under very

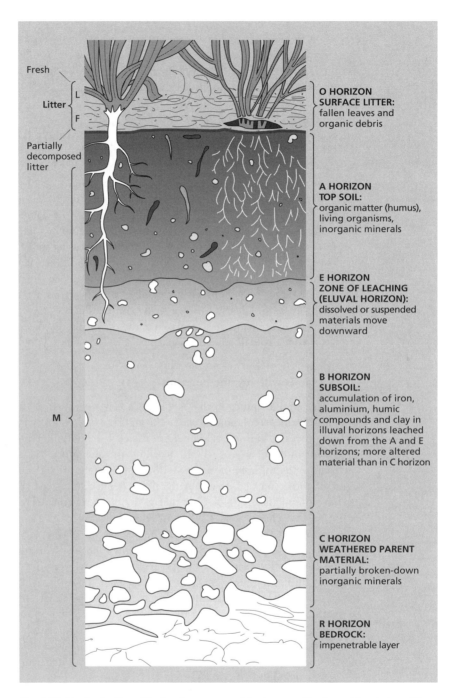

Fig. 4.21 Idealized soil profile showing master soil horizons and horizon abbreviations. The O, A, E and B master horizons can be further subdivided into subordinate horizons depending on composition (see Figs 4.23–4.25). Note that the O horizon is composed of fresh (L) and partially decomposed (F) organic litter. Soil profiles are typically 0.5–1.0 m thick such that master horizons are typically centimetres to tens of centimetres thick. SOM, soil organic matter; CEC, cation exchange capacity.

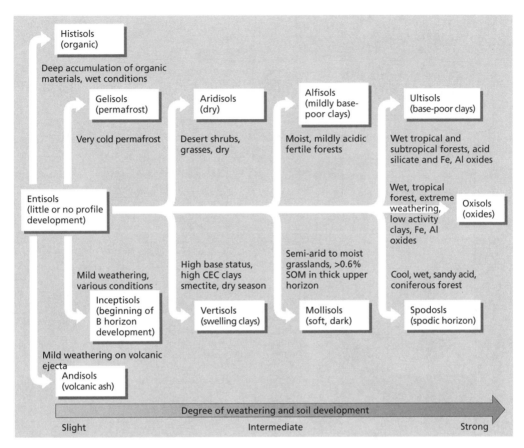

Fig. 4.22 Diagram relating the 12 USDA soil orders to degree of weathering, degree of soil development, climate and vegetation conditions. SOM, soil organic matter. Approximate FAO names where different are: entisol (arenosol, fluvisol, regosol); inceptisol (cambisol); andisol (andosol); ardisol (xerosols); alfisol (luvisols); ultisol (acrisol); spodosol (podzol); oxisol (ferralsol). Gelisol and mollisol have no simple FAO equivalent. Modified from Brady and Weil (2002), reprinted by permission of Pearson Education Inc., Upper Saddle River, NJ.

different conditions to those described for argillic horizons, the materials being removed and deposited are generally metal ions and/or organic matter rather than clay particles. Spodosols are sandy forest soils, forming beneath a horizon of tree litter, (horizon 'L' in Fig. 4.24). In pine forests, for example, large annual inputs of dead pine needles cause an accumulation of organic matter in both the litter horizon 'L' and the organic-rich 'Ah' horizon. The low pH in these horizons, resulting from the release of organic acids, often limits organic matter processing by soil organisms allowing thick accumulations to form.

Spodosols are typified by a prominent eluvated horizon (E) that has lost many of its metal ions, particularly iron, manganese and aluminium. This horizon shows clearly in Plate 4.2 as the grey Ea horizon. Rather insoluble metal ions like iron and aluminium are leached because their solubility has been enhanced by a 'complexing' or chelating agent. In spodosols the 'free' metal ions are chelated (see Box

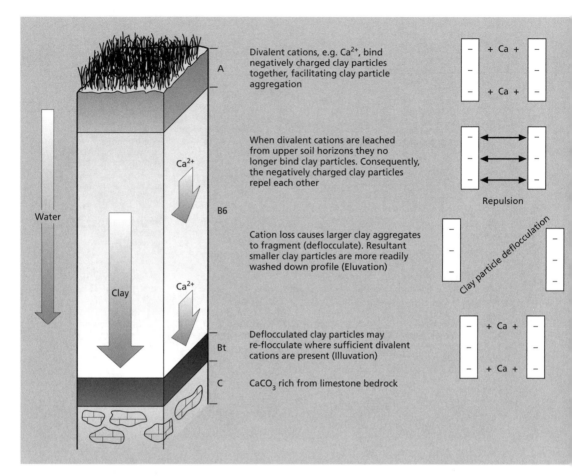

Fig. 4.23 Schematic diagram explaining the formation of argillic horizons in soils. The resulting horizons are comparable to those in the alfisol (luvisol) shown in Plate 4.2(a). The master horizons are typically centimetres to tens of centimetres thick.

6.4) by soluble organic compounds such as fulvic acids, derived from the breakdown of organic matter in the 'L' and 'Ah' horizons (Fig. 4.24). Dissociated functional groups for example -COO⁻ and -O⁻ in fulvic acids, along with lone pairs of electrons (see Box 6.4), for example on the N atom of NH_2, coordinate with metal ions, thereby increasing their solubility. These soluble complexed ions are then translocated downward with percolating water. At depth however, the metal ions reprecipitate to form an illuvial 'Bs' spodic horizon. This metal-rich horizon can also be organic-matter-rich, in which case the horizon is denoted 'Bsh'. Reprecipitation of the metal ions may occur because the metal chelates become unstable due to lowered pH as they move down the soil profile. It is also possible that the organic chelates are degraded by microorganisms as they move down the profile. Clearly if the organic chelates are removed the metal ions will precipitate.

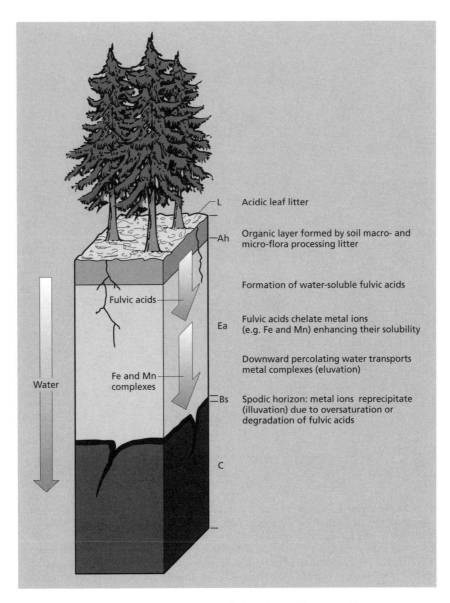

Acidic leaf litter

Organic layer formed by soil macro- and micro-flora processing litter

Formation of water-soluble fulvic acids

Fulvic acids chelate metal ions (e.g. Fe and Mn) enhancing their solubility

Downward percolating water transports metal complexes (eluvation)

Spodic horizon: metal ions reprecipitate (illuvation) due to oversaturation or degradation of fulvic acids

Fig. 4.24 Schematic diagram of a spodosol (podzol) with an explanation of horizon formation. The resulting horizons are comparable to those in Plate 4.2(b). The master horizons are typically centimetres to tens of centimetres thick.

4.9.3 Soils with gley horizons

Gley horizons form when the water table is present within the soil profile. Soils with gley horizons (denoted 'g') are typically aquepts (gleysols), a suborder of the inceptisols (Fig. 4.22). Gley horizons may be either Bg and/or Ag horizons, depending on the height of the water table influence in the soil profile (Fig. 4.25).

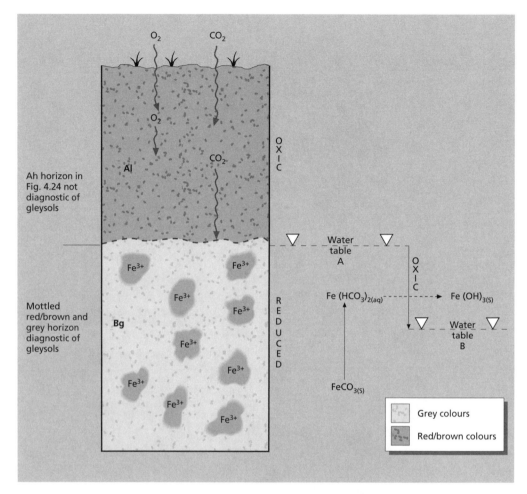

Fig. 4.25 Schematic diagram of a soil with a gley horizon based on the mollisol (mollic gleysol) in Plate 4.2(c). The upper Ah horizon contains oxygen and carbon dioxide in the soil atmosphere, while the lower Bg horizon is very low in oxygen. Under conditions of high water table (A) and low oxygen, Fe^{2+} species are stable and soil colours are grey. A fall in water table (B) allows further ingress of oxygen into the Bg horizon causing oxidation of Fe^{2+} species and precipitation of iron oxides as red/brown patches (mottles). The master horizons are typically centimetres to tens of centimetres thick.

The mollisol (mollic gleysol) in Plate 4.2, for example, has a Bg but no Ag horizon. Gley horizons have a mottled red-brown/grey appearance on account of both reduced (Fe^{2+}, grey) and oxidized (Fe^{3+}, red/brown) iron species being present together. Both iron species are present because of the continuous change in redox conditions (Box 4.3) within the soil profile as the water table rises and falls.

The geochemical cycling of iron species in gley horizons (Fig. 4.25) is mediated strongly by microbiological reactions (Section 4.6.5). Most permanent groundwater has a high Ca^{2+} concentration, near neutral pH and very low dissolved oxygen (Eh near zero). Under these conditions iron is stable in the reduced mineral $FeCO_3$ called siderite (see Box 5.4). However, due to the ingress of atmospheric CO_2 into the soil (or CO_2 that has been generated as the result of biological activity) the rather insoluble $FeCO_3$ can be converted to a reduced aqueous species $Fe(HCO_3)_2$ and mobilized:

$$FeCO_{3(s)} + H_2O_{(l)} + CO_{2(aq)} \rightarrow Fe(HCO_3)_{2(aq)} \qquad \text{eqn. 4.19}$$

As water moves through the soil, either by fluctuating groundwater levels or capillary action, this mobile aqueous iron species is dispersed. If the water table drops, however, the reduced $Fe(HCO_3)_2$ is exposed to more oxygenated condition, resulting in the formation of oxidized (Fe^{3+}) minerals such as $Fe(OH)_3$ (Fig. 4.25).

4.10 Contaminated land

In exceptional cases, the rocks, minerals and soils of the land surface contain compounds that generate natural chemical hazards. Uranium (U) and potassium (K), common elements in granitic rocks, are inherently unstable because of their radioactivity (see Section 2.8) and radioactive decay of isotopes of uranium to form radon (Rn) gas can be a health hazard (Box 4.13). Some chemicals, such as herbicides and pesticides, are present in soils because we put them there intentionally. Other chemicals arrive in soils because of unintentional or unavoidable releases, for example the byproducts of combustion in car engines. Exotic or synthetic chemicals (see Section 1.4) are ubiquitous in the environment today. In the USA alone, 20×10^9 kg of ethylene and 1×10^9 kg of benzene were produced in 1996. These compounds are the feedstock chemicals for synthetic processes that generate a huge array of synthetic organic compounds. Significant proportions of these compounds are released, usually by accident, into the atmosphere, hydrosphere and soils. There are a number of routes through which these compounds might reach soils, including aerial deposition, spillage, leaching and movement in groundwater. Thus, it is possible for some chemicals to reach areas remote from the site of compound production or use (see Section 7.4).

Where a substance is present in the environment at a concentration above natural background levels the term 'contaminant' is used. The term 'pollutant' is used when a contaminant can be shown to have a deleterious effect on the environment. Contaminants are broadly divided into two classes: (i) organic contaminants with chemical structures based on carbon, for example benzene; and (ii) inorganic contaminants, for example asbestos or lead, which may be in compound, molecular or elemental form. In the following section we concentrate on organic contaminants in soils; the inorganic contaminants mercury and arsenic are discussed in the context of water chemistry in Sections 5.6 and 5.7.2. This is an artificial division because organic and inorganic contaminants affect both soils and water. However, the chemical principles involved are similar in each case.

4.10.1 Organic contaminants in soils

Organic contaminants are compounds with a carbon skeleton, usually associated with atoms of hydrogen, oxygen, nitrogen, phosphorus and sulphur (see Section

Box 4.13 Radon gas: a natural environmental hazard

Radon gas (Rn) is a radioactive decay product of uranium (U), an element present in crustal oxides (e.g. uraninite—UO_2), silicates (e.g. zircon—$ZrSiO_4$) and phosphates (e.g. apatite—$Ca_5(PO_4)_3$ (OH, F, Cl)). These minerals are common in granitic rocks, but are also to be found in other rocks, sediments and soils. Uranium decays to radium (Ra), which in turn decays to radon (Rn) (see Section 2.8). The isotope ^{222}Rn exists for just a few days before it also decays, but, if surface rocks and soils are permeable, this gas has time to migrate into caves, mines and houses. Here, radon or its radioactive decay products may be inhaled by humans. The initial decay products, isotopes of polonium, ^{218}Po and ^{216}Po, are non-gaseous and stick to particles in the air. When inhaled they lodge in the lungs' bronchi, where they decay—ultimately to stable isotopes of lead (Pb)—by ejecting α radiation particles (see Section 2.8) in all directions, including into the cells lining the bronchi. This radiation causes cell mutation and ultimately lung cancer. Having said this, radon is estimated to cause only about one in 20 cases of lung cancer in Britain, smoking being a much more serious cause.

Radon gas is invisible, odourless and tasteless. It is therefore difficult to detect and its danger is worsened by containment in buildings. Radon is responsible for about half the annual radiation dose to people in England, compared with less than 1% from fallout, occupational exposures and discharges from nuclear power stations.

In England, about 100 000 homes are above the government-adopted 'action level' of 200 becquerels m^{-3}. Various relatively low-cost steps can be taken to minimize home radon levels, including better underfloor sealing and/or ventilation. Building homes in low-radon areas remains an obvious long-term strategy, but such simple solutions are not always applicable, because of either geographic or economic constraints. For example, bauxite processing in Jamaica produces large amounts of waste red mud. This material binds together strongly when dry, and is readily available as a cheap building material. Unfortunately, the red mud also contains higher levels of ^{238}U than most local soils. These cheap bricks are thus radioactive from the decay of ^{238}U and a potential source of radon. Only careful consideration of the health risks in comparison with the economic benefits can decide whether red mud will be used as a building material.

2.7). These atoms may form an integral part of the molecule, alternatively they may be present in functional groups (see Section 2.7.1). Functional groups impart specific chemical properties on a molecule. Hydroxyl (-OH) and carboxyl (-COOH) functional groups increase polarity, making a molecule more soluble (Box 4.14), while -COOH also makes the molecule acidic due to dissociation of H$^+$ (see Section 2.7.1). The various structural forms of organic molecules, for example saturated and unsaturated chains and rings (see Section 2.7), result in a diverse range of organic contaminants (Fig. 4.26).

Once in the soil environment organic contaminants may move in, or interact with, the soil atmosphere, soil water, mineral fractions and organic matter. Ultimately, however, the organic contaminants will either dissipate or persist (Fig. 4.27). Compounds persist if they are of low volatility, low solubility (Box 4.14) or have a molecular structure that resists degradation. Conversely, if compounds are highly volatile, highly soluble or are easily degraded, they will be destroyed or lost to other environments (e.g. the atmosphere or hydrosphere). The volatility of a compound is controlled by the vapour pressure (Box 4.14), while solubility is governed principally by polarity—a function of molecular structure, molecular weight and functional groups. Degradation results from both biological and abiological mechanisms, although given the profusion of micro-organisms in soil (a gram of soil typically contains 10^6–10^9 culturable micro-organisms) the potential for biodegradation is high. Abiological degradation occurs by hydrolysis, reduction, oxidation and photo-oxidation.

Organic contaminants interact mainly with either the mineral or organic com-ponents of soils. Two types of non-reactive interaction are possible: (i) adsorp-tion (a surface phenomenon); and (ii) entrapment within the soil minerals or components (Fig. 4.28). The nature and extent of interaction is dependent on the properties of the molecule—its aqueous solubility, vapour pressure and hydro-phobicity (Box 4.14)—but also its concentration and the properties of the soil. Soil factors include the amount and type of soil organic matter, the clay content and mineralogy, pore size and pore structure, and the microorganisms present. The net result of both adsorption and entrapment is a decrease in both the bio-logical and chemical availability of soil-associated contaminants with time. Thus, the proportion of compound in the 'available' fraction decreases with time (Fig. 4.29), while the proportion of compound in the 'non-available' fraction increases with time (Fig. 4.29). This process known as 'ageing', may cause a decrease in the rate, and the extent, of degradation an organic contaminant suffers in the soil environment. Organic contaminants attached to soil particles exist in four frac-tions distinguished by the ease with which they can be released or desorbed from

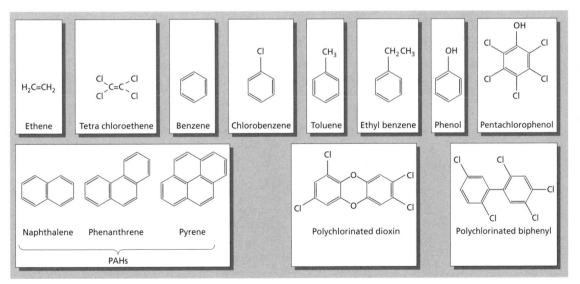

Fig. 4.26 Examples of organic contaminants.

Box 4.14 Physical and chemical properties that dictate the fate of organic contaminants

Polarity

Polarity is a measure of how skewed the electron distributions are within a molecule. We have seen previously that water (H_2O) has an oxygen atom with a negative dipole (imbalance in charge) and two hydrogen atoms with positive dipoles (Box 4.1). These dipoles arise because oxygen is more electronegative (Box 4.2) and therefore draws more of the electron density in the oxygen–hydrogen bonds closer to itself. Conversely, hydrogen being more electropositive assumes a positive dipole in the molecule. Other molecules display polarity as shown below (Fig. 1).

Solubility

The solubility of a compound depends on its polarity. If water is the solvent, the compound will need to be of similar polarity to be dissolved in it, for example ethanol (CH_3CH_2OH). In ethanol the OH-group is polar and therefore the molecule has polar character (Fig. 1b). If the compound is non-polar, for example ethane, it will not dissolve in polar water (Fig. 1c) because opposites do not mix. Ethanol will, however, dissolve in a non-polar solvent, for example hexane. Thus, the term 'solubility' should always be used with clarification of the solvent, for example 'aqueous solubility'. Solubility is expressed as the mass of a substance that will dissolve in a given volume of solvent, for example $mg\,l^{-1}$.

Hydrophobicity

The hydrophobicity of a compound is a measure of its affinity for water. If a compound will not readily partition into the aqueous phase it is known as hydrophobic (fearing water). A converse term, often used when talking about organic compounds, is lipophilicity. If a compound is lipophilic it loves lipid (fat). These terms are used interchangeably when the environmental fate of organic contaminants is discussed. Hydrophobicity is measured as a partition coefficient between octanol/water (K_{OW}). Octanol is chosen to represent lipid (fat) because it is an experimentally reproducible compound. K_{OW} values are determined by allowing the compound of interest to equilibrate between the two phases: water and octanol. After equilibration the concentrations in each phase are determined and their ratio calculated (Fig. 2). Most organic compounds are very hydrophobic such that K_{OW} values are often in the range 10 000 to 1 000 000. In order to work with smaller numbers the log of the K_{OW} value is usually used, such that the values are typically between 4 and 6.

Vapour pressure and volatility

Water vapour condenses to liquid water when cooled. At a constant pressure liquid water appears abruptly at a specific temperature and the pressure is known as the vapour pressure for this temperature. The vapour pressure of a compound is a measure of its volatility, i.e. its tendency to evaporate into the gas phase. A compound with high vapour pressure is described as volatile, while a compound with low vapour pressure is described as non-volatile. The vapour

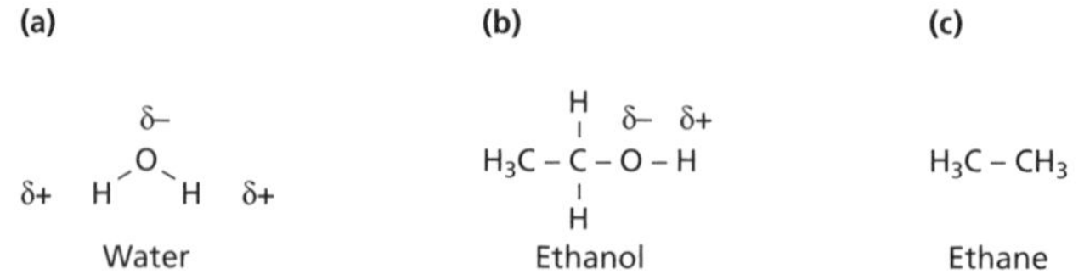

(a)   (b)   (c)

Water   Ethanol   Ethane

Fig. 1 Dipole distribution in (a) water (polar), (b) ethanol (polar) and (c) ethane (non-polar).

(continued)

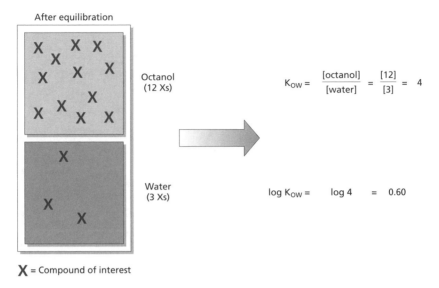

After equilibration

Octanol
(12 Xs)

Water
(3 Xs)

$$K_{OW} = \frac{[octanol]}{[water]} = \frac{[12]}{[3]} = 4$$

$$\log K_{OW} = \log 4 = 0.60$$

X = Compound of interest

Fig. 2 Schematic diagram illustrating experimental equilibration of compound X between water and octanol and determination of the partition coefficient (K_{OW}).

pressure of water can be used as a 'yardstick' for comparison. At 25°C water has a vapour pressure of 3.17 kPa. This means that liquid water is at equilibrium with water vapour at 25°C if the pressure is 3.17 kPa. As temperature changes so too does the equilibrium vapour pressure. At 100°C the vapour pressure of water is 101.3 kPa, equal to normal atmospheric pressure. At this temperature the vapour pressure of water molecules causes rapid formation of bubbles (boiling). Boiling occurs when the vapour pressure of the molecules in water equals (or exceeds) local atmospheric pressure and water vapour escapes to the atmosphere as steam.

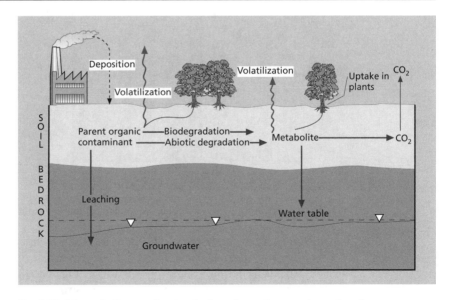

Fig. 4.27 Schematic diagram showing the fate of organic contaminants in soils.

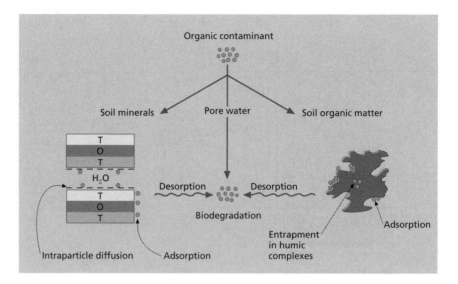

Fig. 4.28 Interaction between organic contaminants and soil components. Of the inorganic components, clay minerals have the most potential to react with organic contaminants. T, tetrahedral sheet; O, octahedral sheet (see Fig. 4.12).

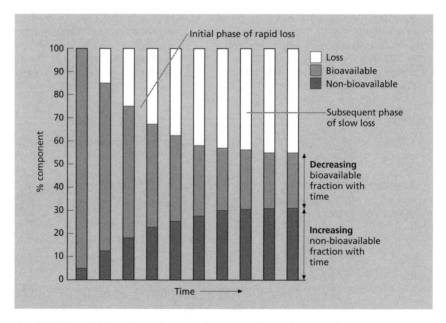

Fig. 4.29 Bioavailability of organic contaminants in soils as a function of time.

the particles: (i) a rapidly desorbable fraction; (ii) a slowly desorbable fraction; (iii) a very slowly desorbable fraction; and (iv) non-extractable (bound) residues. The non-extractable residue is the fraction of an organic compound (or its metabolites) that persists in the matrix following an extraction process that has not substantially changed either the compounds or the matrix. Bound residues usually represent an extreme end-member of ageing. The size of each of these fractions depends mainly on the length of contact time between the soil and contaminant.

Although the attraction between mineral surfaces and organic contaminants in soils can present a problem in cleaning up contaminated land, it can also be put to use by environmental chemists as a way of cleaning up contaminated water (Box 4.15).

4.10.2 Degradation of organic contaminants in soils

Degradation of organic contaminants in soils occurs typically by either chemical or microbiological pathways. The effectiveness of degradation is largely determined by the contaminant availability (Section 4.10.1), although the degree of persistence is influenced by the chemical structure of the contaminant. If the chemical structure of the contaminant is similar to that of a natural substance it is more likely to be degradable. In general, if the structure is complex the rate of degradation is slower and is more likely to be incomplete. Resistance to biodegradation is known as recalcitrance, which is caused by a number of factors:

1 Specific microbes or enzymes required for degradation may not be present in the soil.

2 Unusual or complex substitutions in a molecule (e.g. chlorine (Cl), bromine (Br) or fluorine (F)), or unusual bonds or bond sequences, may 'confuse' microorganisms that would otherwise recognize the molecule as a substrate (Box 4.16).

3 A high degree of aromaticity (see Section 2.7), i.e. strongly bonded structures based on a number of fused benzene rings, results in molecules that are difficult to break down.

4 Large, complex and heavy molecules tend to be less water soluble, and therefore are physically unavailable to microorganisms that use intracellular degradation processes.

Biodegradation of organic contaminants can be carried out by a single microbial species in pure cultures, but in nature the efforts of a mixture of microbes (a consortium) are usually required. The degradation process ranges from only minor structural changes to the parent molecule, known as primary degradation, to complete conversion to mineral constituents, for example CO_2 or H_2O, and termed mineralization:

Phenol → *primary degradation* → Catechol → *mineralization* → $CO_2 + H_2O$

eqn. 4.20

Box 4.15 Use of clay catalysts in clean up of environmental contamination

The interaction between some organic contaminants and mineral surfaces has recently attracted attention as a way of cleaning up contaminants in natural waters. The large cation exchange capacity of smectite clay minerals (Section 4.5.2), in particular, has prompted research into their use as a catalyst, i.e. a substance that alters the rate of a chemical reaction without itself changing. Clay catalysts have potential applications as adsorbents to treat contaminated natural waters or soils.

The compound 2,3,7,8-tetrachlorodibenzo-*p*-dioxin is one of the most toxic priority pollutants on the US Environmental Protection Agency's list. Dioxin compounds act as nerve poisons and are extremely toxic. There is no lower limit at which dioxins are considered safe in natural environments. The destruction of dioxins by biological, chemical or thermal means is costly, not least because their low (but highly significant) concentrations are dispersed in large volumes of other (benign) material. Thus large volumes of material must be treated in dioxin destruction processes.

It is desirable to concentrate soluble contaminants like dioxin by adsorption on to a solid before destruction. The optimal solid adsorbent should be cheap, benign, recyclable and easy to handle and have a high affinity for—and be highly selective to – the contaminant. Finely ground activated carbon and charcoal have been used as adsorbents but they suffer from oxidation during thermal destruction of the contaminant, making them non-recyclable and expensive.

Smectite clay catalysts are potential alternative adsorbents, although some modifications of the natural mineral are necessary. Interlayer sites in smectite dehydrate at temperatures above 200°C, collapsing to an illitic structure. Since the ion-exchange capacity of smectite centres on the interlayer site, collapse must be prevented if clay catalysts are to be used in thermal treatments of chemical organic toxins. The intercalation of thermally stable cations, which act as molecular props or pillars, is one

method of keeping the interlayer sites open in the absence of a solvent like water (Fig. 1). Various pillaring agents can be used, for example the polynuclear hydroxyaluminium cation ($Al_{13}O_4(OH)_{28}^{3+}$), which is stable above 500°C.

The pillar also increases the internal surface area of the interlayer site, making it more effective as an adsorbent. Moreover, by introducing cationic props of different sizes and spacings (spacing is determined by the radius of the hydrated cation and the charge), it is possible to vary the size of spaces between props. It is thus possible to manufacture highly specific molecular sieves, which could be used to trap large ions or molecules (e.g. organic contaminants) whilst letting smaller, benign molecules pass through.

Smectite clays do not have a strong affinity for soluble organic contaminants and this is improved using a surface-active agent (surfactant). A surfactant is a substance introduced into a liquid to affect (usually to increase) its spreading or wetting properties (i.e. those properties controlled by surface tension). Detergents and soaps are examples of surfactants. A soap molecule has two features essential for its cleansing action: a long, non-polar hydrocarbon chain and a

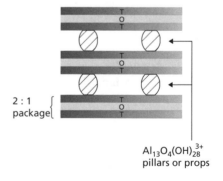

$Al_{13}O_4(OH)_{28}^{3+}$
pillars or props

Fig. 1 Schematic diagram showing props or pillars in the interlayer position in smectite clay. T, tetrahedral sheet; O, octahedral sheet.

(continued)

polar carboxylate group, for example sodium octadecanoate (sodium stearate).

$$CH_3(CH_2)_{16}-COO^-Na^+$$
(non-polar hydrocarbon tail) (carboxylate head)

The polar carboxylate head dissolves in water, while the long, non-polar hydrocarbon-chain tail is hydrophobic (Box 4.14) but mixes well with greasy (lipophilic, Box 4.14) substances, effectively floating the grease into solution in a sheath of COO^- groups.

Surfactant-coated interlayer sites in smectite have similar properties to detergents. A hydrophobic molecule or compound such as dioxin or other chlorinated phenol is attracted to the hydrophobic hydrocarbon tail projecting from the interlayer surface and immobilized.

The modified smectite thus has good affinity and selectivity for its target contaminant and is thermally stable, recyclable and economic to use (Fig. 2).

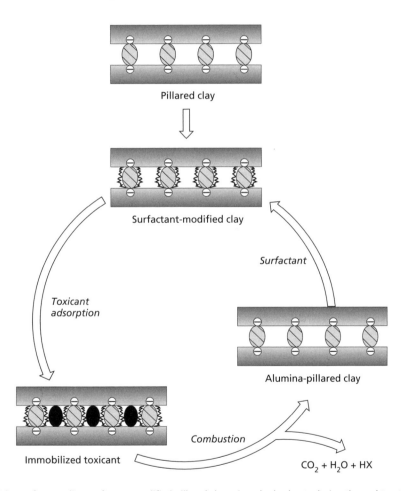

Fig. 2 Scheme for recycling surfactant-modified pillared clay mineral adsorbents during thermal treatment of toxicant. From Michot and Pinnavaia (1991), with kind permission from the Clay Minerals Society.

Box 4.16 Mechanisms of microbial degradation and transformation of organic contaminants

Mineralization

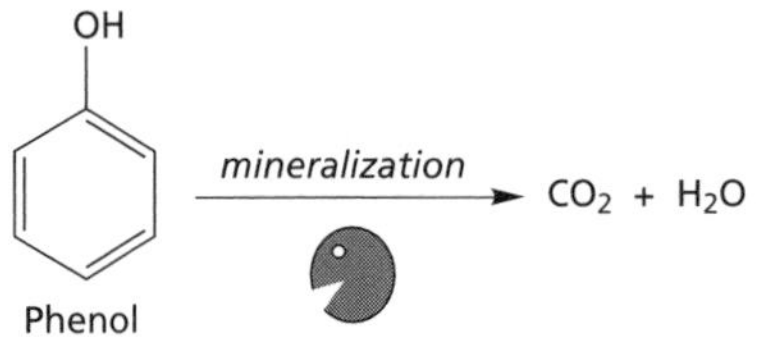

Phenol

$$\text{Phenol} \xrightarrow{\text{mineralization}} CO_2 + H_2O \qquad \text{eqn. 1}$$

Complete biodegradation of an organic contaminant into its inorganic components for example CO_2 and H_2O.

Polymerization

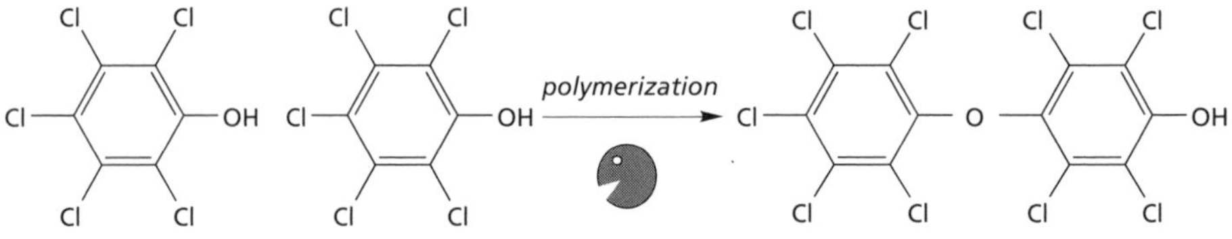

Pentachlorophenol   *Para*-pentachlorophenoxy-tetrachlorophenol

eqn. 2

A metabolic transformation that involves the coupling or bonding of small molecules to form polymers. In equation 2, two pentachlorophenol molecules polymerize to form *para*-pentachlorophenoxy-tetrachlorophenol.

Detoxification

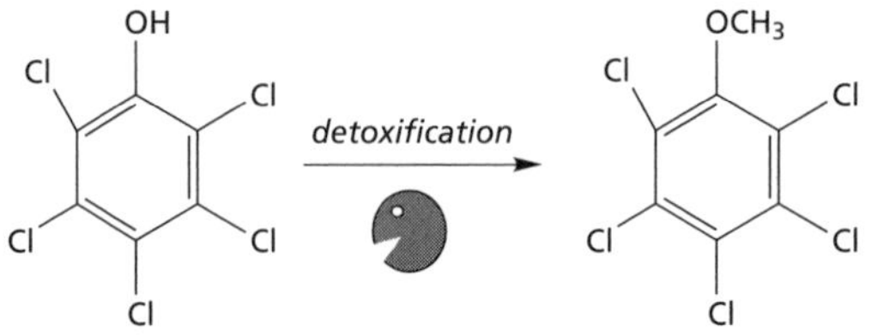

Pentachlorophenol   Pentachloroanisole

eqn. 3

Detoxification involves a chemical change to a molecule. In equation 3, pentachlorophenol (PCP), a powerful biocide used in wood preservative, undergoes *O*-methylation. This transforms PCP to a far less toxic compound pentachloroanisole. Detoxification by one group of microorganisms often allows other organisms to continue biodegradation.

(continued)

Co-metabolism

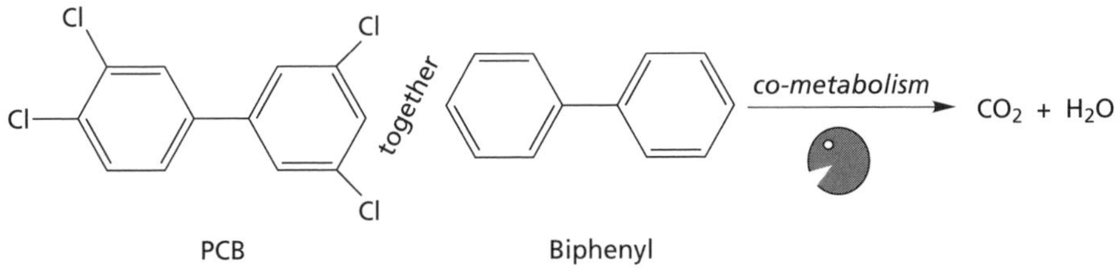

eqn. 4

Although many aromatic compounds are subject to microbial degradation, the presence of substituents (e.g. Cl, Br, CH₃) on the molecule can result in an increase in their resistance to biodegradation. In these instances microbes that would normally recognize the molecule as a substrate fail to do so because of the presence of the novel substituents, and hence do not produce enzymes to destroy them. For example, biphenyl (unsubstituted) is readily degraded. However, the polychlorinated biphenyls (PCBs), for example 2,3,2',4'-tetrachlorinated biphenyl, are recalcitrant. Interestingly where biphenyls and PCBs are present together, both are degraded (eqn. 4). This is because the microbes recognize the biphenyl and produce enzymes to degrade it. The same enzymes degrade the PCBs and thus both contaminants are removed. This is co-metabolism, a process that accounts for the degradation of many xenobiotic (foreign to life) compounds.

Accumulation on or within microbes

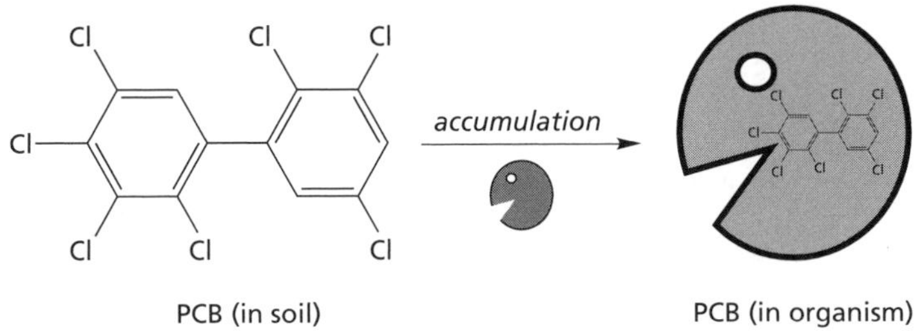

eqn. 5

Certain microorganisms, particularly those with high lipid (fat) contents, can absorb water-insoluble chemicals, such as PCBs. Although this is not strictly biodegradation, it is a process that removes pollutants from the environment.

The main styles of biodegradation and transformation are described in Box 4.16. Even slight structural molecular transformations of the parent contaminant molecule brought about by biodegradation can alter significantly the original contaminant's toxicity, mobility (Fig. 4.30) and its affinity for soil surfaces (Fig. 4.28).

4.10.3 Remediation of contaminated land

In many cases, contaminated land can be treated in order to rehabilitate it for future use. The success of remediation techniques depends on the concentration, type and availability of the contaminants present, and on-site factors, such as soil texture, pH, availability of terminal electron acceptors (Section 4.6.5) and the age

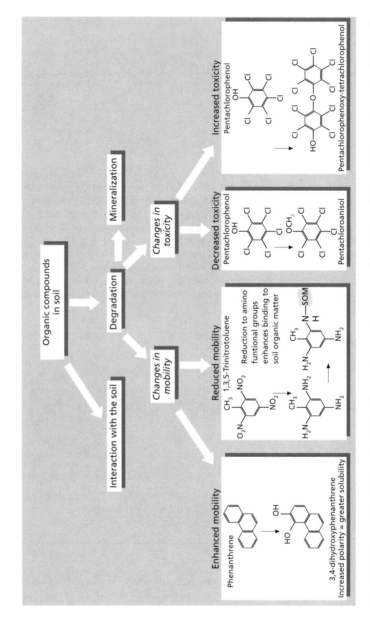

Fig. 4.30 Changes in toxicity and mobility of organic contaminants caused by biodegradation and biotransformation. SOM, soil organic matter.

of the contamination. Social and economic considerations will also affect remediation options. Cost is always a key issue, and this might need to be balanced against the chances of successful remediation. It might be important to know how long the remediation will take or whether the remediation technique is sustainable (see below).

Remediation options are broadly physical, chemical or biological in approach. Physical remediation includes dig-and-dump, incineration or containment of contaminants on site. Generally, these approaches provide a guaranteed 'quick-fix', but at a cost. Some of these options are expensive (incineration), while others simply pass the problem on without addressing the root problem of contamination (dig and dump and containment). Soil washing with, for example, surfactants (Box 4.15) and/or solvents is a chemical remediation option, as is the addition of chemically active reagents to promote contaminant degradation and/or immobilization. Overall, physical and chemical remediation options are non-sustainable because typically they alter a soil's structure, chemistry or biology.

Biological options are described by the term bioremediation, 'the elimination, attenuation or transformation of polluting or contaminating substances by the use of biological processes, to minimize the risk to human health and the environment'. In contrast to physical and chemical methods, biological options generally maintain soil integrity with respect to its chemical and biological elements. However, bioremediation often takes months or years to complete, and success cannot be guaranteed. Although bioremediation is driven by biological vectors, much chemistry (biochemistry) is involved. Biodegradation occurs under both aerobic and anaerobic conditions, although aerobic degradation is generally faster and more extensive for most contaminants (Section 4.6.5). Thus, aerobic conditions are generally promoted during bioremediation strategies, for example air-venting.

The same physical and chemical properties that dictate a contaminant's fate in soils (Box 4.14) also dictate the amenability of contaminants to bioremediation. In general, lighter molecules that are non-halogenated (e.g. without Cl) and with high polarity are more readily biodegraded. This is because lighter and simpler molecules are inherently more biodegradable and because polar molecules are more soluble, and hence available for degradation. The bioavailability of a contaminant may also be influenced by its length of contact with the soil; longer contact usually lowering bioavailability. Toxicity is also an issue because some compounds are highly toxic to microbes (e.g. pentachlorophenol (PCP), the fungicide used in wood preservers).

Bioremediation is either done *in situ*, with contamination treated where it occurs, or *ex situ*, where contaminated soil is removed by excavation prior to treatment. In some cases treatment can be *ex situ* on-site, i.e. where soil is excavated but treated on site in heaps. *In situ* and *ex situ* approaches have their advantages and disadvantages, as outlined in Table 4.11.

In situ *bioremediation by biostimulation*—Exxon Valdez *oil spillage*

Biostimulation is the promotion of favourable conditions to facilitate the degradation of contaminants by *in situ* microorganisms. Stimulation can be achieved

Table 4.11 Consideration of typical factors relating to *in situ* or *ex situ* treatment of contaminated land.

In situ	*Ex situ*
For	*Against*
Less expensive	More expensive
Creates less dust	Creates dust during excavation
Causes less release of contaminants	May disperse contaminants
Treats larger volumes of soil	Limited in scale—batches treated individually
Against	*For*
Slower	Faster
Difficult to manage	Easier to manage—ensure results
Not suited to high clay soils or compacted sites	Suited to a variety of sites including high clay and compacted sites

by addition of nutrients, addition of air/oxygen (a process called 'bioventing'), addition of other terminal electron acceptors (e.g. hydrogen peroxide) or addition of co-metabolic substrates.

On 24 March 1989 the *Exxon Valdez* oil tanker ran aground on Bligh Reef, Prince William Sound, Alaska spilling 37 000 tonnes of oil. Despite efforts to contain the spill, tidal currents and winds caused a significant proportion of the oil to be washed ashore. Approximately 15% (~2000 km) of shoreline in Prince William Sound and the Gulf of Alaska became oiled to some degree. Bioremediation was one of a number of techniques applied in the clean-up operation. Bioremediation was favoured because the majority of molecules in crude oil are biodegradable and because shorelines often support large populations of oil-degrading microorganisms.

Biostimulation of the shoreline microbes was effected through addition of fertilizers. Two products were applied: Inipol EAP22, a urea-based product designed to stick to oil, and Customblen a slow-release granular fertilizer containing ammonium nitrate, calcium phosphate and ammonium phosphate. These products were selected to minimize nutrient losses with the tides and thereby optimize nutrient input to the oiled areas. By late summer 1989 approximately 120 km of shoreline had been treated in this way.

Hydrocarbon degradation by this method starts with attack of the methyl group ($-CH_3$) at the extremity of the hydrocarbon chain (terminal $-CH_3$), a process called methyl-oxidation, resulting in formation of a carboxylic acid group ($-COOH$; Fig. 4.31). The reaction then proceeds by β-oxidation (explained in Fig. 4.31), a process that cleaves C_2 units from the hydrocarbon chain as ethanoic (acetic) acid (CH_3COOH). Ethanoic acid is then utilized in the tricarboxylic acid (TCA) cycle through which the microbes derive energy. This mechanism is limited to straight-chain molecules; branch chains have to be removed by other degradation pathways before β-oxidation can proceed.

Comparison of oil degradation between treated plots and adjacent control plots indicated that after 109 days the treated plots had experienced about 90% consumption of hydrocarbons, whereas no significant changes had occurred

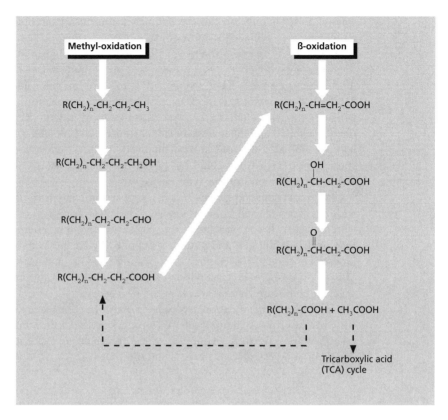

Fig. 4.31 Biodegradation pathways of hydrocarbon. Methyl-oxidation, by attack of the methyl group (–CH₃) at the extremity of the hydrocarbon chain, results in formation of a carboxylic acid group (–COOH). β-oxidation indicates that oxidation occurs at the second carbon atom (counted from the end that bears the –COOH group, the α-carbon atom being immediately adjacent to the –COOH group). β-oxidation continues removing C_2 units, and in effect unzips the hydrocarbon chain until it no longer exists.

on the control plots. Hydrocarbon consumption rate in the control plot was $0.052\% \, d^{-1}$, but this increased to $0.45\% \, d^{-1}$ when the plot was fertilized, a rate enhancement of 8.6 times.

Ex situ *on-site bioremediation by composting — Finnish sawmills*

Gardeners know that compost is added to soil to provide a source of nutrients and to aid soil aeration by creating a more open soil structure. Gardeners might not know, however, that most compost is also a rich source of microorganisms; the compost effectively inoculates the soil with microbes. Under composting conditions heat is generated though the processes of degradation (see eqn. 5.20). This heat changes the microbial community and the rate at which it degrades organic matter, including any bioavailable organic contaminants. Conditions within a soil/compost mixture can thus be optimized in terms of aeration, nutrients and temperature, to achieve the most efficient degradation.

As part of its operations between 1955 and 1977, a Finnish sawmill had been impregnating timber with a preservative to inhibit microbial degradation. This product, called Ky-5, contained a mixture of chlorophenols, namely, 2,4,6-trichlorophenol (7–15%), 2,3,4,6-tetrachlorophenol (~80%) and pentachlorophenol (6–10%). Ky-5 also contained traces of polychlorinated phenoxyphenols and dibenzo-*p*-dioxins as impurities. Over the years this product had contaminated the soils around the sawmill. A cost-effective bioremediation strategy was needed that could be used at this site but also throughout Finland where 800 other sites of this type existed.

The bioremediation strategy used compost and composting materials mixed with the excavated soil in heaps known as biopiles. Two-parts contaminated soil were mixed with one-part inoculant that contained straw-compost, bark chips, lime (to adjust the pH) and nutrients (supplied by a commercial fertilizer). The biopiles contained 7500 kg of material (volume of 13 m^3) and were built on a layer of bark chips to provide insulation. The entire biopile was covered with a plastic sheet and moisture content was adjusted by watering.

Chlorophenol degradation proceeds via dechlorination (removal of Cl-groups) with hydroxylation (addition of OH-groups) at the dechlorinated sites. The microbes are effectively manipulating the molecule to make it susceptible to degradation by cleavage of the benzene ring.

pentachlorophenol

eqn. 4.21

In the case of pentachlorophenol (eqn. 4.21) the Cl-group opposite the OH-group on the benzene ring is replaced by an OH-group first. The next dechlorination/hydroxylation reaction yields a molecule with a total of three OH-groups, two of which are adjacent to each other. At this stage ring cleavage occurs between these adjacent OH-groups (see benzene degradation in Fig. 4.33) and the resultant straight-chain hydrocarbon degrades to derive smaller chlorinated and unchlorinated products.

The extent of degradation in the biopiles was proportional to the starting chlorophenol concentration. Overall, however, chlorophenol loss was between 80 and 90%. Where highly contaminated soil was treated (original chlorophenol concentration 850 mg kg^{-1}) the rate of loss was between 2 and 5 mg $(kg_{dry\ wt})^{-1} d^{-1}$. This is a fast rate of chlorophenol loss and the remediation process was complete within 3 months.

Ex situ *on-site bioremediation using a bioreactor —
polycyclic aromatic hydrocarbons*

A bioreactor is a silo containing a slurry of contaminated soil mixed with water. The slurry is aerated and enriched with nutrients and microorganisms, as necessary, to control and optimize decomposition. Bioreactors are usually small-volume closed systems that allow collection and treatment of volatile components. Their small volume usually limits their use to batches of soil treated individually, such that on large sites bioremediation will be slow and expensive. Bioreactors are particulary useful for the treatment of contaminant hot spots at a site, deployed alongside other bioremediation techniques.

In the 1980s the US Environmental Protection Agency (EPA) embarked on a programme known as Superfund, to clean up abandoned hazardous waste sites. The Superfund site at Burlington Northern (Minnesota) has a historic burden of creosote contamination. Creosote is principally a mixture of polycyclic aromatic hydrocarbons (PAHs). PAHs are a group of compounds based on fused benzene rings resulting in the formation of chains and clusters. The US EPA has listed 16 PAHs as 'priority pollutants' (Fig. 4.32) on account of their toxicity, typically as carcinogens and/or mutagens. PAHs exhibit a wide range of physical and chemical properties, governed principally by the number of fused benzene rings. Naphthalene, for example, the smallest PAH, comprises only two benzene rings (Fig. 4.32), has a molecular weight of 128, an aqueous solubility of $32 \, mg \, l^{-1}$ and a log K_{ow} value of 3.37 (Box 4.14). By contrast, benzo[*a*]pyrene is composed of five benzene rings (Fig. 4.32), has a molecular weight of 252, an aqueous solubility of $0.0006 \, mg \, l^{-1}$ and a log K_{ow} value of 6.04 (Box 4.14). The heavier PAHs with more than four rings are a problem for bioremediation, being relatively insoluble and strongly bonded molecules that are difficult to degrade.

Biodegradation of PAHs is analogous to the biodegradation of benzene (Fig. 4.33). Initial ring oxidation yields 1,2-dihydroxybenzene (commonly known as catechol). The benzene ring is then broken (cleaved). Ring-cleavage occurs either between the –OH groups or adjacent to one of them, known as ortho- and meta-cleavage, respectively (Fig. 4.33). Ring cleavage occurs at these positions because –OH groups are involved as reaction sites. Furthermore, because the process is enzymatically mediated the presence of adjacent –OH groups enables recognition of the molecule by the enzymes responsible for the degradation. Ring cleavage yields straight-chain products, namely *cis,cis*-muconic acid and 2-hydroxymuconic semialdehyde (Fig. 4.33). These products are further degraded to yield simple molecules, such as pyruvate, citrate and acetaldehyde, used in the tricarboxylic acid (TCA) cycle through which the microbes derive energy.

In the case of naphthalene (two-ringed PAH) initial ring oxidation yields 1,2-dihydroxynaphthalene (Fig. 4.34). Ring cleavage then occurs, followed by removal of side-chains to yield salicylic aldehyde. Salicylic aldehyde is then converted to catechol via salicylic acid (Fig. 4.34). Catechol is then degraded as illustrated for benzene in Fig. 4.33. For heavier PAHs the initial phases of degradation yield a catechol analogue to the PAH containing one less benzene ring than the original PAH. By way of illustration phenanthrene is converted to 1,2-

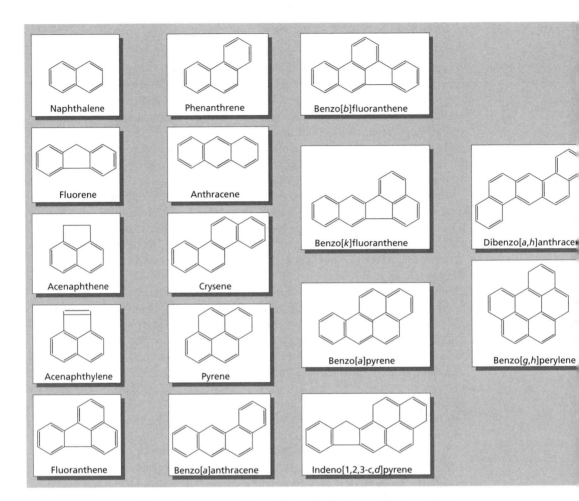

Fig. 4.32 Sixteen PAHs listed as priority pollutants by the US Environmental Protection Agency.

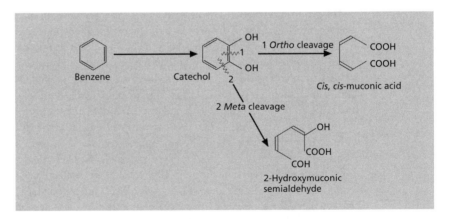

Fig. 4.33 Aerobic biodegradation pathways for benzene.

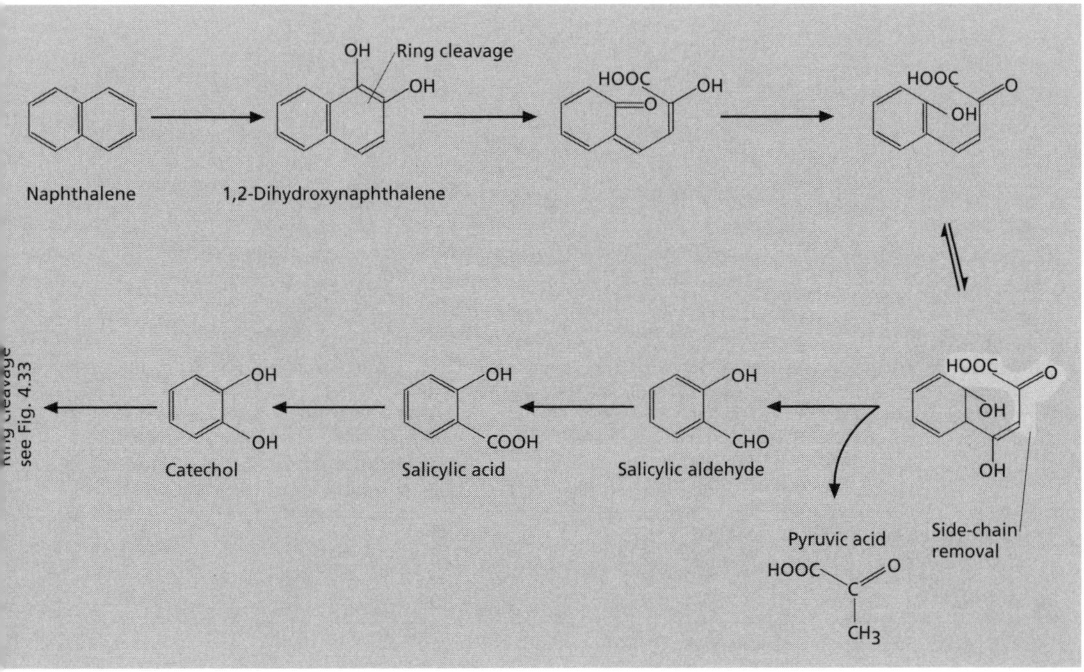

Fig. 4.34 Biodegradation pathways for naphthalene.

dihydroxynaphthalene (Fig. 4.35). Degradation then proceeds as shown for naphthalene (Fig. 4.34).

At the Burlington Northern site the creosote-contaminated soil was sieved and then ball milled to reduce particle size, a process that increases contaminant availability by increasing reactive surface area. The milled creosote-contaminated soil was then slurried with water and placed in five separate (to allow comparisons) 64-litre stainless steel bioreactors, equipped with aeration, agitation and temperature controls. An inoculum of PAH-degrading bacteria was then added, along with an inorganic supplement, containing nitrogen as NH_4, potassium, magnesium, calcium and iron. Conditions within the reactors were controlled to optimize degradation for 12 weeks.

Average initial concentrations in the PAH-contaminated soil and the residual concentrations after 12 weeks in the bioreactor are shown in Table 4.12. Although degradation was clearly greater for the lighter PAHs (98%) it was still extensive for the heavier compounds (70%). Furthermore, these extents of degradation were consistently achieved in the five replicated systems. Thus, while the use of bioreactors is technically more demanding, and more expensive, biodegradation is extensive and reliable.

4.10.4 Phytoremediation

Phytoremediation is the use of plants and trees to clean up metals, pesticides, solvents, explosive hydrocarbons, PAHs and leachates at contaminated sites.

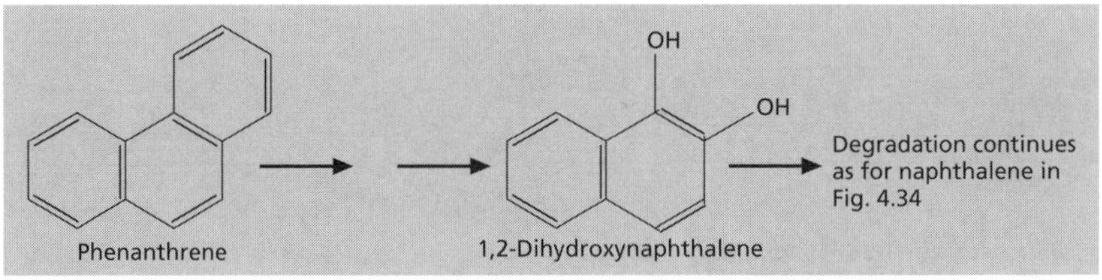

Fig. 4.35 Biodegradation pathway for phenanthrene as an example of a heavier PAH.

Table 4.12 Bioreactor remediation of creosote (PAH) contaminated soil, adapted from US Environmental Protection Agency technology demonstration sheet EPA/540/S5-91/009.

	Initial PAH concentration (mg kg^{-1})	Residual PAH concentration after 12 weeks' treatment (mg kg^{-1})	PAH reduction (%)
Two- and three-ring PAHs	1500	30	98
Four- through six-ring PAHs	960	280	70
Total PAHs*	2460	310	87

*Sixteen PAHs listed as priority pollutants by the US EPA (see Fig. 4.32).

While microbial bioremediation is usually the fastest and most widely applied clean-up technique, phytoremediation can prolong or enhance degradation over longer timescales on sites where microbial techniques have been used first. It is also useful at remote sites missed during the main remediation campaign, and can be aesthetically pleasing.

Plants may accumulate contaminants within their roots, stems and leaves. This is called phytoaccumulation (Fig. 4.36) and is known to remove a variety of heavy metals (see Section 5.6) from soils, including zinc (Zn), copper (Cu) and nickel (Ni). Once the plants have had sufficient time to accumulate contaminants they are harvested and usually incinerated to leave a metal-rich ash. The ash typically represents about 10% of the original mass of the contaminated soil, and is either landfilled or processed as a metal ore (bio-ore) if economically viable. Some plants exude enzymes that are capable of transforming organic contaminants into simpler molecules, used directly by the plants for growth, a process known as phytodegradation (Fig. 4.36). In some plants, degradation of contaminants occurs when root exudates (e.g. simple sugars, alcohols and acids) stimulate proliferation of microbial communities in the soil around the root (rhizosphere). This is known as phyto-enhanced or rhizo-enhanced degradation (Fig. 4.36). Roots also de-aggregate the soil matrix, allowing aeration and promoting biodegradation. Some plants take up volatile and semivolatile compounds from soil and translocate them to their leaves where volatilization to the atmosphere occurs. This phytovolatilization (Fig. 4.36) does not degrade or immobilize the contaminant, but

(a)

(b)

Plate 3.1 (a) Flying into Los Angeles in the early morning before an intense photochemical smog has developed. (b) Later in the day further north on the west coast (USA). Lengthy exposure of primary pollutants to sunlight has induced the photochemical formation of a brown smog. Photographs courtesy of P. Brimblecombe.

facing p. 138

Plate 4.1 Red bauxite-bearing oxisol (ferralsol) overlying Tertiary limestone (white) in south Jamaica. The soil is piped into solution-enlarged hollows of the limestone surface. Cliff face approximately 6 m high. Photograph courtesy of J. Andrews.

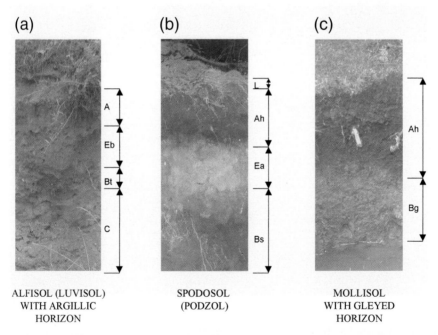

(a)

A
Eb
Bt
C

ALFISOL (LUVISOL)
WITH ARGILLIC
HORIZON

(b)

L
Ah
Ea
Bs

SPODOSOL
(PODZOL)

(c)

Ah
Bg

MOLLISOL
WITH GLEYED
HORIZON

Plate 4.2 Field photographs contrasting three soil profiles. (a) An alfisol (luvisol) with an argillic horizon. Clay material has been washed downward from the E to the B master horizon. Subordinate categories are indicated by lower case letters: the 'b' indicates an eluvation of clay from the E horizon, hence 'Eb'; the 't' indicates a clay-rich or argillic horizon in master horizon B, hence Bt. (b) A spodosol (podzol), where subordinate 'h' indicates illuval accumulation of organic matter, subordinate 'a' indicates highly decomposed organic matter, and subordinate 's' indicates illuvial accumulation of Fe and Al oxides. (c) A mollisol where subordinate 'g' indicates evidence of gleying (see Fig. 4.25). Notice the presence of groundwater in the bottom of the soil pit. Soil master horizons are typically centimetres to tens of centimetres thick (see also Fig. 4.21).

Plate 5.1 Excavations for copper at Parys Mountain, Anglesey, UK. The spoil heaps provide a huge surface area for sulphide oxidation. The lake waters in the pit bottom have pH below 3. Photograph courtesy of J. Andrews.

Plate 5.2 Crusts of iron oxide (goethite) coating stream bed. Drainage from Parys Mountain, copper mine Anglesey, UK. Photograph courtesy of J. Andrews.

Plate 6.1 Black smoker, East Pacific Rise. From Des Oceans aux Continents. *BSGF* 1984, (7), XXVI, no 3. Société Géologique de France, Paris. Reproduced with thanks.

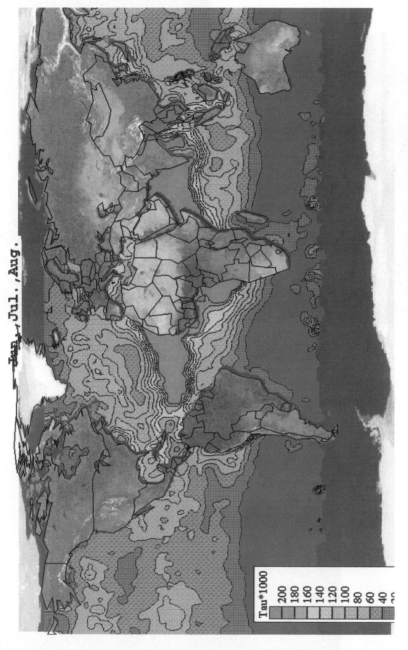

Plate 6.2 Global composite image of optical depth (tau units) in the Earth's atmosphere for June–August 1989–1991. Optical depth data provides a measure of aerosol concentration. Reproduced from Husar RB, Prospero JM and Stowe LL (1997) *Journal of Geophysical Research* 102, 16889–16909, copyright by the American Geophysical Union.

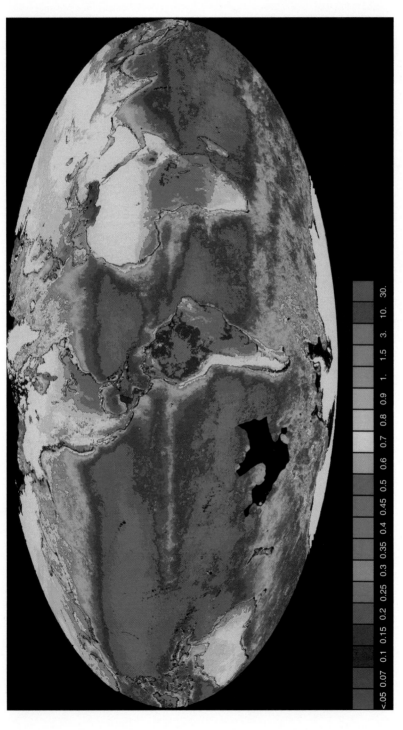

<.05 0.07 0.1 0.15 0.2 0.25 0.3 0.35 0.4 0.45 0.5 0.6 0.7 0.8 0.9 1. 1.5 3. 10. 30.

Plate 6.3 A global view of the Earth's biosphere from multiple satellite images compiled by NASA. Individual images focus on small areas of the Earth concentrating on the light spectrum dominated by green chlorophyll. This image, based on chlorophyll content, provides an estimate of plant standing stock rather than productivity (as in Box 6.6), although the two parameters are closely related. This colour image also demonstrates that there are many small-scale ocean features, such as eddies, which would complicate the simplified map in Box 6.6. The image also highlights the importance of tropical forests as areas of high plant biomass. Estimates of phytoplankton pigment concentrations (mg m^{-2}) shown according to colour code.

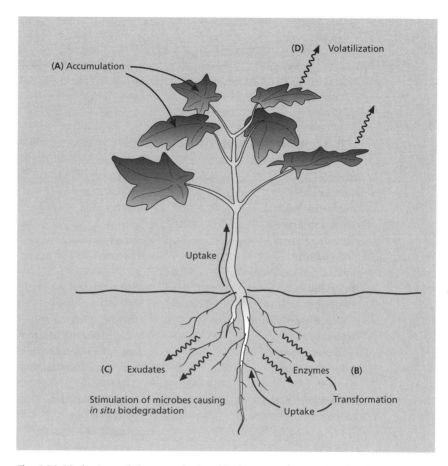

Fig. 4.36 Mechanisms of phytoremediation: (A) phytoaccumulation; (B) phytodegradation: (C) rhizo-enhanced degradation; (D) phytovolatilization.

disperses and dilutes it. Phytovolatilization has also been applied to land contaminated with mercury (see Section 5.6.1).

4.11 Further reading

Alexander, M. (1999) *Biodegradation and Bioremediation*, 2nd edn. Academic Press, San Diego.

Baird, C.B. (1995) *Environmental Chemistry*. WH Freeman, New York.

Berner, K.B. & Berner, R.A. (1987) *The Global Water Cycle*. Prentice Hall, Englewood Cliffs, New Jersey.

Birkeland, P.W. (1999) *Soils and Geomorphology*, 3rd edn. Oxford University Press, New York.

Brady, N.C. & Weil, R.R. (2002) *The Nature and Properties of Soils*, 13th edn. Prentice Hall, New Jersey.

Fitzpatrick, E.A. (1980) *Soils: their Formation, Classification and Distribution*. Longman, London.

Gill, R. (1996) *Chemical Fundamentals of Geology*, 2nd edn. Chapman & Hall, London.

Veizer, J. (1988) The evolving exogenetic cycle. In: *Chemical Cycles in the Evolution of the Earth*, ed. by Gregor, C., Garrels, R.M., Mackenzie, F.T. & Maynard, J.B., pp. 175–220. Wiley, New York.

4.12 Internet search keywords

quartz
feldspar
SiO_4 tetrahedron
radius ratio rule
electronegativity
chain silicate
double chain silicate
sheet silicate
framework silicate
chemical weathering
oxidation
reduction
redox potential
goethite
soil organic matter
kaolinite
illite
smectite
tetrahedral sheet
octahedral sheet
ion exchange clay
cation exchange capacity
soil chemistry
soil catena
oxisol
vertisol
chemical index alteration

soil pH
soil structure
soil horizon
spodosol
aluminium solubility
aquept
gley
clay catalysts
radon
organic contaminant
volatile organic contaminant
biodegradation organic contaminants
remediation contaminated land
PAHs soils
phytoremediation
hydrogen bonds
metastable minerals
reaction kinetics
van der Waals' forces
saturated solution
lignin
base cations
solubility product
solubility
carbonate saturation index
vapour pressure
hydrophobic

The Chemistry of Continental Waters

5

5.1 Introduction

Continental freshwaters are critical to terrestrial life, being the only reliable source of drinking water, and in some cases the medium for life itself. Water is thus also a vital human resource and its quality and quantity are key determinants of human development. Freshwater arrives on the continental land surface from precipitation (rain and snow), mostly derived from evaporation of seawater, and is found in ice-caps, lakes, rivers and groundwater (see Section 1.3.2). The polar ice-caps are long-term storage reservoirs of predominantly pure precipitation. The other reservoirs contain water that is usually much altered from the original precipitation by interactions with mineral weathering and biological processes. The important processes that chemically alter water vary for ground-waters, rivers and lakes. For instance, photosynthetic processes play a major role in regulating lakewater chemistry, but have little direct role in groundwater where there is no light. Similarly, individual rivers, lakes and even aquifers may have quite distinct chemistries depending on the relative importance of the processes that act within them. Of course these three reservoirs are physically interconnected and in general we treat them together in this chapter, although emphasizing clear differences where necessary. Although the chemical composition of rivers, lakes and groundwater varies widely, it is governed predominantly by three factors: element chemistry, weathering regimes and biological processes. In addition, human perturbations may have a major effect on some freshwater systems.

Table 5.1 Comparison of the major cation composition of average upper continental crust (from Wedepohl 1995) and average riverwater (from Berner & Berner 1987); except aluminium and iron from Broecker and Peng (1982).

	Upper continental crust (mg kg^{-1})	Riverwater (mg kg^{-1})
Al	77.4	0.05
Fe	30.9	0.04
Ca	29.4	13.4
Na	25.7	5.2
K	28.6	1.3
Mg	13.5	3.4

5.2 Element chemistry

The 20 largest rivers on Earth carry about 40% of the total continental runoff, with the Amazon alone accounting for about 15% of the total. These rivers give the best indication of global average riverwater chemical composition, which can be compared with average continental crust composition (Table 5.1). Three features stand out from this comparison:

1 Four metals dominate the dissolved chemistry of freshwater, all present as simple cations (Ca^{2+}, Na^+, K^+ and Mg^{2+}).

2 The low concentration of ions in freshwater.

3 The dissolved ionic composition of freshwater is radically different from continental crust, despite the fact that all of the cations in riverwater, with the exception of some of the sodium and chloride (see Section 5.3), are derived from weathering processes.

Although it is not meaningful to derive a single global average composition for groundwater — because of the marked differences in aquifer rocks — it is nonetheless true that most groundwaters share the three features listed above for riverwater (see Table 5.3).

The difference between crustal and dissolved riverwater composition is particularly marked for aluminium and iron relative to other metals (Table 5.1). This difference results from the way specific metal ions react with water.

Ionic compounds dissolve readily in polar solvents like water (see Box 4.1). Once in solution, however, different ions react with water in different ways (Fig. 5.1). Low-charge ions (1+, 2+, 1−, 2−) usually dissolve as simple cations or anions. These ions have little interaction with the water itself, except that each ion is surrounded by water molecules (Fig. 5.1a). In general, for elements with similar atomic number, the smaller the ionic radius the higher the charge on the ion. Small high-charge ions react with water, abstracting OH$^-$ to form uncharged and insoluble hydroxides, liberating hydrogen ions in the process (Fig. 5.1b), for example:

$$Fe^{3+}_{(aq)} + 3H_2O_{(l)} \rightarrow FeOH_{3(s)} + 3H^+_{(aq)}$$

eqn. 5.1

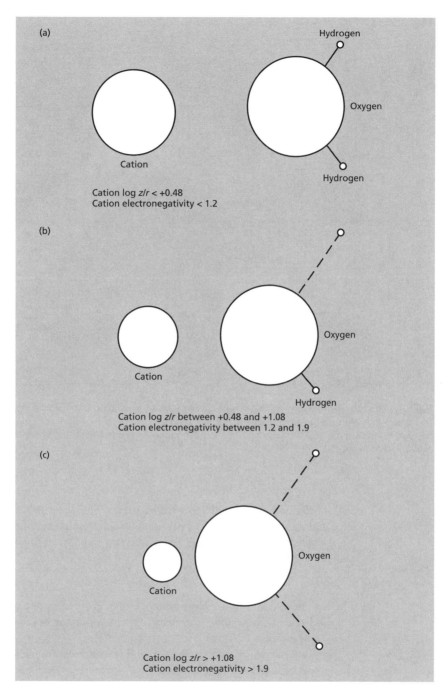

(a)

Hydrogen

Oxygen

Cation

Hydrogen

Cation log z/r < +0.48
Cation electronegativity < 1.2

(b)

Hydrogen

Oxygen

Cation

Hydrogen

Cation log z/r between +0.48 and +1.08
Cation electronegativity between 1.2 and 1.9

(c)

Oxygen

Cation

Cation log z/r > +1.08
Cation electronegativity > 1.9

Fig. 5.1 Relationship between cation properties and force of repulsion between a cation and the hydrogen ions in a water molecule. The log z/r data should be compared to Fig. 5.2. Electronegativity is explained in Box 4.2. After Raiswell *et al.* (1980). (a) Large, low-charge, electropositive ions (e.g. K^+) are surrounded by water molecules such that the centre of the negative charge on the oxygen is aligned toward the cation. (b) Smaller, more highly charged, electronegative cations (e.g. Fe^{3+}) interact more strongly with the water molecule forming a bond to the oxygen and displacing one H^+ to form hydroxide. (c) Small, highly charged, strongly electronegative cations (e.g. P^{5+}) react to displace H^+ ions from the water, forming an oxyanion.

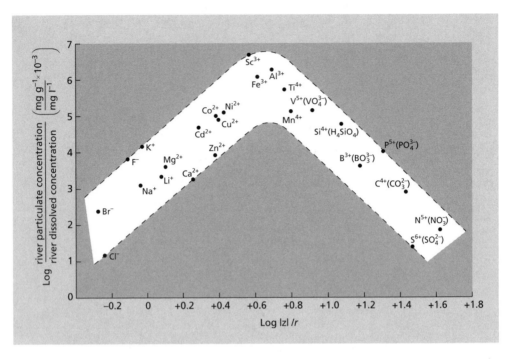

Fig. 5.2 Ratio of average elemental riverine particulate to dissolved concentrations plotted against the ratio of charge to ionic radius for the most abundant ions of those elements. In the case of dissolved oxyanions, the relevant dissolved species are shown in brackets. Concentration data from Martin and Whitfield (1983), other data from Krauskopf and Bird (1995).

Still smaller and more highly charged ions react with water to produce relatively large and stable anions (so called oxyanions), such as sulphate (SO_4^{2-}), by abstracting oxygen ions from water and again liberating hydrogen ions (Fig. 5.1c), for example:

$$S(VI)_{(s)} + 4H_2O_{(l)} \rightarrow SO_{4(aq)}^{2-} + 2H_{(aq)}^{+} \qquad \text{eqn. 5.2}$$

The net effect is to produce large anions, which dissolve readily since the charge is spread over a large ionic perimeter. Other important oxyanions are nitrate (NO_3^-) and carbonate (CO_3^{2-}).

The general pattern of element solubility can be rationalized in terms of charge and ionic radius (z/r) (Fig. 5.2). Ions with low z/r values are highly soluble, form simple ions in solution and are enriched in the dissolved phase of riverwater compared with the particulate phase, for example Na^+. Ions with intermediate z/r values are relatively insoluble and have a relatively high particulate: dissolved ratio in riverwater, for example Fe^{3+}, Al^{3+}. Ions with large z/r values form complex oxyanions and again are soluble, for example SO_4^{2-}, NO_3^-.

Some oxyanions exist in solution as weak acids and will ionize or dissociate (see Box 4.5) depending on the pH, as shown here for phosphorus.

$$H_3PO_{4(aq)} \leftrightharpoons H_2PO_{4(aq)} + H^+_{(aq)} \qquad\qquad \text{eqn. 5.3}$$

$$H_2PO^-_{4(aq)} + H^+_{(aq)} \leftrightharpoons HPO^{2-}_{4(aq)} + 2H^+_{(aq)} \qquad\qquad \text{eqn. 5.4}$$

$$HPO^{2-}_{4(aq)} + 2H^+_{(aq)} \leftrightharpoons PO^{3-}_{4(aq)} + 3H^+_{(aq)} \qquad\qquad \text{eqn. 5.5}$$

At pH 8 the HPO_4^{2-} species predominates. Understanding dissociation behaviour is very important in calculating the solubility of an ion. To do this it is important to know which species is the dominant one. A water chemist, for example, may wish to know if iron phosphate ($FePO_4$) can form in a river, as this species helps regulate phosphorus concentrations in freshwaters. To know this it is necessary to calculate the PO_4^{3-} concentration from the total dissolved inorganic phosphorus concentration that is measured analytically and contains all the species in equations 5.3–5.5. This can be done knowing the pH and dissociation constant (see Box 4.5) for equations 5.3–5.5. A worked example for the carbonate system is given in Section 5.3.1

Silicon is mobilized by the weathering of silicate minerals, mainly feldspars (see eqn. 4.14), and is transported in natural waters at near-neutral pH as undissociated silicic acid (H_4SiO_4), an oxyanion (Fig. 5.2). Silicate minerals weather slowly, such that rates of input—and concentrations—of silicon in most freshwaters are quite low. Despite this, where silicates are the main component of bedrock or soil, H_4SiO_4 can be a significant dissolved component of freshwater.

5.3 Water chemistry and weathering regimes

Comparison of dissolved major ion compositions in four large rivers draining very different crustal areas (Table 5.2) shows the dominance of calcium (Ca), magnesium (Mg), sodium (Na) and potassium (K). Overall, however, the chemistry of each river is different and weathering regimes control most of these variations.

The dissolved ion composition of freshwater depends upon:
1 the varying composition of rainfall and atmospheric dry deposition;
2 the modification of atmospheric inputs by evapotranspiration;

Table 5.2 Dissolved major ion composition ($mmol\,l^{-1}$) of some major rivers. Data from Meybeck (1979); except Rio Grande from Livingston (1963).

	Mackenzie (1)	Orinoco (2)	Ganges (3)	Rio Grande (4)
Ca^{2+}	0.82	0.08	0.61	2.72
Mg^{2+}	0.43	0.04	0.20	0.99
Na^+	0.30	0.06	0.21	5.10
K^+	0.02	0.02	0.08	0.17
Cl^-	0.25	0.08	0.09	4.82
SO_4^{2-}	0.38	0.03	0.09	2.48
HCO_3^-	1.82	0.18	1.72	3.00
SiO_2	0.05	0.19	0.21	0.50

Drainage basin characteristics: (1) northern arctic Canada; (2) tropical northern South America; (3) southern Himalayas; (4) arid southwestern North America.

3 the varying inputs from weathering reactions and organic matter decomposition in soil and rocks;

4 differential uptake by biological processes in soils.

Where crystalline rock or highly weathered tropical soils are present (i.e. where weathering inputs are low or exhausted), dissolved freshwater chemistry is most influenced by natural atmospheric inputs, for example sea spray and dust, as well as by anthropogenic gases, for example SO_2.

Sea-salt inputs are common in coastal regions. These salts have been introduced into the marine atmosphere from bubble bursting and breaking waves and are deposited on land with rain and dust fall. Small amounts of sea-salts are, however, also present in rainwater of central continental areas, thousands of miles from the sea. Sea-salt inputs have broadly similar, predominantly sodium chloride (NaCl), chemistry to the seawater from which they were derived. Thus, sodium or chloride ions can be used as a measure of sea-salt inputs to rainwater. Chloride concentrations in rain falling on oceanic islands are around $200\,\mu mol\,l^{-1}$, rain within 100 km of coastal continental areas contains around $10\text{--}100\,\mu mol\,l^{-1}$, while further inland chloride concentrations fall below $10\,\mu mol\,l^{-1}$, but not to zero.

The importance of a seawater source of ions other than sodium and chloride in rainwater can be assessed by computing their relative abundance with respect to sodium and comparing this with the same ratio in seawater. This comparison can be extended to freshwater, although here there is the complication that some of the ions could be derived from weathering. If we overlook this complication initially, then, where rainwater inputs make a large contribution to the chemistry of freshwater, the dominant cation is likely to be Na^+. If weathering reactions are important, then the major dissolved cations will be those soluble elements derived from local rock and soil. In the absence of evaporite minerals, which are a minor component of continental crust (see Fig. 4.1), the most weatherable rocks are limestones ($CaCO_3$). The calcium ion, liberated by limestone dissolution, is an indicator of this weathering process. This is clearly demonstrated by comparing the Ca^{2+} concentration in groundwater from a limestone aquifer, with groundwater from granites or metamorphic schists (Table 5.3).

Table 5.3 Chemical analyses of US groundwater from various rock types ($mmol\,l^{-1}$). Adapted from Todd (1980). This material is used by permission of John Wiley & Sons, Inc.

	Granite South Carolina	Metamorphic schist Georgia	Limestone Texas
Ca^{2+}	0.3	0.7	1.8
Mg^{2+}	0.2	0.2	0.4
Na^+	0.4	0.7	1.0
K^+	0.1	0.2	0.2
Cl^-	0.1	0.1	0.7
SO_4^{2-}	0.1	0.1	0.2
HCO_3^-	1.2	2.3	4.5
Al	0.004	0	—
Fe	0.003	0.002	0.001
SiO_2	0.6	0.35	0.18
$Na^+/(Ca^{2+} + Na^+)$	0.57	0.50	0.36

The ratio of $Na^+:(Na^+ + Ca^{2+})$ can therefore be used to discriminate between rainwater and weathering sources in freshwaters. When sodium is the dominant cation (sea-salt contribution important), $Na^+:(Na^+ + Ca^{2+})$ values approach 1. When calcium is the dominant ion (weathering contribution important), $Na^+:$ $(Na^+ + Ca^{2+})$ values approach 0.

The composition of dissolved ions in riverwater can be classified by comparing $Na^+:(Na^+ + Ca^{2+})$ values with the total number of ions present in solution (Fig. 5.3). Note that the total dissolved ions or salts can also be expressed as the ionic strength of the water (Box 5.1). Data which plot in the bottom right of Fig. 5.3 represent rivers with low ion concentrations and sodium as the dominant cation. These rivers flow over crystalline bedrock (low weathering rates) or over extensively weathered, kaolinitic, tropical soils (low weathering potential, chemical index of alteration (CIA) *c.* 100 (see Table 4.9)). The Rio Negro, a tributary of the Amazon (Fig. 5.4), draining the highly weathered tropical soils of the central Amazonian region, has low ionic strength (Box 5.1) with weathering-derived sodium as the major cation. The Onyx River in the dry valleys of Antarctica is a better example of a low-ionic-strength, sea-salt-sodium-dominated river. This river has its source as glacial melt water and has a starting chemistry almost totally dominated by marine ions. As it flows over the igneous and metamorphic rocks of the valley floor, its composition evolves to higher ionic strength with an increasing proportion of calcium (Fig. 5.3).

Major river systems flow over a wide range of rock types, acquiring the dissolved products of weathering reactions. Freshwaters originating in areas with active weathering processes will have higher ion concentrations and an increasing predominance of calcium over sodium. These rivers plot along a trend from A to A′ on Fig. 5.3. The Mackenzie and Ganges (Table 5.2) fall within this group, despite very different geomorphological settings.

The Amazon and its tributaries are a good example of a river system where the chemistry of the lower reaches integrates the products of differing soil and bedrock weathering regimes (Fig. 5.4). Rivers draining the intensely weathered soils and sediments of the central Amazonian region, such as the Rio Negro, have low total cation concentrations of less than $200\,\mu eq\,l^{-1}$ (i.e. sum of all major cations concentrations × charge; see also footnote to Table 4.10). The Rio Negro has water relatively enriched in sodium, silica, iron, aluminium and hydrogen ions, because of the limited supply of other cations from weathering reactions. By contrast, rivers draining easily erodible sedimentary rocks (including carbonates) of the Peruvian Andes are characterized by high total cation concentrations of $450–3000\,\mu eq\,l^{-1}$, including abundant calcium, magnesium, alkalinity (see below and Box 5.2) and sulphate. Between these two extremes in water composition are rivers with quite low total cation concentrations, with sodium enriched relative to calcium and magnesium, but also with high concentrations of silica, consistent with the weathering of feldspars (e.g. albite (see eqn. 4.14)). These rivers drain areas without large amounts of easily weatherable rock, but drain soils not so completely degraded as the lowest concentration group characterized by the Rio Negro.

In arid areas, evaporation may influence the major dissolved ion chemistry of rivers. Evaporation concentrates the total amount of ions in riverwater.

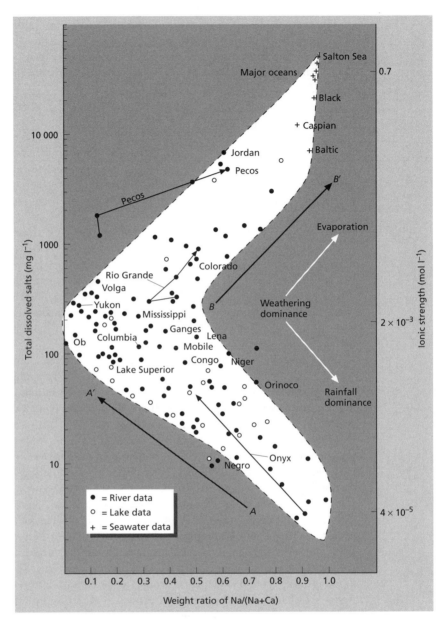

Fig. 5.3 Variation of the weight ratio of Na/(Na + Ca) as a function of the total dissolved salts and ionic strength for surface waters. Arrows represent chemical evolution of rivers from source downstream. Modified from *Science* **170**, 1088–1090, Gibbs, R.J. Copyright (1970) by the AAAS.

Evaporation also causes $CaCO_3$ to precipitate from water before NaCl, the latter being a more soluble salt. The formation of $CaCO_3$ removes calcium ions from the water, increasing the $Na^+:(Na^+ + Ca^{2+})$ value. Data for rivers influenced by evaporation plot along a diagonal B to B' on Fig. 5.3, evolving toward a sea-

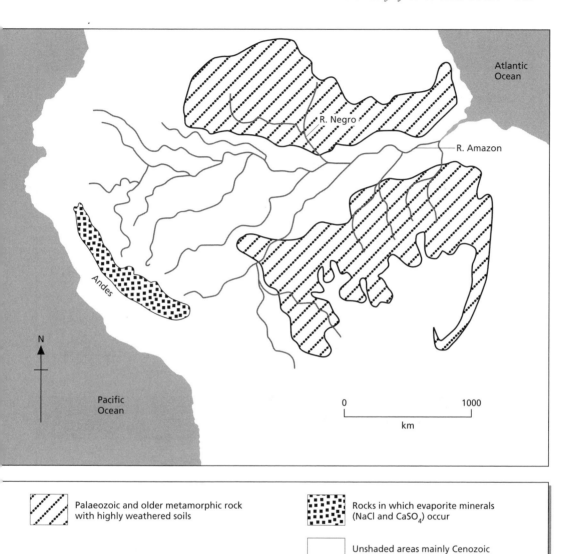

Fig. 5.4 Simplified geological map of the Amazon Basin showing major river systems. After Stallard and Edmond (1983), copyright (1983) American Geophysical Union modified by permission of American Geophysical Union.

water composition in the upper right (see also Table 6.1). The Colarado River in the arid southwest of the USA is an interesting example. Construction of the Hoover Dam on the Colorado created Lake Mead upstream of the dam, increasing evaporation from the catchment. The concentration of dissolved solids increased by 10–20% in outflowing water below the dam, while $CaCO_3$ precipitation occurred in Lake Mead. A more extreme example is seen in the Salton Sea

Box 5.1 Ionic strength

The concentration of an electrolyte solution can be expressed as ionic strength (I), defined as:

$$I = \tfrac{1}{2}\Sigma c_i z_i^2 \qquad \text{eqn. 1}$$

where c is the concentration of ion (i) in mol l^{-1}, z_i is the charge of ion (i) and Σ represents the sum of all ions in the solution.

As a measure of the concentration of a complex electrolyte solution, ionic strength is better than the simple sum of molar concentrations, as it accounts for the effect of charge of multivalent ions.

For example, the Onyx River in Antarctica (Section 5.3), 2.5 km below its glacier source, has the following major ion composition (in μmol l^{-1}): Ca^{2+} = 55.4; Mg^{2+} = 44.4; Na^+ = 125; K^+ = 17.6; H^+ = 10^{-3}; Cl^- = 129; SO_4^{2-} = 32.2, HCO_3^- = 136 and OH^- = 10 (rivers in this region have pH around 9, which means that HCO_3^- is the

dominant carbonate species). Putting these ions in equation 1 gives:

$$
\begin{aligned}
I = \tfrac{1}{2}\Sigma c\,\mathrm{Ca}.4 &+ c\mathrm{Mg}.4 + c\mathrm{Na}.1 + c\mathrm{K}.1 \\
&+ c\mathrm{H}.1 + c\mathrm{Cl}.1 + c\mathrm{SO}_4.4 \\
&+ c\mathrm{HCO}_3.1 + c\mathrm{OH}.1 \qquad \text{eqn. 2}
\end{aligned}
$$

Substituting the μmol l^{-1} values (and correcting to mol l^{-1} with the final 10^{-6} term in eqn. 3) gives:

$$
\begin{aligned}
I = \tfrac{1}{2}[&(55.4 \times 4) + (44.4 \times 4) + 125 + 17.6 \\
&+ 10^{-3} + 129 + (32.2 \times 4) + 136 + 10] \times 10^{-6} \\
& \qquad\qquad\qquad\qquad\qquad\qquad \text{eqn. 3}
\end{aligned}
$$

$$I = 4.73 \times 10^{-4}\,\mathrm{mol}\,l^{-1} \qquad \text{eqn. 4}$$

Freshwaters typically have ionic strengths between 10^{-3} and 10^{-4} mol l^{-1}, whereas seawater has a fairly constant ionic strength of 0.7 mol l^{-1}.

Box 5.2 Measuring alkalinity

Alkalinity is measured by adding acid to a water sample until the pH falls to 4. At this pH, HCO_3^- and CO_3^{2-} alkalinity (known as the carbonate alkalinity (A_c)) will have been converted to carbon dioxide (CO_2), i.e.:

$$HCO_{3(aq)}^- + H_{+(aq)} \rightleftharpoons H_2O_{(l)} + CO_{2(g)} \qquad \text{eqn. 1}$$

$$CO_{3(aq)}^{2-} + 2H_{(aq)}^+ \rightleftharpoons H_2O_{(l)} + CO_{2(g)} \qquad \text{eqn. 2}$$

Note that twice as much H^+ is used up neutralizing the CO_3^{2-} (eqn. 2) relative to the HCO_3^- (eqn. 1). This is expressed in the formulation that expresses carbonate alkalinity, by the 2 in front of the cCO_3^{2-} term, i.e.:

$$A_c = cHCO_3^- + 2cCO_3^{2-} \qquad \text{eqn. 3}$$

This formulation is expressed in concentration because the carbonate species are measured values (see Section 2.6). The volume of acid used is a measure of the alkalinity, which is usually expressed as milliequivalents per litre (see footnote to Table 4.10). These units account for the difference in H^+ neutralizing power between CO_3^{2-} and HCO_3^-. Note, however, that at pH around 7.5–8, monovalent HCO_3^- accounts for almost all alkalinity (Fig. 5.5) such that at these pH values milliequivalents are essentially the same as millimoles.

on the border between the USA and Mexico. This very large lake was created in 1905 by floodwaters of the Colorado, forming a closed basin lake with no outflow to the sea. The Salton Sea began as a freshwater lake with total dissolved salt concentrations around 3.5 g l^{-1} in 1907. Evaporation since this time means that the lakewaters are now saltier than seawater. In 1997, total dissolved salt concentrations stood at $47\,gl^{-1}$. Rivers draining into the Salton Sea have a Na^+/Ca^{2+} ratio of about $5:1$ (on an atomic basis) but as a result of evaporation and $CaCO_3$ precipitation, this ratio increases to $27:1$ in the Salton Sea itself. Other examples of rivers in which evaporation plays an important role include the Jordan and Rio Grande. In all these arid areas, dissolution of evaporite minerals in the catchments may also contribute to the increasing dominance of NaCl.

The classification of riverwater composition in Fig. 5.3 is simplified and does not always work. For example, weathering of feldspars (see Section 4.4.4) can produce solutions of low ionic strength, but rich in sodium and silica, which plot in the bottom right of Fig. 5.3. This effect probably influences the classification of the Rio Negro. Weathering of evaporite minerals will also affect the composition of rivers. For example, in the Amazon catchment there are a small number of tributaries draining areas of predominantly evaporite rock. These have very high total cation concentrations and are characterized by high sodium, chloride, calcium and sulphate concentrations from the weathering of the evaporite minerals, halite and gypsum. Despite these complications, Fig. 5.3 remains a useful way to compare factors controlling riverwater chemistry. Indeed, it is remarkable that most of the world's major rivers can be rationalized in this straightforward way.

5.3.1 Alkalinity, dissolved inorganic carbon and pH buffering

In Section 4.4 we saw that most soilwaters that feed rivers and groundwater have near-neutral pH, with HCO_3^- as the major anion. This results from the dissolution of CO_2 in water (see eqn. 4.7) and from the acid hydrolysis of silicates and carbonates. The total concentration of weak acid anions like HCO_3^- in water is referred to as alkalinity. These anions are available to neutralize acidity (H^+) in natural waters, consequently it is important to understand their chemical behaviour.

In continental waters, bicarbonate (HCO_3^-) and carbonate (CO_3^{2-}) ions are the most important components of alkalinity, although in seawater other ions also contribute to alkalinity. The relative importance of HCO_3^- and CO_3^{2-} depends on the pH of the solution and can be calculated from the known dissociation constants (see Box 4.5) of these ions and the solution pH.

The first dissociation of dissolved carbon dioxide (expressed here as carbonic acid),

$$H_2CO_{3(aq)} \rightleftharpoons H^+_{(aq)} + HCO^-_{3(aq)} \qquad \text{eqn. 5.6}$$

has a dissociation constant,

$$K_1 = \frac{aH^+ \cdot aHCO_3^-}{aH_2CO_3} = 10^{-6.4} \qquad \text{eqn. 5.7}$$

The 'a' denotes activity, the formal thermodynamic representation of concentration (see Section 2.6).

Similarly, for the second dissociation of carbonic acid,

$$HCO_{3(aq)}^- \rightleftharpoons H_{(aq)}^+ + CO_{3(aq)}^{2-}$$ eqn. 5.8

the dissociation constant is,

$$K_2 = \frac{aH^+.aCO_3^{2-}}{aHCO_3^-} = 10^{-10.3}$$ eqn. 5.9

It is simple to demonstrate that the alkalinity in most continental waters is dominated by HCO_3^- by rearranging the above equations at a typical pH value for these waters. For example, many mature rivers have pH values around 8.

Rearranging equation 5.9 to solve for $aHCO_3^-$ gives,

$$aHCO_3^- = \frac{aH^+.aCO_3^{2-}}{K_2}$$ eqn. 5.10

And substituting values for pH 8 (pH = $-\log10aH^+$, see Box 3.5) and K_2 (eqn. 5.9) gives,

$$aHCO_3^- = \frac{10^{-8}.aCO_3^{2-}}{10^{-10.3}} = \frac{1 \times 10^{-8}.aCO_3^{2-}}{5 \times 10^{-11}} = 200\,aCO_3^{2-}$$ eqn. 5.11

This shows that at typical pH values for continental waters the HCO_3^- anion is 200 times more abundant than the CO_3^{2-} anion. Repeating this exercise for a range of pH values results in the graphical relationship shown in Fig. 5.5. Note that when pH falls below 5 on Fig. 5.5, almost all of the weak acid anions (HCO_3^- and CO_3^{2-}) have disappeared and at pH of 4 only undissociated acid (H_2CO_3) remains. This relationship is used as the basis for measuring alkalinity (Box 5.2).

The pH of natural waters is in fact *controlled* mainly by the relative concentrations of dissolved inorganic carbon (DIC) species, i.e. H_2CO_3, HCO_3^- and

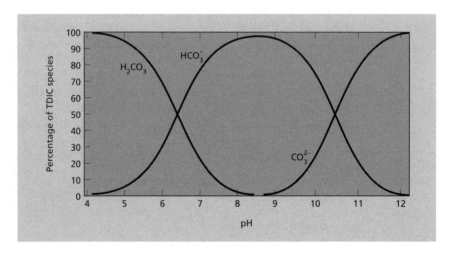

Fig. 5.5 Relationship between the total dissolved inorganic carbon species (TDIC; i.e. H_2CO_3 + HCO_3^- + CO_3^{2-}) and pH. Most natural waters have pH between 7 and 9 where the HCO_3^- anion is abundant (>80% of the TDIC). In highly alkaline waters (pH > 10.3) the CO_3^{2-} anion becomes dominant, while in acidic waters (pH <6.4) the undissociated acid (H_2CO_3) is the dominant TDIC species.

CO_3^{2-} (Fig. 5.5). These species react to maintain the pH within relatively narrow limits. This is known as buffering the pH and the principles, using worked examples, are demonstrated in Box. 5.3. The relationship between pH, $CaCO_3$ weathering and alkalinity is nicely illustrated by real data (Fig. 5.6) from a small stream in North Yorkshire (UK). The stream begins as drainage from an organic peat bog. The initial pH is about 4 and hence the alkalinity is zero (Fig. 5.6). The stream then flows from the bog over siliceous mudrocks with very limited potential for weathering such that the water chemistry changes little. After this the stream flows onto limestone, the acidic water reacting rapidly with the $CaCO_3$ to release Ca^{2+} and HCO_3^- ions:

$$CaCO_{3(s)} + H_2CO_{3(aq)} \leftrightharpoons Ca_{(aq)}^{2+} + 2HCO_{3(aq)}^- \qquad \text{eqn. 5.12}$$

In less than 100 m of distance flowing on the limestone the pH rises sharply to values of 6 or 7 and it continues to increase to a value of about 8 at which it stabilizes due to the buffering action of the alkalinity (Fig. 5.6). The alkalinity is strongly buffering the system at values above $2\,meq\,l^{-1}$ where the relationship between pH and alkalinity is asymptotic (Fig. 5.6). The relationship between Ca^{2+}

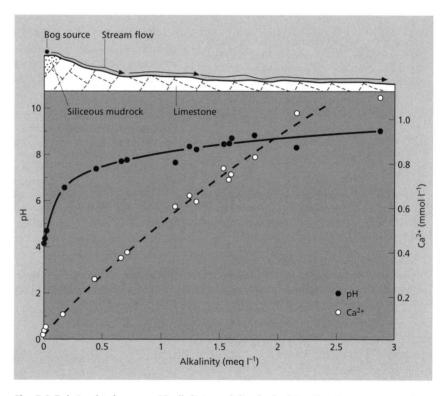

Fig. 5.6 Relationship between pH, alkalinity and dissolved calcium ions in stream waters in the Malham Tarn area of northern England, flowing from bog on siliceous mudrock to limestone. Note that pH is buffered around 8 once limestone weathering begins. Data from Woof and Jackson (1988).

Box 5.3 Worked examples of pH buffering

The principle of pH buffering can be illustrated by considering the simple case of acetic acid, CH_3COOH (abbreviated here to HA) and sodium acetate, CH_3COONa (abbreviated here to NaA). Acetic acid partially dissociates in water (H_2O) while the sodium salt completely dissociates.

$$HA \rightarrow H^+ + A^- \qquad \text{eqn. 1}$$

$$NaA \rightarrow Na^+ + A^- \qquad \text{eqn. 2}$$

and

$$K_{HA} = \frac{aH^+ . aA^-}{aHA} = 10^{-4.75} \, mol \, l^{-1} \qquad \text{eqn. 3}$$

Rearranging gives:

$$aH^+ = 10^{-4.75} \frac{aHA}{aA^-} \qquad \text{eqn. 4}$$

For a 0.1 molar (M) solution of HA and NaA (for simplicity we will assume that activity (a) and concentration (c) are the same) and assuming very little HA dissociates, then:

$$aH^+ = 10^{-4.75} \times \frac{0.1}{0.1} = 10^{-4.75} \, mol \, l^{-1} \qquad \text{eqn. 5}$$

We know that pH $= -\log_{10} aH^+$ (see Box 3.5), so in our example:

$$pH = -\log_{10} 10^{-4.75} = 4.75 \qquad \text{eqn. 6}$$

To illustrate the principle of buffering, consider what happens when 0.005 moles of NaOH (sodium hydroxide, a strong base) are added to 1 litre of 0.1 M HA and NaA. The added base reacts with the hydrogen ions (H^+), causing an amount of the HA equivalent to the added NaOH to dissociate (eqn. 1); the HA concentration decreases and the A^- concentration increases by this amount.

$$aH^+ = 10^{-4.75} \times \frac{(0.1 - 0.005)}{(0.1 + 0.005)} = 1.61 \times 10^{-5} \, mol \, l^{-1}$$
$$\text{eqn. 7}$$

Now pH becomes:

$$pH = -\log_{10} 1.61 \times 10^{-5} = 4.79 \qquad \text{eqn. 8}$$

The pH is barely altered because the excess of undissociated HA dissociates to neutralize the added OH^-. Buffering will continue if an excess of HA is available. If acid is added to the solution, H^+ will react with the excess of A^- to increase the HA and decrease the A^- concentration by an amount equivalent to the added H^+, resulting in a similar buffering effect. The HA and NaA solution is an effective buffer because it can react to neutralize either added acid or base.

By contrast, if 0.005 moles of NaOH are added to 1 litre of water, the pH will rise to 11.7, as illustrated below.

$$K_w = aOH^- . aH^+ = 10^{-14} \, mol^2 \, l^{-2} \qquad \text{eqn. 9}$$
(see Box 4.1)

Thus:

$$aH^+ = K_w / aOH^- \qquad \text{eqn. 10}$$

and so:

$$pH = -\log_{10} aH^+ = -\log_{10}(K_w / aOH^-) \qquad \text{eqn. 11}$$

So, if 0.005 moles of OH^- are added to 1 litre of water (again assuming that activity and concentration are the same), then:

$$pH = -\log_{10}(10^{-14} / 0.005) = 11.7 \qquad \text{eqn. 12}$$

In natural waters the buffering system involves the weak acid, carbonic acid (H_2CO_3), and the associated anions, bicarbonate (HCO_3^-) and carbonate (CO_3^{2-}). At pH 4–9, HCO_3^- is the major anion. In the following example we ignore CO_3^{2-} (and again assume that activity and concentration are the same). First, we can rewrite equation 4 for the HCO_3^- system:

$$aH^+ = K_{HCO_3^-} \times \frac{aH_2CO_3}{aHCO_3^-} \qquad \text{eqn. 13}$$

and define the terms used (eqns 14–17). At 25°C the equilibrium constant for equation 1 in Box 5.2 is defined as:

$$K_{HCO_3^-} = aHCO_3^- . aH^+ = 4 \times 10^{-7}$$
$$= 10^{-6.4} \, mol \, l^{-1} \qquad \text{eqn. 14}$$

(see also Box 3.7), whilst the relationship between partial pressure of carbon dioxide (pCO_2) and H_2CO_3 is:

$$CO_{2(g)} + H_2O_{(l)} \rightleftharpoons H_2CO_{3(aq)} \qquad \text{eqn. 15}$$

Thus:

(continued)

$$K_{CO_2} = \frac{aH_2CO_3}{aCO_2 \cdot aH_2O} = \frac{aH_2CO_3}{pCO_2} = 0.04$$

$$= 10^{-1.4}\, mol\, l^{-1}\, atm^{-1} \qquad \text{eqn. 16}$$

(see also Box 3.7), and so:

$$aH_2CO_3 = 10^{-1.4} \times pCO_2 \qquad \text{eqn. 17}$$

Now equation 13 can be rewritten:

$$aH^+ = 10^{-6.4} \times \frac{10^{-1.4} \times pCO_2}{aHCO_3^-}\, mol\, l^{-1} \qquad \text{eqn. 18}$$

Consider the case of the Mackenzie River, where $HCO_3^- = 1.8\, mmol\, l^{-1}$ (Table 5.2) and atmospheric $pCO_2 = 3.6 \times 10^{-4}\, atm$:

$$aH^+ = 10^{-6.4} \times \frac{10^{-1.4} \cdot 3.6 \times 10^{-4}}{1.8 \times 10^{-3}}$$

$$= 10^{-6.4} \times 0.008 = 3.2 \times 10^{-9}\, mol\, l^{-1} \qquad \text{eqn. 19}$$

$$pH = -\log_{10} aH^+ = -\log_{10} 3.2 \times 10^{-9} = 8.49 \qquad \text{eqn. 20}$$

Although this treatment is simplified, it serves to illustrate the way in which pH can be calculated. In practice the pH of most natural water containing HCO_3^- and CO_3^{2-} is buffered between pH 7 and 9.

and alkalinity is, in this example, approximately linear, although for every mmol of Ca^{2+}, 2 mmol of HCO_3^- are released, as predicted in equation 5.12.

5.4 Aluminium solubility and acidity

Aluminium is largely insoluble during weathering processes (Table 5.1), but becomes soluble when pH is both low and high. At the simplest level, three aluminium species are identified; soluble Al^{3+}, dominant under acid conditions, insoluble aluminium hydroxide ($Al(OH)_3$), dominant under neutral conditions, and $Al(OH)_4^-$, dominant under alkaline conditions.

$$Al(OH)_{3(s)} + OH^-_{(aq)} \leftrightharpoons Al(OH)_4^- \qquad \text{eqn. 5.13}$$

$$Al(OH)_{3(s)} \leftrightharpoons Al^{3+}_{(aq)} + 3OH^-_{(aq)} \qquad \text{eqn. 5.14}$$

Aluminium solubility is therefore pH-dependent and aluminium is insoluble in the pH range 5–9, which includes most natural waters. The details of aluminium solubility are also complicated by the formation of partially dissociated $Al(OH)_3$ species and complexing between aluminium and organic matter (see Box 6.4). Although aluminium is soluble at high pH, alkaline waters are uncommon because they absorb acid gases, for example CO_2 and SO_2, from the atmosphere. However, alkaline rivers with aluminium mobility are known. For example, the industrial process for abstracting aluminium from bauxite involves leaching the ore with strong sodium hydroxide (NaOH) solutions. In Jamaica, discharge of wastes from bauxite processing produces freshwater streams with high pH (>8) in addition to high sodium and aluminium concentrations. As these streamwaters evolve, the pH falls to about 8 and the dissolved aluminium:sodium ratio declines as the aluminium precipitates.

It is, however, *acidification* of freshwaters that commonly results in aluminium mobility resulting in ecological damage. This acidification is typically caused by two anthropogenic processes, acid rain and acid mine drainage.

5.4.1 Acidification from atmospheric inputs

Acidification of soilwater occurs if the rate of displacement of soil cations by H^+ exceeds the rate of cation supply from weathering. Ion-exchange reactions in soils (e.g. eqn. 4.18) help to buffer pH in the short term (see Section 4.8), but over longer periods cation supply to soils from the underlying bedrock is very slow. Rainwater is naturally acidic (see Box 3.7) and soilwaters are further acidified by the production of H^+ from the decomposition of organic matter (see eqns 4.6–4.10). Thus, acidification can be a natural process, although acid rain (see Section 3.9) has greatly increased the rate of these processes in many areas of the world.

Acidification of freshwater is most marked in upland areas with high rainfall (hence high acid flux), steep slopes (resulting in a short residence time for water in the soil) and crystalline rocks (which weather, and supply cations, slowly). Thus, while acid rain is a widespread phenomenon, acidified freshwaters are less common and are controlled both by rates of atmospheric input and by rock types (Fig. 5.7). All weathering processes, except sulphide oxidation (see Sections 4.4.2 & 5.4.2), consume hydrogen ions, driving pH toward neutrality. Hence, mature rivers, which drain deeper, cation-rich lowland soils, have higher pH and lower aluminium concentrations.

The effects of upland acidification of freshwaters can be dramatic. Between 1930 and 1975 the median pH of lakes in the Adirondack Mountains of north-eastern USA decreased from 6.7 to 5.1, caused by progressively lower pH in rain-water (Fig. 5.7). The acidified lakewater killed fish and other animals by several mechanisms. The problem for fish is that the dissolved Al^{3+} in the acidic water precipitates as an insoluble $Al(OH)_3$ gel on the less acidic gill tissues, preventing normal uptake of oxygen and suffocating the animal. Similar problems have occurred in Scandinavia and Scotland. In addition to problems in freshwaters, the loss of forests in high-altitude areas has been linked to acid leaching, which leads to impoverishment of soils coupled with direct loss of cations from plant leaves.

5.4.2 Acid mine drainage

Acidification of surface and groundwaters in mining regions is a worldwide problem. In a mid-1980s survey, 10% of streams fed by groundwater springs in the northern Appalachians (USA) were found to be acidified by mine drainage. The acidity is caused by oxidation of sulphide minerals (see eqn. 4.4), common in sedimentary mudrocks, mineral veins and coal deposits. When mineral and coal deposits are mined, sulphide is left behind in the waste rock, which is piled in heaps. These waste heaps have large surface areas exposed to the atmosphere (Plate 5.1, facing p. 138), allowing extensive and rapid oxidation of the sulphide. The problem is long-lived, and often intensifies once mining has finished, because abandoned mines are rapidly flooded by groundwater once the pumps are switched off.

While the oxidation reaction is a natural one (see Section 4.4.2), the mining activities increase the scale and rate of the reaction (see eqn. 4.4). The resulting

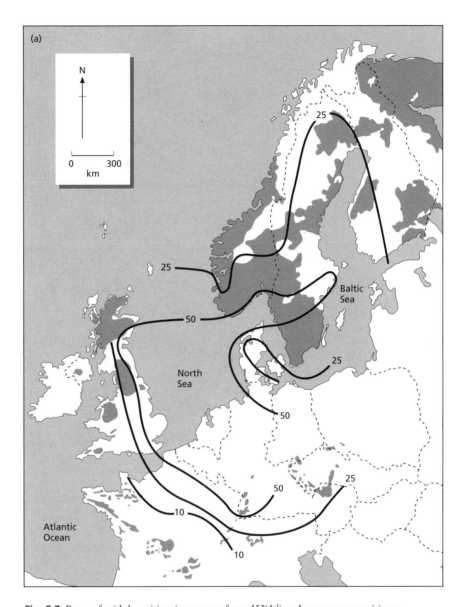

Fig. 5.7 Rates of acid deposition (contours of $\mu mol\,H^+l^{-1}$) and areas most sensitive to acidification (shaded) based on their rock type: (a) Europe, (b) North America. Modified from Likens *et al.* (1979), with kind permission of A.M. Tomko III.

H_2SO_4 makes drainage from abandoned mines strongly acidic (pH as low as 1 or 2). This acidity increases the solubility of aluminium and other metals, causing toxicity in aquatic ecosystems. Microorganisms are closely involved in sulphide oxidation, which can be modelled by a series of reactions:

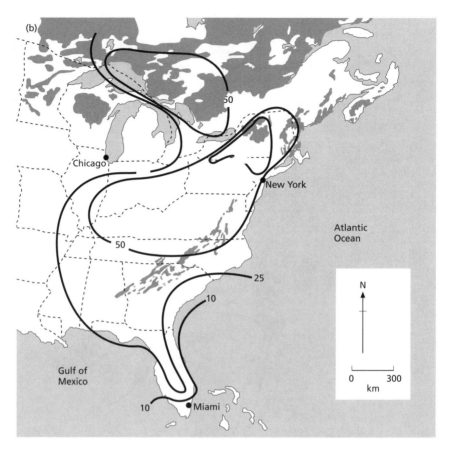

Fig. 5.7 *Cont.*

$$2FeS_{2(s)} + 2H_2O_{(l)} + 7O_{2(g)} \longrightarrow 4H^+_{(aq)} + 4SO^{2-}_{4(aq)} + 2Fe^{2+}_{(aq)}$$
(pyrite oxidation)

eqn. 5.15

followed by the oxidation of ferrous iron (Fe(II)) to ferric iron (Fe(III)):

$$4Fe^{2+}_{(aq)} + O_{2(g)} + 10H_2O_{(l)} \longrightarrow 4Fe(OH)_{3(s)} + 8H_{+(aq)}$$

(Fe(II))   (Fe(III))   eqn. 5.16

Oxidation happens very slowly at the low pH values found in acid mine waters. However, below pH 3.5 iron oxidation is catalysed (see Box 4.4) by the iron bacterium *Thiobacillus thiooxidans*. At pH 3.5–4.5 oxidation is catalysed by *Metallogenium*. Ferric iron may react further with pyrite:

$$FeS_{2(s)} + 14Fe^{3+}_{(aq)} + 8H_2O_{(l)} \longrightarrow 15Fe^{2+}_{(aq)} + 2SO^{2-}_{4(aq)} + 16H^+_{(aq)}$$

eqn. 5.17

At pH values much above 3 the iron(III) precipitates as the common iron(III) oxide, goethite (FeOOH):

$$Fe^{3+}_{(aq)} + 2H_2O_{(l)} \rightarrow FeOOH_{(s)} + 3H^+_{(aq)}$$

eqn. 5.18

The precipitated goethite coats stream beds and brickwork as a distinctive yellow-orange crust (Plate 5.2, facing p. 138), a very visible manifestation of the problem.

Bacteria use iron compounds to obtain energy for their metabolic needs (e.g. oxidation of ferrous to ferric iron). Since these bacteria derive energy from the oxidation of inorganic matter, they thrive where organic matter is absent, using carbon dioxide (CO_2) as a carbon (C) source. Iron oxidation, however, is not an efficient means of obtaining energy; approximately $220\,g$ of Fe^{2+} must be oxidized to produce $1\,g$ of cell carbon. As a result, large deposits of iron(III) oxide form in areas where iron-oxidizing bacteria survive.

We should note that the common sulphide of iron (pyrite—FeS_2) often contains significant amounts of the toxic semimetal arsenic, as impurities. As a result, when iron sulphides are oxidized (eqn. 5.15) arsenic is released along with the dissolved iron and sulphate. In very rare circumstances this arsenic release can result in groundwater contamination (see also Section 5.7.2).

The acidity caused by mining operations can be treated (neutralized) by adding crushed $CaCO_3$ to the system and by removing dissolved trace metals. At active mines this is the responsibility of the mine operators. Abandoned mines, however, create a bigger problem because the source of leakage from the mine area is unpredictable, flowing out of various fissures and fractures in the rock as the mine fills with water. Furthermore, in abandoned mines there is often no operator to take responsibility for treatment. Moreover, as developed countries move away from coal as an energy source, more coal mines are being abandoned increasing the risk of acid drainage. There are various strategies being developed to create passive, low-cost treatments for acid waste including phytoremediation (see Section 4.10.4), where reed-beds are used to encourage oxidation and trap the solid iron oxides that precipitate.

5.4.3 Recognizing acidification from sulphate data—ternary diagrams

We have already seen how the factors regulating river chemical composition can be summarized using simple cross-plots of weathering and atmospheric (sea-salt) inputs (Fig. 5.3). Recognizing the significance of acidification either from acid rain or from acid mine drainage is aided using a ternary (triangular) diagram to allow for three inputs—weathering, sea-salt and sulphuric acid (Fig. 5.8). The diagram plots alkalinity, chloride and sulphate data to trace weathering, sea-salt and sulphuric acid inputs respectively for river systems discussed elsewhere in this chapter. Sulphate is a good tracer of acid mine drainage (eqns. 5.15 & 5.17), although not totally unique to this system (see below).

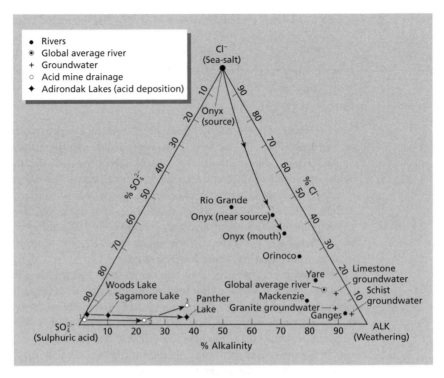

Fig. 5.8 Ternary plot of alkalinity, chloride and sulphate data to trace weathering, sea-salt and sulphuric acid (acid deposition/mine drainage) inputs to river- and groundwaters. River and groundwater data are from Tables 5.1–5.3 and Box 5.1. River Yare (eastern England) is representative of a river system developed on limestone (chalk) bedrock. The Adirondak Lakes data (from Galloway *et al.* 1983) represent response of nearby lakes to acid deposition. Woods Lake has pH of 4.7 and zero alkalinity; it cannot buffer effects of acid input. As alkalinity increases, due to increased carbonate weathering in the bedrock, the effects of acid deposition are progressively buffered (Sagamore (pH 5.6) to Panther Lake (pH 6.2)). Acid mine drainage (AMD) data (from Herlihy *et al.* 1990) are all averaged from streams in the eastern USA where (1) are strongly impacted acidic streams, (2) are strongly impacted non-acidic streams and (3) are weakly impacted non-acidic streams.

The Onyx River in Antarctica plots in the top corner of Fig. 5.8 as a pure chloride (sea-salt-dominated) system. As this river evolves and starts to weather rocks in the catchment, it picks up alkalinity moving on a trajectory toward the bottom right corner (weathering dominated). Upland systems heavily impacted by acid mine drainage or acid rain plot toward the bottom left corner of Fig. 5.8, their anion chemistry is dominated by sulphate with little or no alkalinity present. Again as these systems mature and acquire weathering products including alkalinity, they too move in trajectories toward the bottom right corner (weathering dominated). As noted earlier, HCO_3^- alkalinity is the dominant anionic component of most mature rivers and groundwaters, explaining their position at the bottom right of Fig. 5.8. The only exception is the Rio Grande that plots well away from most mature rivers. Although this river has the highest HCO_3^- alkalinity of those plotted on Fig. 5.8 (3.0 mmol l^{-1} HCO_3^-; Table 5.2), the weathering of evaporite minerals ($NaCl$ and $CaSO_4$) in the catchment, and the concentration of dissolved salts by evaporation, leads to much increased dissolved chloride and sulphate compared to all the other mature rivers, explaining its near central position on the diagram.

5.5 Biological processes

In streams and small rivers, biological activity in the water has little influence on water chemistry because any effects are diluted by the rapid flow. Conversely, in large slow-flowing rivers and in lakes, biological activity can cause major changes in water chemistry.

All photosynthetic plants absorb light and convert this to chemical energy within a chlorophyll molecule. The liberated energy is then used to convert CO_2 (or HCO_3^-) and water into organic matter. This complex biochemical process is crudely represented by the familiar equation:

$$CO_{2(g)} + H_2O_{(l)} \xrightarrow{\text{light}} CH_2O_{(s)} + O_{2(g)} \qquad \text{eqn. 5.19}$$

CH_2O represents organic matter in a simplified way as carbohydrate. The reaction depicted by equation 5.19 requires the input of energy ($\Delta G° = +475\,kJ\,mol^{-1}$) (see Box 4.8) to proceed and this is provided by light. In shallow freshwater, large plants and drifting microscopic algae (phytoplankton) are responsible for photosynthesis, while in deep lakes (and in the oceans) phytoplankton account for almost all photosynthesis. The reverse of equation 5.19 is organic matter decomposition, i.e. oxidation or respiration, which liberates the energy that sustains most life:

$$CH_2O_{(s)} + O_{2(s)} \rightarrow CO_{2(g)} + H_2O_{(l)} \quad \Delta G° = -475\,kJ\,mol^{-1} \qquad \text{eqn. 5.20}$$

Since photosynthesis requires light, it is confined to the surface layers of waters — the euphotic zone (the region receiving >1% of the irradiance arriving at the water surface). The depth of the euphotic zone varies with the angle of the sun, the amount of light absorbed by suspended matter (including phytoplankton) and the presence of dissolved coloured compounds in the water.

The decomposition of organic matter, which is almost always bacterially mediated, can occur at any depth in the water column. Decomposition consumes oxygen (eqn. 5.20), which is supplied to the water largely by gas exchange at the water/air interface and partly as a byproduct of photosynthesis. Temperature influences the amount of oxygen that can dissolve in water. Oxygen-saturated freshwater contains about $450\,\mu mol\,l^{-1}$ oxygen at 1°C and $280\,\mu mol\,l^{-1}$ at 20°C.

In summer, the surface layers of many lakes are warmed by insolation. The warmer surface water is less dense than the cold deep water, causing a stable density stratification. Stratification limits exchange of oxygenated surface water with the deeper waters. Organic matter produced in surface waters sinks into the deeper waters where it is oxidized, consuming and depleting dissolved oxygen. In some cases, oxygen levels fall below those needed to support animal life. This

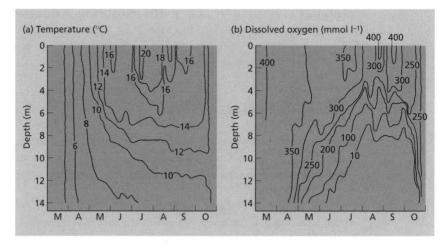

Fig. 5.9 Variation in (a) temperature and (b) dissolved oxygen for lakewater (Esthwaite in northwest England) between March and October 1971. Modified from Heany *et al.* (1986). © 1986. This material is used by permission of John Wiley & Sons, Inc.

process is illustrated in Fig. 5.9 using data from Esthwaite, a lake in northwest England. In March the water column is well mixed and oxygen concentrations are uniformly high, around 350-400 µmol l^{-1} at all depths. By May, however, the water column is stratifying (warmer above 7 metres water depth, cooler below) and oxygen concentrations begin to fall in water depths below 8 metres. By the end of summer a large area of very low oxygen concentrations (<10 µmol l^{-1}) has formed below 8 metres depth. In the autumn the stratification is destroyed by cooling and by strong winds that mix oxygenated surface waters with the deep water.

The rate of oxygen consumption usually increases as the supply of organic matter increases, either due to enhanced photosynthesis in the surface waters, or due to direct discharge of organic waste, for example sewage. The Thames estuary (UK), Chesapeake Bay (USA) (see Section 6.2.4) and the Baltic Sea (see Section 6.8.1) are all examples of water bodies affected by low oxygen concentrations, and similar processes occur in groundwater when oxygen consumption exceeds supply.

Once oxygen has been used up, bacteria use alternative oxidizing agents (electron acceptors, see Box 4.3) to consume organic matter. These alternative oxidants are used in an order that depends on energy yields (see Table 4.7). Nitrate reduction (denitrification) is energetically favourable to bacteria, but is often limited in natural freshwaters by low nitrate concentrations. Anthropogenic inputs, however, have resulted in increased nitrate concentrations in rivers and groundwater (Section 5.5.1), increasing the availability of nitrate for bacterial reduction. One of the byproducts of denitrification is nitrous oxide (N$_2$O; Fig. 5.12), a powerful greenhouse gas which is increasing in concentration, probably because of human perturbation of the nitrogen cycle.

Iron (Fe) and manganese (Mn), both potential electron acceptors, are common as insoluble Fe(III) and Mn(IV) oxides. In reducing environments (at about the same redox potential as nitrate reduction), these oxides may be reduced to soluble Fe(II) and Mn(II) (see Table 4.7). Indeed, iron is soluble only under low redox or acidic conditions (Box 5.4).

Sulphate reduction (see Table 4.7) is not an important mechanism of organic matter respiration in freshwaters because dissolved sulphate levels are usually low. In seawater, however, sulphate is abundant and sulphate reduction is very important (see Section 6.4.6). Methanogenesis (see Table 4.7) can be an important respiration process in some organic-rich freshwater lake and swamp sediment. The reduced reaction product, methane (CH_4), a greenhouse gas (see Section 7.2.4), is known to bubble out of some wetlands, including rice paddy fields, contributing significantly to atmospheric CH_4 budgets (see Section 3.4.2).

5.5.1 Nutrients and eutrophication

In addition to CO_2, water and light, ions (or nutrients) are needed for plant growth. Some of these ions, for example Mg^{2+}, are abundant in freshwater, but other essential nutrients, for example nitrogen (N) and phosphorus (P), are usually present at low concentrations in natural systems. If light availability does not limit algal growth, chemical limitation is likely to occur when demand for nitrogen and phosphorus exceeds their availability. Consequently, a great deal of attention has been focused on the behaviour of nitrogen and phosphorus in natural waters and their role as potential, or actual, limiting nutrients. In seawater, the ratio of nitrogen to phosphorus required for optimal growth is quite well known, being 16:1 on an atomic basis. In freshwater, the required nitrogen:phosphorus ratio (N:P) is more variable. If, however, either nitrogen or phosphorus are in excess of the ratio required for optimal growth, it follows that the less abundant nutrient may be totally consumed and become limiting. Of course, nutrient elements can be available in excess when artificially introduced into the environment, for example as nitrate- and phosphate-based fertilizers. Nitrogen and phosphorus that leach from fertilizer applications often stimulate exessive algal growth (biomass) in water courses, a problem referred to as *eutrophication*. The exess algal biomass can cause toxicity, clogging of water filters, unsightly water bodies, reduced biodiversity and low oxygen concentrations in stratified waters.

Phosphorus

In natural waters, dissolved inorganic phosphorus (DIP) exists predominantly as various dissociation products of phosphoric acid (H_3PO_4) (eqns 5.3–5.5). Phosphorus is usually retained in soils by the precipitation of insoluble calcium and iron phosphates, by adsorption on iron hydroxides or by adsorption on to soil particles. As a result, DIP in rivers is derived mainly from direct discharges, for example sewage. DIP concentrations vary inversely with river flow (Fig. 5.10), the input being diluted under higher flow conditions. Since phosphate is usually in sediments as insoluble iron(III) phosphate ($FePO_4$), under reducing conditions (such as occur in sediments when oxygen consumption exceeds supply (Section

Box 5.4 Eh-pH diagrams

Acidity (pH) and redox potential (Eh) (see Box 4.3) may determine the chemical behaviour of elements or compounds in an environment. In theory, an infinite valley of Eh-pH combinations is possible, although the pH of most environments on Earth is between 0 and 14 and more usually between 3 and 10. Redox potential is constrained by

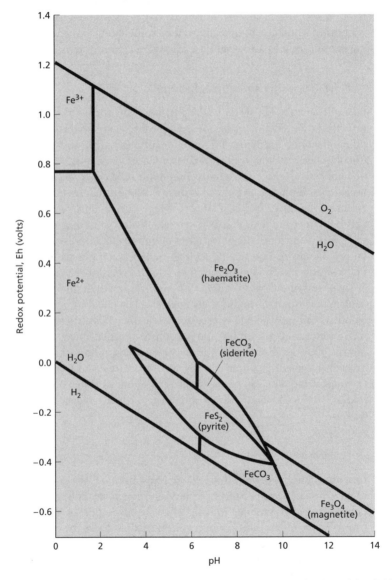

Fig. 1 Simplified Eh-pH diagram showing stability fields of common iron minerals. The stability fields change position slightly depending on the activity of the components. In this case Fe^{2+}, Fe^{3+} and $S = 10^{-6}\,mol\,l^{-1}$ and dissolved inorganic carbon $= 1\,mol\,l^{-1}$. After Garrels and Christ (1965).

(continued)

the existence of water. Under very oxidizing conditions (Eh 0.6 to 1.2 V) water is broken into oxygen and hydrogen ions and under highly reducing conditions (Eh 0.0 to –0.6 V) water is reduced to hydrogen. Eh-pH diagrams are used to visualize the effects of changing acidity and/or redox conditions. The diagram for iron minerals is typical (Fig. 1). The lines represent conditions under which species on either side are present in equal concentrations. The exact position of the lines varies depending on the activities of the various species.

From the diagram it is clear that the mineral haematite (Fe_2O_3) is usually the stable iron species under oxidizing conditions with pH above 4. Soluble Fe^{3+} is only present under very acidic conditions because of its tendency to form insoluble hydroxides (Section 5.2). This tendency is only overcome under very acid conditions when hydroxide ion (OH^-) concentrations are low. Fe^{2+} is less prone to form insoluble hydroxides because of its small z/r value (Section 5.2). Fe^{2+} is thus soluble at higher pH, but can only persist under low Eh conditions, which prevent its oxidation to Fe^{3+}. The small stability field for the common iron sulphide, pyrite (FeS_2), shows that this mineral only forms under reducing conditions, usually between pH 6 and 8. Iron carbonate, (siderite $FeCO_3$) is typically stable at either slightly higher or slightly lower Eh than pyrite.

5.5)), DIP can be returned to the water column in association with iron(III) reduction to iron(II). This process of DIP release from sediments can confound efforts to control eutrophication in lakes that are based on reducing the direct DIP inputs.

Increased riverine NO_3^- concentrations due to human activity (see below) mean that DIP is now the main limiting nutrient for plant growth in many freshwaters. The consequent relationship between DIP and chlorophyll levels (a measure of algal biomass) (Fig. 5.11) makes the management of phosphorus inputs to rivers and lakes very important. In some areas DIP inputs are consequently stringently

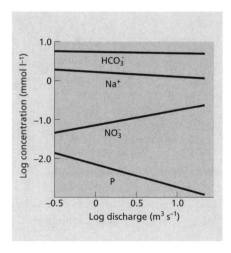

Fig. 5.10 Relationship between dissolved ion concentration and river discharge in the River Yare (Norfolk, UK). Direct discharges of phosphorus from sewage and products of weathering (HCO_3^- and Na^+) decline in concentration (are diluted) as discharge increases. By contrast, heavy rainfall leaches NO_3^- from soil, causing NO_3^- concentrations to rise as discharge increases. After Edwards (1973).

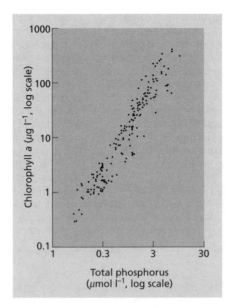

Fig. 5.11 Relationship between summer particulate chlorophyll *a* levels (as a measure of phytoplankton abundance) and total phosphorus concentrations in various lakes, with data plotted on log scales. After Moss (1988).

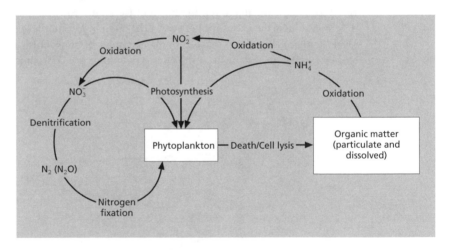

Fig. 5.12 Nitrogen cycling in natural waters.

controlled to limit algal growth in the receiving waters, even though the need for this strategy arises from nitrate enrichment.

Nitrogen

Nitrogen (N) chemistry and cycling (Fig. 5.12) is complex because nitrogen exists in several oxidation states (see Box 4.3), of which N(0) nitrogen gas (N_2), N(3−)

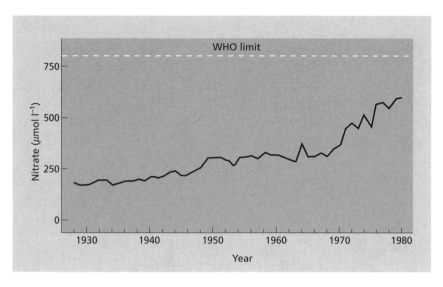

Fig. 5.13 Mean annual dissolved nitrate concentrations in the River Thames, UK, 1930–1980. WHO limit refers to the World Health Organization recommended maximum safe nitrate concentration in drinking water. Data courtesy of UK Department of the Environment National Water Council, Crown Copyright 1984.

ammonium (NH_4^+) and N(5+) nitrate (NO_3^-) are the most important. Nitrogen gas dissolved in natural waters cannot be utilized as a nitrogen source by most plants and algae because they cannot break its strong triple bond (see Section 2.3.1). Specialized 'nitrogen-fixing' bacteria and fungi do exist to exploit N_2, but it is not an energetically efficient way of obtaining nitrogen. Hence, these microorganisms are only abundant when N_2 is the only available nitrogen source. Nevertheless, along with fixation of N_2 by lightning, nitrogen-fixing microorganisms provide the major natural source of nitrogen for rivers. Increases in both the area and the intensity of agricultural activity are probably responsible for the increased NO_3^- concentrations seen in British (Fig. 5.13), other European and North American rivers. Globally, human activity has doubled the natural riverine transport of reactive nitrogen, mostly as NO_3^-.

Biological processes use nitrogen in the 3- oxidation state, particularly as amino functional groups (see Table 2.1) in proteins. This is the preferred oxidation state for algal uptake (although NO_3^- and NO_2^- can be taken up) and also the form in which nitrogen is released during organic matter decomposition, largely as NH_4^+. Once released into soils or water, NH_4^+ being cationic may be adsorbed on to negatively charged organic coatings on soil particles or clay mineral surfaces. Ammonium is also taken up by plants or algae, or oxidized to NO_2^- and ultimately NO_3^-, a process that is usually catalysed by bacteria.

In contrast to NH_4^+, NO_3^-—the thermodynamically favoured species in oxygenated waters—is anionic, soluble and not retained in soils. Therefore, NO_3^- from rainwater or fertilizers, or derived from the oxidation of soil organic matter

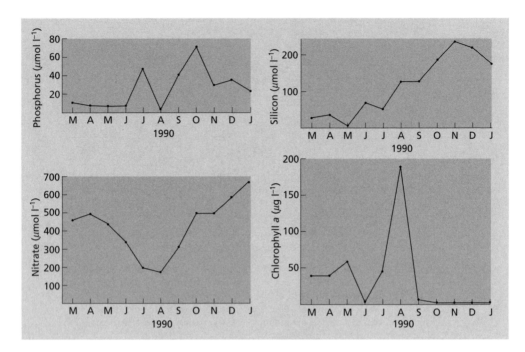

Fig. 5.14 Seasonal variations in dissolved phosphorus, silicon, nitrate and particulate chlorophyll *a* (as a measure of phytoplankton abundance) in the River Great Ouse (eastern England). Data from Fichez *et al.* (1992).

and animal wastes, will wash out of soils and into rivers. The seasonal variation of NO_3^- concentrations in many temperate rivers is caused mainly by fluctuation in supply of NO_3^- from soil. In summer, NO_3^- concentrations are low because soil-water flushing by rainfall is low. In the autumn, soil moisture content increases, allowing nitrate to wash out of the soil into rivers (Fig. 5.14). In nitrate-rich rivers like the Great Ouse in eastern England (N:P about 30:1 in winter), biological production makes little initial impact on NO_3^- levels until later in the season as the NO_3^- supply decreases due to reduced runoff. A NO_3^- minimum is reached in summer as a result of the reduced supply from soils and increased biological uptake, before rising again in the autumn (Fig. 5.14). DIP concentrations, in contrast, show more erratic behaviour (Fig. 5.14), reflecting the influences of biological and dilution control working out of phase, but are generally higher during low-flow conditions in summer.

Apart from biological uptake, denitrification in low-oxygen environments is the most important way that NO_3^- is removed from soil, rivers and groundwater. It has been estimated that, in the rivers of northwest Europe, half of the total nitrogen input to the catchment is lost by denitrification before the waters reach the sea. Thus, under low redox conditions, DIP is mobilized during iron(III) reduction and NO_3^- is lost, again emphasizing the importance of redox processes in environmental chemistry.

Phosphorus and nitrogen in groundwater

The very different chemistry of DIP and NO_3^- is illustrated by their behaviour in groundwater. On the limestone island of Bermuda there is little surface water because rainfall drains rapidly through the permeable rock to form groundwater. Almost all of the sewage waste on Bermuda is discharged to porous walled pits that allow effluent to gradually seep into the groundwater. The sewage has a N : P ratio of about 16:1, and yet Bermudian groundwater is characterized by very low DIP concentrations (average $3.5\,\mu mol\,l^{-1}$) and very high NO_3^- concentrations (average $750\,\mu mol\,l^{-1}$). The N:P ratio of 215:1 for the groundwater implies removal of more than 90% of the DIP, by natural precipitation of calcium phosphate minerals. This process is now used as a sewage treatment stage for DIP removal from some waste waters.

High nitrate concentrations are characteristic of groundwater in areas of intensive agriculture and can compromise its use as a drinking water supply (Section 5.7). The main potential human health hazard is a condition called methaemoglobinaemia, where NO_3^- combines with and oxidizes haemoglobin in the blood, robbing the cells of oxygen. This condition affects some adults with specific enzyme deficiencies, but also newborn babies, resulting in the name 'blue-baby syndrome'. To safeguard drinking water supplies in agricultural areas such as southeast England, farmers are required to control fertilizer inputs in areas of groundwater recharge.

Silicon

One other important nutrient, silicon, is used by diatoms (a group of phytoplankton) to build their exoskeletons. Diatoms are capable of rapid and prolific growth in nutrient-rich conditions. In temperate rivers, diatom blooms occur early in the year. For example, in the River Great Ouse in eastern England, silicon levels fall in early spring as diatom growth begins and rise again in summer as diatoms are displaced by other algal groups (Fig. 5.14). Since silicon is derived entirely from weathering reactions, its naturally low concentrations may be drastically reduced by diatom blooms, such that further diatom growth is limited, particularly where NO_3^- and DIP have been enriched by human activity. Thus silicon availability limits species diversity but not total phytoplankton biomass.

The presence of lakes and dams in river catchments has an important effect on nutrient transport. Increased water residence times and improved light conditions promote algal (particularly diatom) blooms that are very effective at removing dissolved silicon (DSi). In Scandinavia, for example, catchments without lakes have DSi concentrations around $164\,\mu mol\,l^{-1}$, almost four times higher than those in catchments where lakes and reservoirs cover more than 10% of the area (DSi = $46\,\mu mol\,l^{-1}$). A particularly striking example of this effect is seen in the Danube, the largest river draining to the Black Sea, which was dammed by the 'Iron Gates' on the Yugoslavian/Romanian border in the early 1970s. DSi concentrations fell three-fold as a result of the damming, and the resulting decrease in Si:N ratios in water draining to the Black Sea (exacerbated by increases in nitrate inputs) has resulted in large-scale changes in phytoplankton ecology in the Black Sea itself. Large numbers of the world's rivers are now extensively dammed for flood control and to provide hydroelectric power. These numbers are expected to increase in the near future and the resulting effects on riverine nutrient fluxes will also grow.

5.6 Heavy metal contamination

Heavy metals such as mercury (Hg) and lead (Pb) are so called because of their very high densities ($Hg = 13.5\,g\,cm^{-3}$, $Pb = 11.3\,g\,cm^{-3}$) when compared to other common metals, for example $Mg = 1.7\,g\,cm^{-3}$. Heavy metals are of concern because of their toxicity to humans and other animals. However, other elements, for example the semimetal arsenic ($As = 5.7\,g\,cm^{-3}$), are also toxic. It can be argued that the term heavy metals should be more encompassing to include the toxicity of the element. Toxicity depends on an element's chemistry, the mode of contact with the host organism, the concentration of the element and the host organism's biochemistry. Indeed some substances that are toxic at high concentrations may be essential to life at low concentration (Box 5.5). Heavy metals are of concern because of their toxicity to humans and other animals, but also because they are non-biodegradable. All heavy metals occur naturally, but industrial activity can markedly increase their concentrations in natural waters. These contamination maxima tend to be localized rather than widespread. In Section 5.6.1 we highlight the specific case of mercury associated with gold mining. Mercury is particularly interesting as its toxicity is intimately related to biogeochemical processes and redox conditions. In Section 5.7.2 we discuss the case of natural rather than anthropogenic arsenic contamination of groundwater.

5.6.1 Mercury contamination from gold mining

Mercury has had a host of industrial and commercial uses, ranging from use in batteries, for the production of commercial chlorine, as a fungicide on seeds and in the mining of precious metals such as silver (Ag) and gold (Au). In the latter context the unusual property of metallic mercury (Hg^0)—being liquid at 'room temperature'—is exploited to help isolate particulate gold from mined gravel slurries. When mercury is added to the slurry it readily forms an amalgam with the gold:

$$Au_{(metal)} + Hg^0 \rightarrow Au\text{-}Hg_{(amalgam)} \qquad \text{eqn. 5.21}$$

This amalgam or alloy has a higher density than the surrounding gravel slurry allowing easy separation. While this method of separation has long been used by miners, for example in the Californian gold rush of the late 19th century, it is currently an environmental concern in many developing countries. In the early 1980s, for example, the Amazon Basin experienced its own 'gold rush'—some claim the largest single gold rush in history—and mercury is still used extensively as an amalgam in the myriad small-scale gold mines (garimpos). The environ-

Box 5.5 Essential and non-essential elements

Many substances that might be considered toxic are in fact essential to life, for example the heavy metals Co, Cu, Fe, Se, Zn. Thus, there is a relationship between concentration of the substance and the health response it invokes. An element is essential to life when a deficient intake of the element consistently results in impairment of a life function from optimal to suboptimal (Fig. 1a). Moreover, when physiological levels of this element, but not of others, are supplemented or restored, the impairment is cured and optimal health is restored (Fig. 1a). By contrast, non-essential elements do not produce a positive health response. An organism may tolerate low concentrations of some non-essential elements (Fig. 1b). Even dangerous poisons such as arsenic oxide (As_2O_3) can be tolerated

to some extent if the amount taken (dose) is increased in small degrees. The so-called 'arsenic eaters' of Styria, a mountainous Austrian state, are said to have taken six times the minimum fatal dose of arsenic oxide without ill effect. These people maintained that their curious diet supplement improved their personal appearance and increased their energy for ascents of their Alpine homeland! Of course if concentration exceeds the organism's tolerance threshold, impairment is observed (Fig. 1b). In both essential and non-essential elements, excessive concentrations will result in toxicity and ultimately death (Fig. 1). The 16th century Swiss scientist Paracelsus recognized this with his comment 'Tis the dose that maketh the poison'.

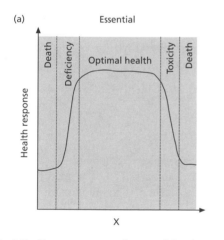

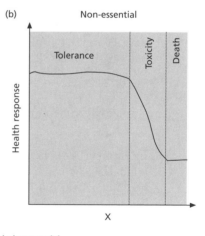

Fig. 1 Health–response curves for essential and non-essential elements (X).

mental concerns revolve around both the amounts of mercury used and its accidental release into the environment. On average 1.3 kg of Hg^0 is used for each kilogram of gold recovered, although much of this mercury is contained and reused. However, many hundreds of tonnes of mercury have been released into the Amazon Basin during the 1980s and 1990s. This practice is widespread not only in the Amazon, but in Latin America as a whole, the Pacific Rim countries, and Africa, which together produce thousands of tonnes of gold each year.

During the amalgamation process Hg^0 is easily introduced into the local soils, rivers and atmosphere. Although the effects of leaching of Hg^0 from waste tips and direct spillages into water courses are reduced by the low solubility of Hg^0 in water (solubility of $0.12 \times 10^{-6} \, mol \, l^{-1}$), the mercury is potentially subject to a number of chemical transformations that make it harmful to organisms. During the final stages of gold processing, burning of the Au-Hg amalgam in open pans allows evaporation of Hg^0 as vapour to the local atmosphere. The annual flux to the atmosphere by this process is around 80 tonnes of Hg^0, approximately 67% of Brazil's total mercury emission. In the atmosphere Hg^0 can be dispersed globally (atmospheric residence time around 1 year) but much is readily oxidized to Hg^{2+}. The reaction occurs in water droplets, the oxidant is ozone (see also Section 3.9) and experiments suggest that mercuric oxide is produced, for example:

$$Hg^0 + O_3 \rightarrow HgO + O_2 \qquad \text{eqn. 5.22}$$

The oxide might then be ionized, for example by:

$$2H^+ + HgO \rightarrow Hg^{2+} + H_2O \qquad \text{eqn. 5.23}$$

This ionic mercury (HgII) adheres to aerosols and thus has a short (days to weeks) residence time in the atmosphere; rainfall delivers it to the local soils and rivers. Ionic mercury is readily methylated (eqn. 5.24) by both abiotic and biotic pathways. However, most scientists now agree that methylation by anaerobic sulphate reducing bacteria (SRB) is most important.

$$Hg^{2+} + 2CH_3^- \xrightarrow{\hspace{4cm}} Hg(CH_3)_2$$

(methylcobalamin)

SRB $\qquad \qquad \qquad$ eqn. 5.24

In equation 5.24 the Hg^{2+} ion forms a covalent compound dimethylmercury ($Hg(CH_3)_2$) by bonding with the methyl anion (CH_3^-). The methyl anion is a derivative of vitamin B_{12} called methylcobalamin, a common constituent of the bacteria themselves. At low to neutral pH, methyl mercury (CH_3Hg^+) forms in a similar way to dimethylmercury (Fig. 5.15). Methyl mercury is a potent toxin because it is soluble in the fatty tissues (lipophylic, see Box 4.14) of animals, ultimately attacking the central nervous system. It enters the food chain either directly absorbed from the water (e.g. into fish) or when plankton, a food source for fish, feed on the methyl mercury-rich bacteria. Each step in the food chain increases the concentration of mercury in the organic tissues (biomagnification) because organisms cannot eliminate mercury as fast as they ingest it (Fig. 5.15). As a result, mercury-contaminated fish are the main cause of mercury poisoning in humans. This is a particularly alarming situation for the Amazon region where fish are the staple food of the poor. Unfortunately, this pattern of mercury poisoning is very well known. Between 1953 and 1975 in the now infamous Japanese fishing village of Minamata, over 600 people were poisoned (115

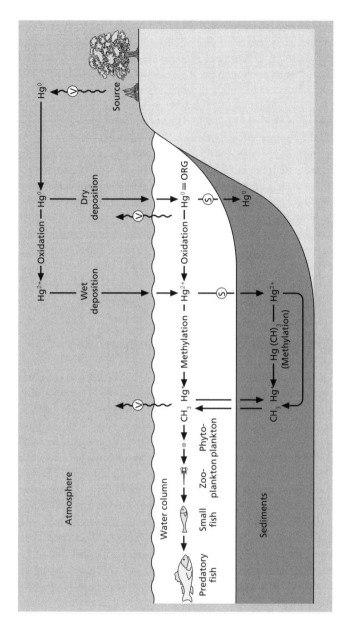

Fig. 5.15 Cycling of mercury in lakes. Ionic mercury (Hg^{2+}) is readily methylated by bacteria, especially sulphate-reducing bacteria in reducing sediments. Methyl mercury (CH_3Hg^+) enters the food chain through phytoplankton and is biomagnified in fish. Elemental mercury (Hg^0) is quite insoluble except when complexed by dissolved organic ligands ($Hg^0 \equiv$ org), making it more mobile and susceptible to oxidation in the water column. Pathways labelled S represent sedimentation, pathways labelled V represent volatilization.

fatally), by methylated mercury in fish. The methyl mercury, a waste product of the plastics industry, had been discharged into coastal waters near the village.

The bioavailability of mercury is complex, controlled by fundamental biogeochemical parameters such as type of microbial communities and Eh/pH (as discussed above), but also by its partitioning between solid and liquid phases and its complexation particularly with dissolved organic ligands (see Box 6.4). Laboratory studies show that humic acids acting as ligands (see Box 6.4) drastically increase the solubility of otherwise rather insoluble Hg^0 at near-neutral pH. Moreover, the soluble complex formed is much less reactive than Hg^0 to mineral surfaces in river sediments. This finding is highly significant for tropical rivers like the Amazon that contain high levels of dissolved organic matter and humic acids. It implies that mercury will remain mobile rather than fixed to mineral surfaces in river sediments, and may be transported to sites where oxidation and methylation is occurring, increasing the risk of toxicity. Once in local soils and water, mercury is also difficult to remove. In the Sierra Nevada region of the western USA, millions of kilograms of mercury were lost to the environment during the late 19th century gold mining. Much of this legacy remains today, slowly releasing mercury into surface and groundwaters.

5.7 Contamination of groundwater

Although aspects of groundwater chemistry have been discussed elsewhere in this chapter, this section highlights issues relating to the contamination of groundwater. Groundwater is critically important to humans since it is a major source of drinking water. For example, in the USA over 50% of the population rely on groundwater as a source of drinking water. Groundwater quality is therefore very important and, in most developed countries, water must conform to certain standards for human consumption. Groundwater may fail to meet water quality standards because it contains dissolved constituents arising from either natural or anthropogenic sources. Typical anthropogenic mechanisms of groundwater contamination are shown in Fig. 5.16. In the USA, major threats to groundwater include spillage from underground storage tanks, effluent from septic tanks and leachate from agricultural activities, municipal landfills and abandoned hazardous waste sites. The most frequently reported contaminants from these sources include nitrates, pesticides, volatile organic compounds, petroleum products, metals and synthetic organic chemicals.

The chemistry of contaminated groundwater is little different from that of surface waters, except that most groundwaters are anaerobic. This has posed a major problem for remediation of benzene (C_6H_6), one of the most prevalent organic contaminants in groundwater and of concern because of its toxicity. Although much is known regarding aerobic degradation of benzene (see Section 4.10), until recently no pure culture of an organism capable of anaerobic degradation of benzene existed, which made remediation almost impossible. It is now known, however, that the bacterium *Dechloromonas aromatica* strain RCB oxidizes benzene anaerobically, using nitrate as the electron acceptor.

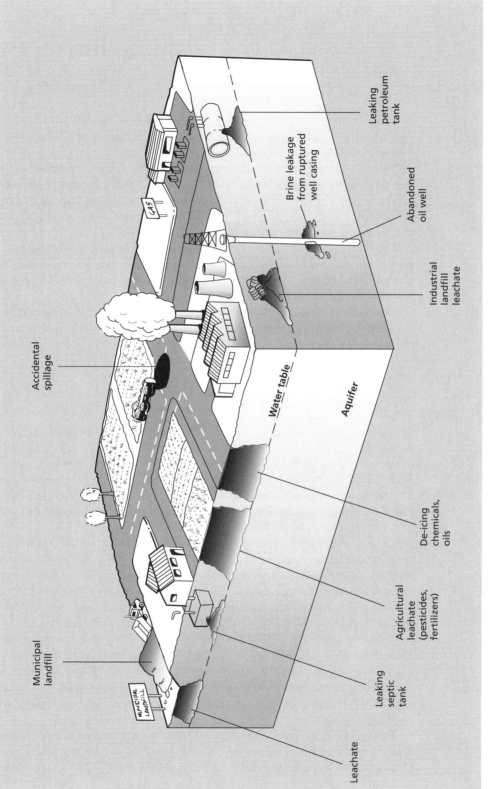

Fig. 5.16 Typical anthropogenic mechanisms of groundwater contamination.

$$C_6H_{6(aq)} + 6NO_{3(aq)}^- + 6H_{(aq)}^+ \longrightarrow 6CO_{2(aq)} + 3N_{2(g)} + 6H_2O_{(l)}$$

(benzene)

eqn. 5.25

In addition to benzene this bacterium also oxidizes toluene, ethylbenzene and xylene and thus offers great potential for the treatment of petroleum-contaminated aquifers (Section 5.7.1).

Natural degradation processes that occur in days or weeks in surface waters may take decades in groundwater, where flow rates are slow and microbiological activity is low. This limits the potential for natural purification through flushing or biological consumption. Once contaminated, groundwater is difficult and expensive—in many cases impossible—to rehabilitate.

The location of older sites of contamination may be imprecisely known, or even unknown, and hydrogeological conditions may dictate that contaminated groundwater discharges at natural springs into rivers or lakes, spreading contamination to surface waters.

The following sections highlight different styles of groundwater contamination where chemical considerations have proved important. Nutrient element contamination of groundwater was discussed in Section 5.5.1.

5.7.1 Anthropogenic contamination of groundwater

Landfill leakage—Babylon, Long Island, New York, USA

At Babylon landfill site, New York, shallow groundwater contamination of a surface sand aquifer has resulted from leakage of leachate rich in Cl⁻, nitrogen compounds, trace metals and a complex mixture of organic compounds. Landfilling began in the 1940s with urban and industrial refuse and cesspool waste. The refuse layer is now about 20 m thick, some of it lying below the water table. Chloride behaves conservatively (see Section 6.2.2) and is thus an excellent tracer of the contaminant plume, which is now about 3 km long (Fig. 5.17).

Close to the landfill, most nitrogen species are present as NH_4^+, indicating reducing conditions resulting from microbial decomposition of organic wastes. With increasing distance from the landfill, NO_3^- becomes quantitatively important due to the oxidation of NH_4^+ (Fig. 5.12), brought about by mixing of the leachate plume with oxygenated groundwater. This demonstrates how nitrogen speciation can be used to assess redox conditions in a contaminant plume.

Reducing conditions within the leachate plume also cause metal mobility, particularly of manganese and iron. The plume near the landfill has a pH of 6.0–6.5 and is reducing (−50 mV), making Fe^{2+} stable (Box 5.4). The transition to oxidizing conditions down gradient in the aquifer allows solid iron oxides (e.g. FeOOH) to precipitate, dramatically, reducing the mobility of metals which co-precipitate with iron.

This relatively inoffensive example illustrates the importance of redox conditions in contaminated groundwater. Worse scenarios are known where toxic chlorophenolic compounds in very alkaline groundwaters (pH 10) ionize to neg-

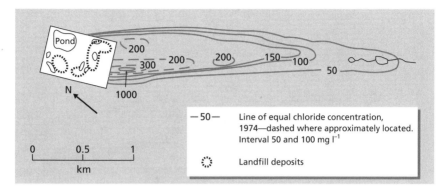

Fig. 5.17 Map of Cl⁻ plume at 9–12 m depth below the water table, Babylon landfill site, 1974, showing the extent of groundwater contamination. After Kimmel and Braids (1980).

atively charged species and become much more mobile than in the neutral conditions generally typical of groundwater.

Petroleum contamination—Bowling Green, Kentucky, USA

Although groundwater flow rates are generally low when compared to surface waters, large cracks and conduits in some contaminated aquifers cause specific problems. The US city of Bowling Green, Kentucky, is built on limestone bedrock (Ste Geneviève limestone), with underground drainage through the Lost River Cave (Fig. 5.18). Limestone bedrock is often heavily fissured and joints in the rock are enlarged by dissolution, resulting in interconnected caves. Sinkholes may divert surface streams into these fissures and caves, resulting in a subsurface drainage system. Accidental spillage of toxic chemicals or any other contaminant is rapidly dispersed in these conduits, making remediation particularly difficult.

In the 1970s and 1980s up to 22 000 litres of petroleum leaked from storage tanks at auto service stations in Bowling Green into the subsurface water. Petroleum, being a non-aqueous phase liquid (NAPL), floats on the surface of groundwater and its volatile components (see Box 4.14), for example benzene, rapidly fill air spaces with explosive fumes, particularly at sumps or traps in the cave system (Fig. 5.18). The trapped fumes then escape into the basements of buildings, water wells and storm drains.

In addition, leaking tanks at a chemical company are believed to have delivered benzene, methylene chloride (CH_2Cl_2), toluene ($C_6H_5CH_3$), xylene ($C_6H_4(CH_3)_2$) and aliphatic hydrocarbons (see Section 2.7) to the subsurface water. These toxic (some carcinogenic) chemicals vaporize in the cave atmosphere, collect at traps and then rise into homes in a similar way to petroleum fumes.

The potential explosive/toxicity risk in Bowling Green has resulted in a number of evacuations of homes in the last 20 years. Remediation measures have

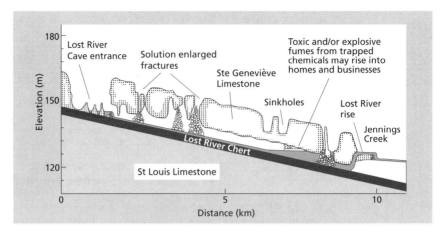

Fig. 5.18 Cross-section through the Lost River Cave drainage system underlying the city of Bowling Green, Kentucky, showing potential trap for floating or gaseous contaminants. After Crawford (1984). With permission from Swets & Zeitlinger Publishers.

included better storage tank containment, regular monitoring of cave conduit outlets and ventilation of basements in homes at risk. It is hoped that these will prevent a disaster such as occurred in nearby Louisville, Kentucky, where an underground sewer explosion travelled 11 blocks, causing damage estimated at over 43 million dollars.

5.7.2 Natural arsenic contamination of groundwater

Those who have seen Victorian melodramas will know that arsenic is a common and effective poison; murderers have used the oxide (As_2O_3) successfully for the last two millennia. Ingestion of just 20 mg of the oxide is said to be lethal (but see Box 5.5), caused through damage to the stomach and intestines; lower exposures cause cancers. Given this macabre background it is perhaps obvious that exposure to arsenic in foodstuffs and drinking water should be low. Although tiny amounts of arsenic occur in some foods, it is typically a water-soluble form of organic arsenic that is easily excreted. Moreover, most drinking water contains much less than $5\,\mu g\,l^{-1}$ of inorganic arsenic, such that typical daily intakes are about $4\,\mu g$.

Arsenic contamination of drinking waters by industrial and commercial activities is not particularly commonplace, although arsenic is still used in pesticides in some underdeveloped countries. Surprisingly, however, natural arsenic contamination of groundwater is now well known, and today affects areas of Argentina, Taiwan, Vietnam, Cambodia, China, Hungary, Bangaladesh, India and the USA. The symptoms of low-level arsenic poisoning are dependent on dose received, but can take up to 10 years to develop, while cancers may take 20 years. It is therefore important to identify arsenic contamination of water supplies quickly.

Nearly all cases of large-scale arsenic contamination in groundwater are caused by reduction of iron oxides in aquifer sediments.

Microbial reduction of arsenic-bearing iron oxides

Reduction of iron oxide (FeOOH) is a potential source of arsenic to groundwater because it releases arsenic adsorbed to the oxide surfaces. A representation of this process might be:

$$4FeOOH_{(s)} + CH_2O_{(s)} + 7H_2CO_{3(aq)} \longrightarrow 4Fe^{2+}_{(aq)} + 8HCO^-_{3(aq)} + 6H_2O_{(l)}$$

eqn. 5.26

The reaction requires anoxic conditions, since iron is barely soluble in oxic water (Box 5.4), and is fuelled by the microbial metabolism of organic matter (depicted as CH_2O in eqn. 5.26).

This mechanism of arsenic release to groundwater has recently been proposed as the cause of high arsenic concentrations in millions of deep drinking water wells across Asia, the example in the Ganges–Meghna–Bramaputra delta plain of Bangladesh and West Bengal being much publicized. The original source of the arsenic in these aquifer sediments is not known with much certainty, but probably comes from weathering of arsenic-rich coals and sulphide ores in the upstream drainage basin. The arsenic was then transported downstream in solution, adsorbing to clay-rich and organic-rich sediments that accumulated in the delta over the last 2 million years or so.

Many of these drinking water wells, which provide 20 million people with almost all their drinking water, exceed both the World Health Organization (WHO) guideline value of $10\,\mu g\,l^{-1}$ As and the Bangladesh drinking water maximum of $50\,\mu g\,l^{-1}$ As. The world's press have recently picked up on this problem, first discovered in the mid 1990s, with headlines claiming 'the largest mass poisoning of a population in history'. It is feared that up to 20 000 people could soon die each year as a delayed result of accumulating arsenic in their bodies from wells sunk up to 25 years ago.

The highest levels of contamination ($>250\,\mu g\,l^{-1}$ As) typically occur between 25 and 45 m depth—too deep to implicate sulphide oxidation (Section 5.4.2)—and concentrations in excess of $50\,\mu g\,l^{-1}$ As occur down to 150 m. The requirement for microbial metabolism of organic matter has recently led scientists to link the presence of highly organic peat beds in the deltaic aquifer sediments as the driver for equation 5.26. Clearly other sedimentary aquifers that host peats or organic-rich muds might be vulnerable to arsenic contamination, including other large deltas such as the Mekong and Irrawaddy.

As the release of arsenic requires reducing conditions, a logical application of environmental chemical principles suggests that aeration of the water might reverse the effects of equation 5.26, for example:

$$4Fe^{2+}_{(aq)} + O_{2(g)} + 10H_2O_{(l)} \rightarrow 4Fe(OH)_{3(s)} + 8H^+_{(aq)} \qquad \text{eqn. 5.27}$$

This would force iron oxide ($Fe(OH)_3$) to precipitate, but also remove some of the dissolved arsenic as it adsorbs to the precipitating oxide surface. Trials at water-treatment plants in the area have shown that water with $220\,\mu g\,l^{-1}$ As, can be reduced to around $40\,\mu g\,l^{-1}$ As (below the Bangladesh drinking water maximum) by this method. Longer-term mitigation strategies might include the construction of shallower wells in the oxidized upper zone of the aquifer. Whatever the long-term mitigation strategies are, it is clear that these wells are currently critical for the water supply of millions of people. As abandoning the wells is not an option in many cases, short-term mitigation methods, even imperfect ones such as aeration with a stick in a bucket, seem preferable to continued exposure to high levels of arsenic.

It is worth noting that an 'early warning' of potential arsenic contamination by reduction of iron oxide is given by high values of dissolved iron in reducing drinking water wells. This element is simple to analyse, and routinely measured during water quality analysis. It could therefore be used as a sign that arsenic contamination is possible, prompting further analysis for dissolved arsenic.

5.8 Further reading

Baird, C.B. (1995) *Environmental Chemistry*. WH Freeman, New York.
Berner, K.B. & Berner, R.A. (1987) *The Global Water Cycle*. Prentice Hall, Englewood Cliffs, New Jersey.
Harrison, R.M., deMora, S.J., Rapsomanikas, S. & Johnston, W.R. (1991) *Introductory Chemistry for the Environmental Sciences*. Cambridge University Press, Cambridge.
Moss, B. (1998) *Ecology of Freshwaters: Man and Medium*, 3rd edn. Blackwell Science, Oxford.
Stumm, W. & Morgan, J.J. (1996) *Aquatic Chemistry*, 3rd edn. Wiley, New York.

5.9 Internet search keywords

water chemistry	acid mine drainage
freshwater chemistry	landfill leachate
chemistry river water	electron acceptors
chemistry groundwater	nitrate water
oxyanions	phosphate water
phosphorous natural water	eutrophication
silica natural water	heavy metal contamination
weathering regimes	mercury contamination
geochemistry Amazon	groundwater contamination
alkalinity	arsenic contamination
pH buffering	ionic strength
aluminium solubility	Eh-pH
acid deposition	redox potential
acid rain	

The Oceans 6

6.1 Introduction

The oceans are by far the largest reservoir of the hydrosphere (see Fig. 1.4) and have existed for at least 3.8 billion years. Life on Earth probably began in seawater and the oceans are important in moderating global temperature changes. Rivers draining continental land areas carry both dissolved and particulate matter to the oceans and the average input of dissolved major ions can be estimated by considering the input from some of the rivers with the largest discharges (see Section 5.2). The transport of particulate matter to the oceans depends on both the discharge and the supply of suspended sediment. Some large rivers such as those in central Africa carry rather small amounts of sediment because of the relatively low relief and dry climates in the catchments (Fig. 6.1). By contrast, rapidly eroding areas of South East Asia carry a disproportionate volume of sediment compared to the volume of water, due to high relief and heavy rainfall in the catchments (Fig. 6.1). Much of this sediment falls to the seafloor, usually in estuaries and on continental shelves, although in some parts of the ocean where the shelf area is small, this material may reach deep-sea environments. Most riverwater enters the oceans through estuaries and here freshwater mixes with seawater. The chemical composition of seawater is quite different from that of freshwater, a difference that affects the transport of some dissolved and particulate components. In addition, humans often perturb the natural chemistry of coastal areas, either through contamination of the freshwater runoff, or due to activities located close to estuaries and shallow seas.

We begin this chapter by examining the chemistry of seawater close to continental areas, in the transition zone between terrestrial and open-ocean environments, before moving on to discuss open-ocean environments.

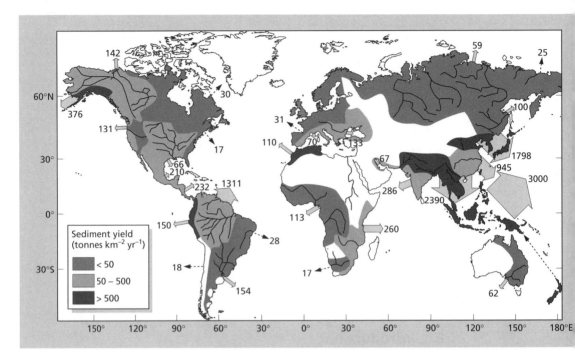

Fig. 6.1 Annual river sediment flux from large drainage basins to the oceans. Numbers in 10^6 tonnes yr^{-1} and arrow size proportional to the numbers. After Milliman and Meade (1983), with permission from the University of Chicago Press.

6.2 Estuarine processes

There are many differences between the chemistry of continental surface waters and seawater. In particular, seawater has a much higher ionic strength than most continental water (see Fig. 5.3) and seawater has a huge concentration of sodium and chloride ions (Na^+ and Cl^-) (Table 6.1), in contrast with calcium bicarbonate-dominated continental waters (see Section 5.3). Seawater is a high-concentration chemical solution, such that mixing only 1% (volume) of seawater with average riverwater produces a solution in which the ratio of most ions, one to another, is almost the same as in seawater. Thus, the chemical gradients in estuaries are very steep and localized to the earliest stages of mixing. In addition to the steep gradient in ionic strength, in some estuaries there is also a gradient in pH.

Unidirectional flow in rivers is replaced by tidal (reversing) flows in estuaries. At high and low tide, water velocity drops to zero, allowing up to 95% of fine-grained suspended sediment (mainly clay minerals and organic matter) to sink and deposit. The efficiency of estuaries as sediment traps has probably varied over quite short geological timescales. For example, over the last 11 000 years as sealevel has risen following the last glaciation, estuaries seem to have been filling with sediment reworked from continental shelves. We might regard estuaries as

Table 6.1 Major ion composition of freshwater and seawater in mmol l^{-1}. Global average riverwater data from Berner and Berner (1987); seawater data from Broecker and Peng (1982).

	Riverwater	Seawater
Na$^+$	0.23	470
Mg^{2+}	0.14	53
K$^+$	0.03	10
Ca^{2+}	0.33	10
HCO$_3^-$	0.85	2
SO$_4^{2-}$	0.09	28
Cl$^-$	0.16	550
Si	0.16	0.1

temporary features on a geological timescale, but this does not reduce their importance as traps for riverine particulate matter today.

6.2.1 Aggregation of colloidal material in estuaries

In estuarine water the steep gradient in ionic strength destabilizes colloidal material (i.e. a suspension of very fine-grained (1 nm to 1 μm) material), causing it to stick together (flocculate) and sink to the bed. We can better understand this by considering clay minerals, the most abundant inorganic colloids in estuarine waters. Clay minerals have a surface negative charge (see Section 4.5) that is partly balanced by adsorbed cations. If surface charges are not neutralized by ion adsorption, clay minerals tend to remain in suspension, since like charges repel. These forces of repulsion are strong relative to the van der Waals' attractive forces (see Box 4.7) and prevent particles from aggregating and sinking. It follows that anything which neutralizes surface charges will allow particles to flocculate. Many colloids flocculate in an electrolyte, and seawater—a much stronger electrolyte than riverwater—fulfils this role in estuaries. The cations in seawater are attracted to the negative charges on clay surfaces. The cations form a mobile layer in solution adjacent to the clay surface (Fig. 6.2) and the combined 'electrical double layer' is close to being electrically neutral. Adjacent particles can then approach each other and aggregate. In nature, this simple explanation is vastly complicated by the presence of organic and oxyhydroxide coatings on particles.

Sedimentation in estuaries is localized to the low-salinity region by the physical and chemical effects discussed above. The sediment is, however, continuously resuspended by tidal currents, moving upstream on incoming tides and downstream on the ebb. The net effect is to produce a region of high concentration of suspended particulate matter, known as the turbidity maximum. The turbidity maximum is an important region because many reactions in environmental chemistry involve exchange of species between dissolved and particulate phases. Clearly, these reactions occur most where particle concentrations are high and decrease as particle concentrations decline away from the turbidity maximum.

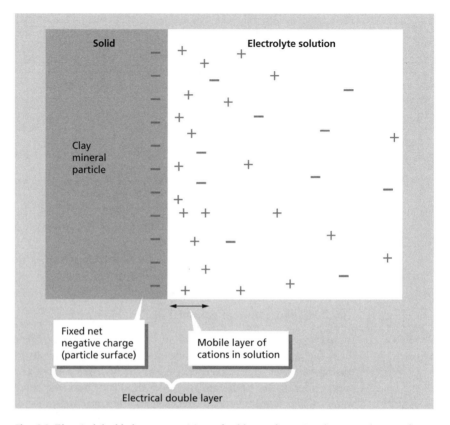

Fig. 6.2 Electrical double layer, comprising a fixed layer of negative charge on the particle and a mobile ionic layer in solution. The latter is caused because positive ions are attracted to the particle surface. Note that with increasing distance from the particle surface the solution approaches electrical neutrality. After Raiswell *et al.* (1980).

These reactions can significantly affect fluxes of riverine material to the oceans and therefore must be quantified in order to understand global element cycling.

6.2.2 Mixing processes in estuaries

Water flow in estuaries is not unidirectional; it is subject to the reversing flows of tides. There is therefore no constant relationship between a fixed geographic point and water properties (e.g. calcium ion (Ca^{2+}) concentration). For this reason, data collected in estuaries are usually compared with salinity (Box 6.1) rather than location. The underlying assumption is that salinity in an estuary is simply the result of physical mixing and not of chemical changes. If the estuary has just one river entering it and no other inputs, the behaviour of any component can be assessed by plotting its concentration against salinity.

Box 6.1 Salinity

Salinity is defined as the weight in grams of inorganic ions dissolved in 1 kg of water. Seven ions constitute more than 99% of the ions in seawater and the ratios of these ions are constant throughout the world oceans. Consequently, the analysis of one ion can, by proportion, give the concentration of all the others and the salinity. The density of seawater and light and sound transmission all vary with salinity.

Salinity is measured by the conductance of electrical currents through the water (conductivity). Measured values are reported relative to that of a known standard; thus salinity has no units—although, in many older texts, salinities are reported in units of parts per thousand (ppt or ‰) or grams per litre.

Open-ocean waters have a narrow range of salinities (32–37) and most are near 35. In estuaries, values fall to less than 1 approaching the freshwater end-member. In hypersaline environments salinities can exceed those of seawater, reaching values greater than 300.

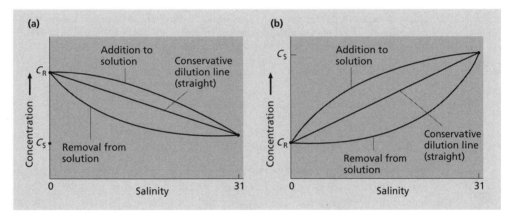

Fig. 6.3 Idealized plots of estuarine mixing illustrating conservative and non-conservative mixing. C_R and C_S are the concentrations of the ions in river and seawater respectively. After Burton and Liss (1976), with permission from Elsevier Science.

If the concentration of the measured component is, like salinity, controlled by simple physical mixing, the relationship will be linear (Fig. 6.3). This is called conservative behaviour and may occur with riverine concentrations higher than, or lower than, those in seawater (Fig. 6.3). By contrast, if there is addition of the component, unrelated to salinity change, the data will plot above the conservative mixing line (Fig. 6.3). Similarly, if there is removal of the component, the data will plot below the conservative mixing line (Fig. 6.3). In most cases, removal or input of a component will occur at low salinities and the data will approach the conservative line at higher salinity (Fig. 6.4). Extrapolation of such a 'quasi-conservative line' back to zero salinity can provide, by comparison with the measured zero salinity concentration, an estimate of the extent of removal (Fig. 6.4a) or release (Fig. 6.4b) of the component.

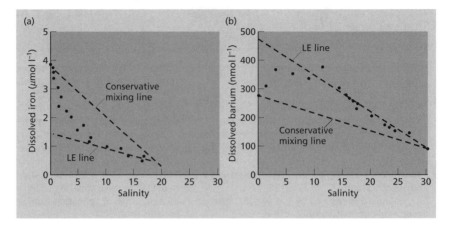

Fig. 6.4 (a) Dissolved iron versus salinity in the Merrimack Estuary (eastern USA), illustrating non-conservative behaviour. Linear extrapolation (LE) of high-salinity iron data to zero salinity gives an estimate of 60% low-salinity removal of iron (after Boyle *et al.* 1974). (b) Dissolved barium versus salinity in the Chesapeake Bay (eastern USA). In this case linear extrapolation (LE) of high-salinity barium data to zero salinity indicates low-salinity release of barium (after Coffey *et al.* 1997), with permission from Elsevier Science.

6.2.3 Halmyrolysis and ion exchange in estuaries

The electrochemical reactions that impinge on soil-derived river-borne clay minerals carried into seawater do not finish with flocculation of particles and sedimentation of aggregates. The capacity for ion exchange in clay minerals (see Section 4.8) means that their transport from low-ionic-strength, Ca^{2+}- and HCO_3^- -dominated riverwater, to high-ionic strength, sodium chloride (NaCl)-dominated seawater demands reaction with the new solution to regain chemical equilibrium (see Box 3.2). The process by which terrestrial materials adjust to marine conditions has been called 'halmyrolysis', derived from Greek roots hali (sea) and myros (unguent), literally 'to anoint with the sea'. Halmyrolysis is imprecisely defined but we will consider it to encompass all those reactions that affect a particle in seawater before burial in sediment.

Various measurements of cation exchange on river clays in seawater have shown that clay minerals exchange adsorbed Ca^{2+} for Na^+, potassium ions (K^+) and magnesium ions (Mg^{2+}) from seawater (see Section 4.8), consistent with the differences in ionic composition between river and seawater. In general, components with a high affinity for solid phases, such as dissolved phosphorus (P) or iron (Fe) (Fig. 6.4a), are removed from solution. Thus the rules of ionic behaviour arising from consideration of charge/ionic radius (z/r) ratios (see Section 5.3) are helpful in understanding chemical behaviour in estuarine environments, as well as in weathering.

6.2.4 Microbiological activity in estuaries

As in most environments, biological, particularly microbial, processes are important in estuaries; these can include both primary production by phytoplankton and organic matter decomposition by heterotrophic bacteria. In many estuaries the high particulate concentrations make waters too turbid to allow phytoplankton growth. However, in shallow or low turbidity estuaries, or at the seaward end of estuaries where suspended solid concentrations are low due to sedimentation of flocculated particles, sunlight levels may be sufficient to sustain phytoplankton growth. Estuaries frequently provide safe, sheltered harbours, often centres of trade and commerce. As a result, in developed and developing countries estuary coasts are often sites of large cities. Discharge of waste, particularly sewage, from the population of these cities increases nutrient concentrations and, where light is available, large amounts of primary production occur (see Section 5.5). In the dynamic environment of an estuary, dilution of phytoplankton-rich estuarine water with offshore low-phytoplankton waters occurs at a faster rate than cells can grow (phytoplankton populations under optimum conditions can double on timescales of a day or so). Thus, phytoplankton populations are often limited by this dilution process, rather than by nutrient or light availability.

The extent of nutrient removal that can occur in estuaries is illustrated for silicate (SiO_2) and phosphorus (P) in the estuary of the River Great Ouse in eastern England (Fig. 6.5). In this example there are a number of parameters whose values point to the role of phytoplankton in nutrient removal. Most importantly, silicate removal is related to high particulate chlorophyll concentrations, oxygen (O) supersaturation (arising from photosynthesis—see eqn. 5.19) and removal of other nutrients. During winter, silicate removal ceases when chlorophyll levels are low, allowing oxygen concentrations to fall to lower levels.

In Chesapeake Bay, a large estuary on the east coast of the USA, phytoplankton blooms in the high-salinity part of the estuary generate large amounts of organic matter which sink into the deep waters. The deep waters are isolated

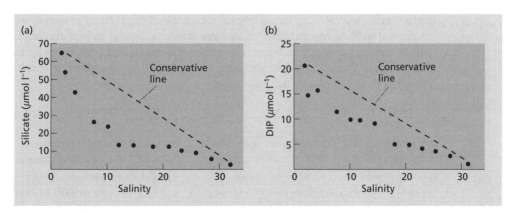

Fig. 6.5 (a) Dissolved silicate and (b) dissolved inorganic phosphorus (DIP) plotted against salinity in the Great Ouse Estuary (eastern England), illustrating non-conservative removal.

from surface waters by thermal stratification and the breakdown of the phyto-plankton debris results in occasional, seasonal, dissolved oxygen depletion and consequent death of fish and invertebrates.

The processes that concentrate sediments in estuaries also concentrate particulate organic matter. If large amounts of organic matter are present in an estuary, oxygen consumption rates, resulting from aerobic bacterial consumption of organic matter, can exceed the rate at which oxygen is supplied. This results in decreasing dissolved oxygen concentrations.

The discharge of sewage from cities often causes low or zero dissolved oxygen concentrations. A particularly well-documented example is the River Thames in southern England. London is built around the Thames estuary and throughout the 18th century, the wastes of the population were dumped in streets and local streams. During the early 19th century, public health improvements led to the development of sewers and the discharge of sewage directly into the Thames. The result was an improvement in local sanitation but massive pollution of the Thames. Although there was no systematic environmental monitoring at that time, historic evidence reveals the scale of the problem. Salmon and almost all other animals disappeared, the river was abandoned as a water supply and the literature of the time refers to the foul smells. Public concern prompted the development of sewage treatment works and also routine monitoring of environmental conditions in the estuary. The sewage treatment allowed dissolved oxygen concentrations to rise and fish returned to the estuary. However, as the population of London grew, the treatment system became overloaded and environmental conditions in the estuary deteriorated once more. The decrease of dissolved oxygen concentrations, and their subsequent increase, arising from improved sewage treatment in the 1950s, are illustrated in Fig. 6.6. The story of the Thames indicates the interaction between public health improvements and pollution; it also shows that some environmental problems are at least in part reversible, given political will and economic resources.

Most of the discussion above has centred on processes occurring in the water column. We should not forget, however, that estuarine sediments and their fringing marshes and wetlands also play a role in trapping sediment, storing organic matter and promoting microbiological reactions. These often nutrient-rich marsh and wetland environments allow large amounts of plant growth and resultant accumulation of organic matter along with nitrogen and phosphorus. The breakdown of this organic matter in water-logged, low-oxygen sediments in turn promotes denitrification which can convert nitrate into nitrogen gases:

$$5CH_2O_{(s)} + 4NO_{3(aq)}^- \rightarrow 2N_{2(g)} + 4HCO_{3(aq)}^- + CO_{2(g)} + 3H_2O_{(l)} \qquad \text{eqn. 6.1}$$

Over the last few hundred years there has been large-scale loss of wetland environments in rivers and estuaries throughout the world due to development pressures and flood control measures. The continued loss of wetlands removes valuable ecological habitats and also reduces the capacity for both carbon storage and nutrient removal by the processes described above. In the Humber estuary (UK), more than 90% of the intertidal marshes and supratidal wetlands have been lost to land reclamation over the last 300 years or so, resulting in a 99% reduc-

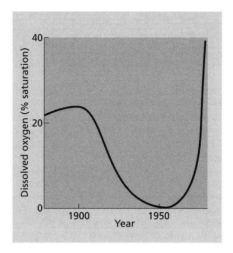

Fig. 6.6 Average autumn dissolved oxygen concentrations in the tidal River Thames. After Wood (1982).

tion of carbon burial. The situation for nitrogen and phosphorus is complicated by the increase in fluxes from human activity that has paralleled the loss of wetlands. However, it is estimated that restoration of all the wetlands in the Humber system could remove 25% of the phosphorus and 60% of the nitrogen currently flowing through the estuary, and preventing it reaching the North Sea where it can cause eutrophication.

6.3 Major ion chemistry of seawater

Having examined the chemistry of estuarine environments, we now turn to global chemical cycling in the open ocean. This chapter began by noting that the major ion chemistry of seawater is different from that of continental surface waters (Table 6.1). Three principal features clearly mark this difference:

1 The high ionic strength of seawater (see Fig. 5.3), containing about $35\,g\,l^{-1}$ of salts (Box 6.1).

2 The chemical composition of seawater, with Na^+ and Cl^- overwhelmingly dominant (Table 6.1).

3 Seawater has remarkably constant relative concentrations of major ions in all the world's oceans. For example the $Na^+:Cl^-$ ratio changes by less than 1% from the Arabian Gulf to the Southern Ocean. In the oceans, bicarbonate ions (HCO_3^-) and Ca^{2+} are biologically cycled (Section 6.4.4), causing vertical gradients in their ratios relative to the other major ions. However, the differences in the ratios to Na^+ are small—less than 1% for calcium.

There is evidence that the major ion composition of seawater has varied over many millions of years, linked to very long term geochemical cycling. Evidence from ancient marine evaporite sequences (Box 6.2) sets limits on the possible extent of that variability.

Box 6.2 Salinity and major ion chemistry of seawater on geological timescales

The evidence that the salinity and major ionic composition of seawater have remained reasonably constant over at least the last 900 million years comes from ancient marine evaporite deposits. Evaporites are salts that crystallize from evaporating seawater in basins largely cut off from the open ocean.

Over the last 900 million years, marine evaporites have normally begun with a gypsum-anhydrite section ($CaSO_4.2H_2O$-$CaSO_4$), followed by a halite (NaCl) sequence. Bittern salts (named due to their bitter taste) have precipitated from the final stages of evaporation and have variable composition, including magnesium salts, bromides, potassium chloride (KCl) and more complex

salts, depending on conditions of evaporation (Fig. 1).

The order of precipitates is the same as that seen in modern marine evaporites and can be reproduced by experimental evaporation of seawater. This sequence of salt precipitation sets limits on the possible changes of major ion compositions in seawater, since changes beyond these limits would have resulted in different sequences of salt formation.

Calculations demonstrating the actual limits on changes in the major ion chemistry of seawater imposed by the evaporite precipitation sequence are beyond the scope of this book. However, some simple

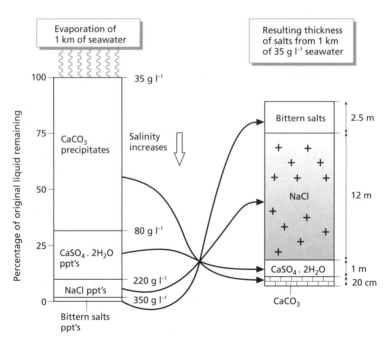

Fig. 1 Successional sequence and approximate thickness of salts precipitated during the evaporation of 1 km of seawater. After Scoffin (1987), with kind permission of Kluwer Academic Publishers.

(continued)

observation indicate the possible variations. For example, doubling calcium ion (Ca^{2+}) seawater concentrations at present sulphate ion (SO_4^{2-}) concentrations would not affect the sequence, whereas tripling the Ca^{2+} concentration would. Similarly, halving or doubling present-day potassium ion (K^+) concentrations would result in the formation of some very unusual bittern salts, not seen in the geological record.

Ideas about variations in sodium ion (Na^+) and chloride ion (Cl^-) concentrations are based on ancient halite inventories. The total volume of known halite deposits amounts to about 30% of the NaCl content of the present oceans. If all of this salt were added to the present oceans, the salinity of seawater would increase by about 30%, setting an upper limit. However, the ages of major halite deposits are reasonably well dispersed through geological time, suggesting that there was never a time when all of these ions were dissolved in seawater.

Setting lower limits on Na^+ and Cl^- concentrations in seawater can be estimated by considering the larger evaporite deposits in the geological record. For example, in Miocene times (5–6 million years ago) about 28×10^{18} mol of NaCl was deposited in the Mediterranean–Red Sea basins. This volume of salt represents just 4% of the present mass of oceanic NaCl. This suggests that periodic evaporate-forming events are only able to decrease Na^+ and Cl^- concentrations of seawater by small amounts. It has been suggested that the salinity of seawater has declined in 'spurts' from 45 to $35\,g\,l^{-1}$ over the last 570 million years. During this time the formation of Permian-aged salts alone (280–230 million years ago) may have caused a 10% decrease in salinity, possibly contributing to the extinction of many marine organisms at the end of this period.

Overall, these types of constraints suggest that the major ion chemistry of seawater has varied only modestly (probably by no more than a factor of 2 for each individual ion) during the last 900 million years or so (slightly less than a quarter of geological time).

6.4 Chemical cycling of major ions

The concept of residence times was introduced when discussing atmospheric gases (see Section 3.3), but it is applicable to most other geochemical systems, including the oceans. Residence times of the major ions in seawater (Box 6.3) are important indicators of the way chemical cycling operates in the oceans. These residence times are all very long (10^4 to 10^8 years), similar to or longer than the water itself (around 3.8×10^4 years) and very much longer than those calculated for atmospheric gases (see Section 3.3). Long residence times mean there is ample opportunity for ocean currents to mix the water and constituent ions thoroughly. This ensures that changes in ion ratios arising from localized input or removal processes are smoothed out. It is the long residence times of the ions that create the very constant ion ratios in seawater. The long residence times result from the high solubility of the ions and hence their z/r ratios (see Section 5.2). Other cations with similar z/r ratios will also have long oceanic residence times (e.g. caesium ion (Cs^+)), but these are not major ions in seawater because of their low crustal abundances. Chloride is an interesting exception as it is abundant in seawater, has a long residence time and yet has a low crustal abundance. Most of this Cl^- was degassed from the Earth's mantle as hydrogen chloride (HCl) very early in Earth history (see Section 1.3.1) and has been recycled in a hydrosphere–evaporite cycle since then (Section 6.4.2).

Box 6.3 Residence times of major ions in seawater

The total volume of the oceans is 1.37×10^{21} l and the annual river discharge to the oceans is 3.6×10^{16} yr^{-1}. The residence time of water in the oceans is therefore:

$$\frac{\text{Inventory}}{\text{Input}} = \frac{1.37 \times 10^{21}}{3.8 \times 10^{16}} = 3.8 \times 10^4 \text{ years}$$

eqn. 1

Applying this approach to the data in Table 6.1, it is straightforward to calculate the residence times of the major ions (Table 1), assuming that:

1 dissolved salts in rivers are the dominant sources of major ions in seawater;
2 steady-state conditions apply (see Section 3.3).

Table 1 Residence times of major ions in seawater.

Ion	Residence time (10^6 yr)
Na$^+$	78
Mg^{2+}	14
K$^+$	13
Ca^{2+}	1.1
HCO$_3^-$	0.09
SO$_4^{2-}$	12
Cl$^-$	131

The first assumption is probably valid, since the other sources listed in Table 6.2 do not greatly alter the results derived by considering rivers alone. The issue of steady state cannot be verified for very long (millions of years) timescales, but the geological evidence does suggest that the concentration of major ions in seawater has remained broadly constant over very long time periods (Box 6.2). As an example of the residence time calculation, consider sodium (Na$^+$):

$$\text{Input} = \text{discharge in rivers} \times \text{river concentration}$$
$$= 3.6 \times 10^{-6} \times 0.23 \times 10^{-3} \text{ mol yr}^{-1}$$
$$= 8.28 \times 10^{12} \text{ mol yr}^{-1}$$

eqn. 2

$$\text{Inventory} = \text{water content of the oceans} \times \text{ocean concentration}$$
$$= (1.37 \times 10^{21}) \times (470 \times 10^{-3}) \text{ mol}$$
$$= 644 \times 10^{18} \text{ mol}$$

eqn. 3

$$\text{Residence time} = \frac{644 \times 10^{18} \text{ mol}}{8.28 \times 10^{12} \text{ mol yr}^{-1}}$$
$$= 78 \times 10^6 \text{ yr}$$

eqn. 4

The residence times in Box 6.3 are based on riverwater being the only input of ions to the oceans. This is a simplification as there are also inputs from the atmosphere and from hydrothermal (hot water) processes at mid-ocean ridges (Fig. 6.7). For major ions, rivers are the main input, so the simplification in Box 6.3 is valid. For trace metals, however, atmospheric and mid-ocean ridge inputs are important and cannot be ignored in budget calculations (Section 6.5).

The long residence times of the major ions compared with the water mean that seawater is a more concentrated solution than riverwater. However, the different ionic ratios of seawater and riverwater show that the oceans are not simply the result of riverwater filling the ocean basins, even if the resulting solution has been concentrated by evaporation. Although major ion residence times are all long, they vary over four orders of magnitude, showing that rates of removal for specific ions are different. Processes other than evaporative concentration must be operating.

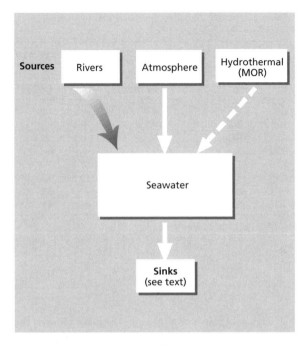

Fig. 6.7 Simple box model summarizing material inputs to seawater. Style of arrow indicates relative importance of input: bold, high; pecked, low. MOR, mid-ocean ridges.

Identifying removal mechanisms for a specific component is difficult because removal processes are usually slow and occur over large areas. Some removal processes are very slow—operating on geological timescales of thousands or millions of years—and impossible to measure in the present oceans. The requirement to study element cycles on geological timescales is further complicated by processes like climate change and plate tectonics which affect the geometry of ocean basins and sealevel. These large-scale geological processes can have significant effects on removal processes of major ions from the oceans.

The effects of the geologically recent glacial–interglacial oscillations during the Quaternary period (the last 2 million years) are particularly relevant. Firstly, the rapid rise in sealevel over the last 11 000 years, following the melting of polar ice accumulated during the last glacial period, has flooded former land areas to create large, shallow, continental shelves, areas of high biological activity and accumulation of biological sediments (Section 6.2.4). Also, the unconsolidated glacial sediments which mantle large areas of northern-hemisphere (temperate-arctic zone) land surfaces are easily eroded. This results in high particulate concentrations in rivers, which carry material to estuaries and continental shelves. This enhanced sediment supply results in correspondingly high detrital sediment–seawater interactions, increasing the importance of removal processes such as ion exchange.

Table 6.2 Simplified budget for major ions in seawater. All values are in $10^{12} \, mol \, yr^{-1}$. Data from Berner and Berner (1987).

Ion	River input	Sea–air fluxes	Evaporites	CEC–clay	CaCO$_3$	Opaline silica	Sulphides	MOR
Cl$^-$	5.8	1.1	4.7	—	—	—	—	—
Na$^+$	8.3	0.9	4.7	0.8	—	—	—	1.6
Mg^{2+}	5.0	—	—	0.1	0.6	—	—	4.9
SO$_4^{2-}$	3.2	—	1.2	—	—	—	1.2	—
K$^+$	1.1	—	—	0.1	—	—	—	−0.8
Ca^{2+}	11.9	—	1.2	−0.5	17	—	—	−4.8
HCO$_3^-$	30.6	—	—	—	34	—	−2.4	—
Si	5.8	—	—	—	—	7.0	—	−1.1

*Minus sign indicates a source.
CEC–clay, cation exchange on estuarine clay minerals; MOR, mid-ocean ridge and other seawater–basalt interactions.

Despite these complications, the main removal mechanisms of major ions from seawater are known (Table 6.2). Quantifying the importance of each mechanism is less easy and the uncertainty of data in Table 6.2 should not be forgotten. The amount of removal is compared with the riverine inputs resulting in a geochemical 'budget' that helps constrain the quality of the data. In the following section we outline the important removal processes for major ions in seawater.

6.4.1 Sea-to-air fluxes

Sea-to-air fluxes of major ions are caused by bubble bursting and breaking waves at the sea surface. These processes eject sea-salts into the atmosphere, the majority of which immediately fall back into the sea. Some of these salts are, however, transported over long distances in the atmosphere and contribute to the salts in riverwater (see Section 5.3). These airborne sea-salts are believed to have the same relative ionic composition as seawater and their flux out of the oceans is estimated by measuring the atmospheric deposition rates on the continents. In terms of global budgets, airborne sea-salts are an important removal process only for Na$^+$ and Cl$^-$ from seawater; removal of other major ions by this route is trivial.

6.4.2 Evaporites

Evaporation of seawater will precipitate the constituent salts, the so-called evaporite minerals, in a predictable sequence (Box 6.2). This sequence starts with the least soluble salts and finishes with the most soluble (see Box 4.12). If approximately half (47%) of the water volume is evaporated, CaCO$_3$ precipitates (see eqn. 6.4). With continued evaporation and an approximately four-fold increase in salinity, $CaSO_4.2H_2O$ (gypsum) precipitates:

$$Ca^{2+}_{(aq)} + SO^{2-}_{4(aq)} + 2H_2O_{(l)} \leftrightharpoons CaSO_4.2H_2O_{(s)} \qquad \text{eqn. 6.2}$$

Once 90% of the water (H_2O) has been evaporated, at dissolved salt concentrations around $220\,g\,l^{-1}$, NaCl precipitates:

$$Na^{+}_{(aq)} + Cl^{-}_{(aq)} \leftrightharpoons NaCl_{(s)} \qquad \text{eqn. 6.3}$$

and, in addition, some magnesium (Mg) salts begin to crystallize; if evaporation continues, highly soluble potassium (K) salts precipitate (Box 6.2).

The problem with invoking evaporation as a removal mechanism for ions in seawater is that there are currently very few environments in which evaporite salts are accumulating to a significant extent. This is because enormous volumes of seawater need to be evaporated before the salts become concentrated enough to precipitate. Clearly this cannot occur in the well-mixed open oceans, where net evaporative water loss is roughly balanced by resupply from continental surface waters (river flux). This implies that evaporative concentration of seawater can only occur in arid climatic regions within basins largely isolated from the open ocean and other sources of water supply. There are no modern examples of such basins; modern evaporite deposits are two to three orders of magnitude smaller than ancient deposits and are restricted to arid tidal flats and associated small salt ponds; for example, on the Trucial Coast of the Arabian Gulf. Note, however, that large evaporite deposits do exist in the geological record, the most recent example resulting from the drying out of the Mediterranean Sea in late Miocene times (about 5–6 million years ago).

As Cl^- has a very long oceanic residence time, the sporadic distribution of evaporate-forming episodes (Box 6.2), integrated over million-year timescales, results in only quite small fluctuations in the salinity of seawater. However, the lack of major evaporate-forming environments today suggests that both Cl^- and sulphate (SO^{2-}_4) are gradually accumulating in the oceans until the next episode of removal by evaporite formation.

In Table 6.2 the amount of Cl^- removed from seawater by evaporation has been set to balance the input estimate. This is acceptable because there are no other major Cl^- sinks, after allowing for sea-to-air fluxes and burial of pore water (Section 6.4.8). The Cl^- removal term dictates that the same amount of Na^+ is also removed to match the equal ratios of these ions in NaCl. The figure for SO^{2-}_4 removal by evaporation (Table 6.2) is plausible, albeit poorly constrained. Again, the SO^{2-}_4 estimate dictates an equal removal of Ca^{2+} ions to form $CaSO_4.2H_2O$.

6.4.3 Cation exchange

Ion-exchange processes on clay minerals moving from riverwater to seawater (Section 6.2.3) remove about 26% of the river flux of Na^+ to the oceans and are significant removal processes for K^+ and Mg^{2+} (Table 6.3). To balance this removal, clay minerals exchange a significant amount of Ca^{2+} to the oceans, adding an extra 8% to the river flux (Table 6.3). These modern values are,

Table 6.3 Additions to the river flux from ion exchange between river-borne clay and seawater. From Drever *et al.* (1988).

	Laboratory studies (10^{12} mol yr^{-1})	Amazon (10^{12} mol yr^{-1})	Average* (10^{12} mol yr^{-1})	Percentage of river flux[†]
Na$^+$	−1.58	−1.47	−1.53 ± 0.06	26
K$^+$	−0.12	−0.27	−0.20 ± 0.08	17
Mg^{2+}	−0.14	−0.49	−0.31 ± 0.08	7
Ca^{2+}	0.86	1.05	0.96 ± 0.10	8

*The averages are based on modern suspended sediment input to the oceans, which is probably double the long-term input rate (see Section 4.4). These values are thus halved when used in Table 6.2.
[†] Corrected for sea-salt and pollution inputs.
Minus sign indicates removal from seawater.

however, thought to be double the long-term values due to the effects of abnormally high postglacial suspended-solid input rates (see introductory text to Section 6.4). The modern values are halved in Table 6.2 to account for this.

6.4.4 Calcium carbonate formation

It is surprisingly difficult to calculate whether seawater is supersaturated or undersaturated with respect to $CaCO_3$. There are a number of different approaches to the problem all based on equilibrium relationships that describe the precipitation (forward reaction) or dissolution (back reaction) of $CaCO_3$.

$$Ca^{2+}_{(aq)} + 2HCO^-_{3(aq)} \leftrightharpoons CaCO_{3(aq)} + CO_{2(g)} + H_2O_{(l)} \qquad \text{eqn. 6.4}$$

We have already noted the importance of this reaction in buffering the pH of continental waters (see eqn. 5.12), and it behaves in exactly the same way in seawater. The Le Chatelier Principle (see Box 3.2) predicts that any process that decreases the concentration of HCO^-_3 in equation 6.4, will encourage dissolution of $CaCO_3$ to restore the amount of HCO^-_3 lost (Fig. 6.8). The oceans contain an effectively infinite amount of $CaCO_3$ particles suspended in surface waters and in bed sediments (see below). This $CaCO_3$ helps maintain (buffer) the pH of the oceans at values between 7.9 and 8.1, in much the same way as pH in continental waters is buffered by reacting with limestone (see Section 5.3.1 & Fig. 5.5). If, for example, more atmospheric CO_2 is taken up by the oceans as a result of global warming (see Sections 7.2.2 & 7.2.4), the acidity added to the oceans is largely neutralized by dissolution of $CaCO_3$, such that the HCO^-_3 ion concentration is unchanged (Fig. 6.8), maintaining the pH around 8.

Seawater is a concentrated and complex solution in which the ions are close together, compared with those in a more dilute solution. Electrostatic interaction occurs between closely neighbouring ions and this renders some of these ions 'inactive'. We are interested in the available or 'active' ions and we correct for this effect, using activity coefficients, denoted γ (see Section 2.6).

Activity coefficients are notoriously difficult to measure in complex solutions like seawater, but are thought to be around 0.26 for Ca^{2+} and around 0.20 for car-

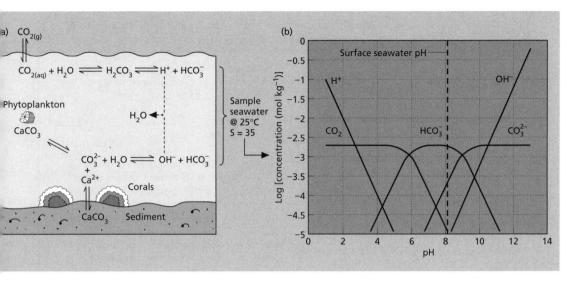

Fig. 6.8 (a) Schematic diagram to illustrate the buffering effect of $CaCO_3$ particles (suspended in the water column) and bottom sediments on surface seawater HCO_3^- concentrations (after Baird 1995, with permission from W.H. Freeman and Company.). (b) A sample of the seawater in (a) will have a pH very close to 8 because of the relative proportions of CO_2, HCO_3^- and CO_3^{2-}, which in seawater is dominated by the HCO_3^- species. Increased CO_2 concentrations in the atmosphere from anthropogenic sources could induce greater dissolution of $CaCO_3$ sediments including coral reefs.

bonate (CO_3^{2-}). Measured Ca^{2+} and CO_3^{2-} ocean surface concentrations are 0.01 and 0.000 29 mol l^{-1} respectively and thus the ion activity product (IAP) (see Box 4.12) can be calculated:

$$IAP = aCa^{2+} \times aCO_3^{2-} = cCa^{2+} \cdot \gamma \times cCO_3^{2-} \cdot \gamma$$
$$= 0.01 \times 0.26 \times 0.00029 \times 0.2$$
$$= 1.5 \times 10^{-7} \, mol^2 \, l^{-2} \hspace{3cm} \text{eqn. 6.5}$$

This value is much greater than the solubility product (see Box 4.12) of calcite ($CaCO_3$), which is $3.3 \times 10^{-9} \, mol^2 \, l^{-2}$ at 25°C and 1 atmosphere pressure, a suitable choice for tropical surface seawater. The degree of saturation (see Box 4.12) is:

$$\text{Degree of saturation} = \Omega = \frac{IAP}{K_{sp}} \hspace{3cm} \text{eqn. 6.6}$$

An Ω value of 1 indicates saturation, values greater than 1 indicate supersaturation, and values less than 1 indicate undersaturation. Using the values for calcite above we get:

$$\Omega = \frac{1.5 \times 10^{-7}}{3.3 \times 10^{-9}} = 45.4 \hspace{3cm} \text{eqn. 6.7}$$

a value that implies surface seawater in the tropics is highly supersaturated with respect to calcite. This approach neglects the effects of ion pairing (Box 6.4).

Box 6.4 Ion interactions, ion pairing, ligands and chelation

Ions and water

When ionic salts dissolve in water, the salts dissociate to release the individual ions. The charged ions attract the polar water molecules such that a positively charged ion will be surrounded most closely by oxygen atoms of the water molecules (Fig. 1). Thus, ions are not free in solution, but interacting, or coordinating, with water molecules. The water molecules can be considered to be bound to the ion by so called coordinate bonds. For example, the hydrogen ion (H^+) is hydrated to form H_3O^+. For simplicity in chemical equations the simple H^+ notation is used.

In addition to interaction with water molecules, individual ions may interact with other ions to form *ion pairs*. For example, dissolved sodium and sulphate ions can interact to form a sodium sulphate ion pair:

$$Na^+_{(aq)} + SO^{2-}_{4(aq)} \rightleftharpoons NaSO^-_{4(aq)} \qquad \text{eqn. 1}$$

The extent of this interaction varies for different ions and is measured by an equilibrium constant. In this case:

$$K = \frac{aNaSO^-_4}{aNa^+ \cdot aSO^{2-}_4} \qquad \text{eqn. 2}$$

If all the relevant equilibrium constants are known, together with the amounts of the ions present, the proportions of the various ions associated with each ion pair can be calculated. The results of such an analysis for seawater (Table 1) show that ion pairing is common. A full analysis of the properties of seawater requires that these species are taken into account.

Since anions are present at lower concentrations than cations (except for chloride (Cl^-)), ion pairing has a proportionately greater effect on anions relative to cations. The extent of ion pairing is dependent on temperature, pressure and salinity.

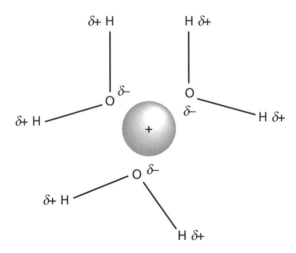

Fig. 1 Ionic salts dissolved in water.

(continued)

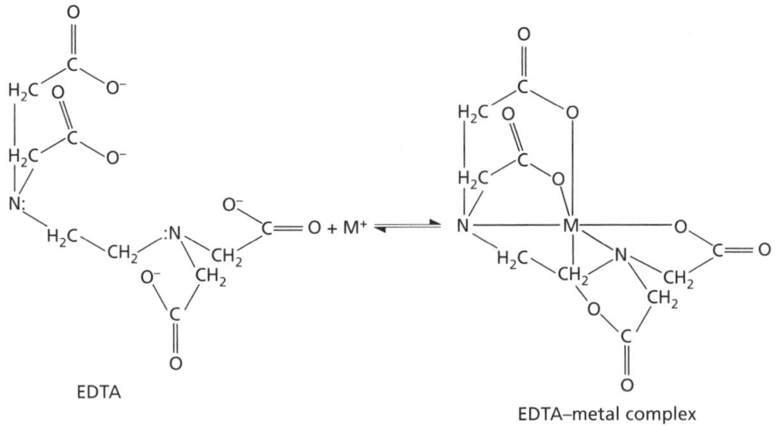

Fig. 2 Chelation between a metal ion (M+) and ethylenediaminetetra-acetic acid (EDTA), a well-known laboratory ligand. Note that the six coordination bonds (white) are made between dissociated -O− groups and lone pairs of electrons (denoted by double dots) on the N atoms of the EDTA. Similar coordination bonds are made in fulvic acid chelates.

Table 1 Percentage of each major ion in seawater present in various ion pairs.

	Na^+	Mg^{2+}	Ca^{2+}	K^+
Free ion	98	89	89	99
MSO_4	2	10	10	1
$MHCO_3$	−	1	1	−
MCO_3	−	−	−	−

	Cl^-	SO_4^{2-}	HCO_3^-	CO_3^{2-}
Free ion	100	39	81	8
NaX	−	37	11	16
MgX	−	20	7	44
CaX	−	4	4	21
KX	−	1	−	−
Mg_2CO_3	−	−	−	7
$MgCaCO_3$	−	−	−	4

M, cation; X, anion; −, species present at less than 1%.

Ligands and chelation

Beside water, other polar molecules and some anions coordinate to metal ions; such firmly bonded molecules are known as ligands. Each ligand contains at least one atom that bears a lone pair of electrons. Lone pairs of electrons can be envisaged as non-bonded electrons that are grouped at the extremities of an atom's electron orbital, and these 'plug-in' to gaps in the electron orbital of the metal ion forming a coordinate bond. Most ligands form only one coordinate bond with a metal ion, for example the Cl^- anion which helps keep metal ions in solution in mid-ocean ridge hydrothermal systems. In some cases, however, a ligand may form more than one link with the metal ion. The complex ions that form between these ligands and cations are known as chelates or chelated complexes from the Greek *chelos* meaning 'a crabs claw'. The ligand is envisioned to have a claw-like grip on the metal ion, binding it firmly and forming a stable complex ion (Fig. 2), which is much more soluble than the metal ion itself. Natural chelates include fulvic and humic acids formed by the degradation of soil organic matter.

Table 6.4 Total activity coefficients (γ_t) for selected ions in surface seawater at 25°C and salinity of 35. After Millero and Pierrot (1998) with kind permission of Kluwer Academic Publishers.

Ion	Measured γ_t	Calculated γ_t*
Cl$^-$	0.666	0.666
Na$^+$	0.668	0.664
H$^+$	0.590	0.581
K$^+$	0.625	0.619
OH$^-$	0.255	0.263
HCO$_3^-$	0.570	0.574
B(OH)$_4^-$	0.390	0.384
Mg^{2+}	0.240	0.219
SO$_4^{2-}$	0.104	0.102
Ca^{2+}	0.203	0.214
CO$_3^{2-}$	0.039	0.040
H$_2$PO$_4^-$	0.453	0.514
HPO$_4^{2-}$	0.043	0.054
PO$_4^{3-}$	0.00002	0.00002

*Values recommended for use in calculations.

Allowance for ion pairing indicates that about 90% of the calcium in seawater is present as the free ion with the remainder present as CaSO$_4^0$ and CaHCO$_3^+$ ion pairs. For CO$_3^{2-}$ it is calculated that only about 10% exists as the free ion with the remainder in ion pairs with Mg^{2+}, Ca^{2+} and Na$^+$. Applying these corrections decreases the IAP in the following way:

$$IAP = 1.5 \times 10^{-7} \times 0.9 \times 0.1 = 1.35 \times 10^{-8} \, mol^2 \, l^{-2} \qquad \text{eqn. 6.8}$$

Now the calculation of saturation state becomes:

$$\Omega = \frac{1.35 \times 10^{-8}}{3.3 \times 10^{-9}} = 4 \qquad \text{eqn. 6.9}$$

On this basis, surface seawater in the tropics is about four times supersaturated with respect to calcite. In practice, calculating seawater Ω values has been simplified for the environmental chemist because chemical oceanographers have measured and modelled the combined effects of ion activity and ion pairing over a range of pressures, temperatures and salinities (ionic strength) to yield total activity coefficients, denoted γ_t (Table 6.4). Using total activity coefficients for Ca^{2+} and CO$_3^{2-}$ from Table 6.4, the IAP calculations (eqns. 6.5 & 6.8) now simplify to:

$$\begin{aligned} IAP &= a\text{Ca}^{2+} \times a\text{CO}_3^{2-} \\ &= c\text{Ca}^{2+} \cdot \gamma_t \times c\text{CO}_3^{2-} \cdot \gamma_t \\ &= 0.01 \times 0.214 \times 0.00029 \times 0.04 \\ &= 2.48 \times 10^{-8} \, mol^2 \, l^{-2} \end{aligned} \qquad \text{eqn. 6.10}$$

The calculation of saturation state is now:

$$\Omega = \frac{2.48 \times 10^{-8}}{3.3 \times 10^{-9}} = 7.51$$

eqn. 6.11

The difference between the Ω values in equations 6.9 and 6.11 is mainly due to the effects of ionic strength, which are properly compensated for using total activity coefficients. Overall, the data show that surface seawater is about six to seven times supersaturated with respect to calcite. We might then reasonably expect $CaCO_3$ to precipitate spontaneously in the surface waters of the oceans.

The evidence from field studies is somewhat contrary to the predictions based on equilibrium chemistry. Abiological precipitation of $CaCO_3$ seems to be very limited, restricted to geographically and geochemically unusual conditions. The reasons why carbonate minerals are reluctant to precipitate from surface seawater are still poorly understood, but probably include inhibiting effects of other dissolved ions and compounds. Even where abiological precipitation is suspected—for example, the famous ooid shoals and whitings of the Bahamas (Box 6.5)—it is often difficult to discount the effects of microbial involvement in the precipitation process.

Volumetrically, the biological removal of Ca^{2+} and HCO_3^- ions, built into the skeletons of organisms, is much more important. In the modern oceans, the large continental shelf areas created by sealevel rise in the last 11 000 years probably account for about 45–50% of global carbonate deposition. Moreover, about half of this sink for Ca^{2+} and HCO_3^- ions occurs in the massive coral reefs of tropical and subtropical oceans (e.g. the Australian Great Barrier Reef). It is tempting to assume that coral reefs have always represented a major removal process for Ca^{2+} and HCO_3^- ions. However, over the last 150 million years it can be shown that it is carbonate sedimentation in the deep oceans which has been volumetrically more important, accounting for between 65 and 80% of the global $CaCO_3$ inventory. These deep-sea deposits, which average about 0.5 km in thickness, mantle about half the area of the deep ocean floor (Fig. 6.9). The ultra fine-grained calcium carbonate muds (often referred to as oozes) are composed of phytoplankton (coccolithophores) and zooplankton (foraminifera) skeletons (Fig. 6.10). Although these pelagic organisms live in the ocean surface waters, after death their skeletons sink through the water column, either directly or within the faecal pellets of zooplankton.

The controls on the distribution of pelagic oozes are partly related to the availability of nutrients, which must be capable of sustaining significant populations of phytoplankton (see Section 5.5). More important, however, is the dissolution of $CaCO_3$ as particles sink into ocean deep waters. In the deep ocean, carbon dioxide (CO_2) concentrations increase, particularly in the deep Pacific, as a result of the decomposition of sedimenting organic matter. Decreased temperature and increased pressure also promote dissolution of $CaCO_3$, favouring the reverse reaction in equation 6.4.

By mapping the depth at which carbonate sediments exist on the floors of the oceans, it is possible to identify the level where the rate of supply of biogenic $CaCO_3$ is balanced by the rate of solution. This depth, known as the calcite compensation depth (CCD), is variable in the world's oceans, depending on the

Box 6.5 Abiological precipitation of calcium carbonate

Where a skeletal source cannot be identified, calcium carbonate ($CaCO_3$) grains and fine-grained muds may be of abiological origin. The most famous occurrences occur in shallow, warm, saline waters of the Bahamas and the Arabian Gulf. In these areas two distinctive morphologies are present, ooids and needle muds (Fig. 1).

Ooids are formed by aggregation of aragonite* crystals around a nucleus, usually a shell fragment or pellet. Successive layers of aragonite precipitation build up a concentric structure, which may vary in size from about 0.2 to 2.0 mm in diameter. Needle muds are also aragonitic; typically each needle is a few micrometres wide and tens of micrometres long.

It has long been thought that the warm,

(a)

(b)

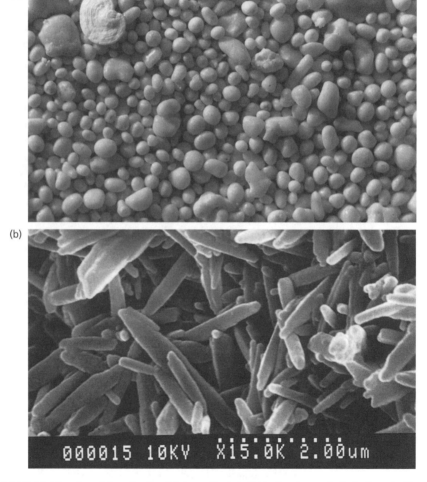

Fig. 1 (a) Ooid-rich sediment from the Great Bahama Bank. Individual grains are typically 1 mm in diameter (photograph courtesy of J. Andrews). (b) Scanning electron microscope photograph of aragonitic needles from the Great Bahama Bank. Scale bar = 1 µm (photograph courtesy of I.G. Macintyre, from Macintyre and Reid 1992).

(continued)

shallow, saline waters where these deposits are found favour increased concentrations of carbonate ions (CO_3^{2-}), increasing the ion activity product of $aCa^{2+}.aCO_3^{2-}$ such that precipitation of $CaCO_3$ occurs. The formation of ooids probably requires fairly agitated, wind- or wave-stirred waters, allowing periodic suspension of the grains into the $CaCO_3$-saturated water, whereas aragonitic needles may precipitate as clouds of suspended particles, known as whitings.

There has been, and still is, much debate about the origins of these particles. Firstly, it is difficult to disprove the effects of microbial mediation in their formation. Thus we might regard the grains as non-skeletal, while accepting a possible microbial influence. Secondly, various geochemical and mineralogical studies have produced equivocal results in attempting to demonstrate an abiological origin. Having said this, recent work based on crystal morphology and strontium substitution lends support for inorganic precipitation of needle muds.

Despite the considerable interest these phenomena provoke, we should remember that they are of minor significance to the modern oceanic $CaCO_3$ budget. The relative importance of inorganic $CaCO_3$ in the geological past is more difficult to assess, but may have been more significant before the evolution of shelly organisms about 570 million years ago.

* Aragonite and calcite are known as polymorphs of $CaCO_3$. Both minerals have the formula $CaCO_3$ but they differ slightly in the structural arrangement of atoms.

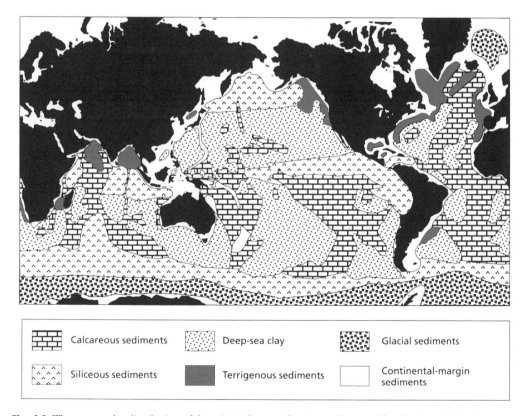

Calcareous sediments   Deep-sea clay   Glacial sediments

Siliceous sediments   Terrigenous sediments   Continental-margin sediments

Fig. 6.9 The present-day distribution of the principal types of marine sediments. After Davies and Gorsline (1976), with kind permission from R. Chester.

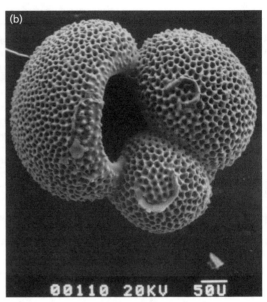

Fig. 6.10 (a) Planktonic coccolithophore *Emiliania huxleyi*, a very common species in the modern oceans. This specimen has a diameter of 8 μm. The skeleton is clearly composed of circular shields packed around a single algal cell. After death the coccosphere breaks down, releasing the shields to form the microscopic particles of deep-sea oozes and chalks (photograph courtesy of D. Harbour). (b) Skeleton of modern planktonic foraminifer *Globigerinoides sacculifer*, common in tropical oceans. Scale bar = 50 μm (photograph courtesy of B. Funnell).

degree of $CaCO_3$ undersaturation in the deep waters (Fig. 6.11). In the Atlantic Ocean the CCD is at about 4.5 km depth; above the CCD, at about 4 km depth in the Atlantic, there is a critical depth, known as the lysocline (Fig. 6.11). Here the rate of calcite dissolution increases markedly and all but the most robust particles dissolve rapidly. It is estimated that about 80% of the $CaCO_3$ settling into deep waters is dissolved, either during transit through the water column or on the seabed. As a consequence, pelagic carbonate deposits are most common on the shallower parts of the deep ocean floors (Fig. 6.11) or on topographic highs that project above the CCD.

Planktonic coccolithophores and foraminifera did not evolve until the mid-Mesozoic (about 150 million years ago), whereas shallow-water shelly organisms are known to have existed throughout Phanerozoic time (570 million years to the present day). This means that the locus of carbonate deposition has shifted to the deep oceans only in the last quarter of Phanerozoic time.

The removal of Ca^{2+} by $CaCO_3$ precipitation can be estimated directly from the abundance of $CaCO_3$-rich ocean sediments and their sedimentation rates (Table 6.2). From equation 6.4, we see that two moles of HCO_3 are removed with each mole of Ca^{2+}, a process that releases dissolved CO_2 into seawater, ultimately to be returned to the atmosphere. Calcium carbonate also incorporates a small but significant amount of Mg^{2+} by isomorphous substitution for Ca^{2+} (see Box 4.6) and this is used to derive the Mg^{2+} removal in Table 6.2.

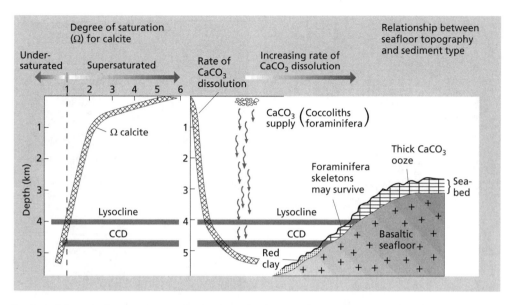

Fig. 6.11 Schematic diagram showing depth relationship between degree of saturation for calcite in seawater and rate of CaCO₃ dissolution. At 4 km depth, as seawater approaches undersaturation with respect to calcite, rate of dissolution of sinking calcite skeletons increases. The lysocline marks this increased rate of dissolution. Below the lysocline only large grains (foraminifera) survive dissolution if buried in the seabed sediment. Below the calcite compensation depth (CCD; see text) all CaCO₃ dissolves, leaving red clays.

Modern surface seawater is demonstrably supersaturated with respect to CaCO₃, and the presence of carbonate sediments in rocks of all ages suggests that it has been throughout much of Earth history. Oceanic pH is unlikely to have fallen below 6 as such a shift requires a 1000-fold increase in atmospheric pCO₂ over its present value of $10^{-3.6}$ atm. An atmospheric pCO₂ of this magnitude may have occurred very early in Earth history, but the existence of 3.8 billion-year-old CaCO₃ sediments shows that seawater was still close to saturation with calcite. Achieving calcite saturation at these high pCO₂ values implies that oceanic Ca²⁺ concentrations were at that time 10 times larger than modern values. Seawater pH has probably never exceeded 9 in the geological past, because at that pH sodium carbonate would be a much more common mineral precipitate than calcite. There is no evidence in ancient marine sediments that sodium carbonate has ever been a common marine precipitate.

6.4.5 Opaline silica

Opaline silica (opal) is a form of biologically produced silicon dioxide (SiO₂.nH₂O) secreted as skeletal material by pelagic phytoplankton (diatoms) and one group of pelagic zooplankton (radiolarians) (Fig. 6.12). Opaline silica-rich sediments cover about one-third of the seabed, mainly in areas where sedimen-

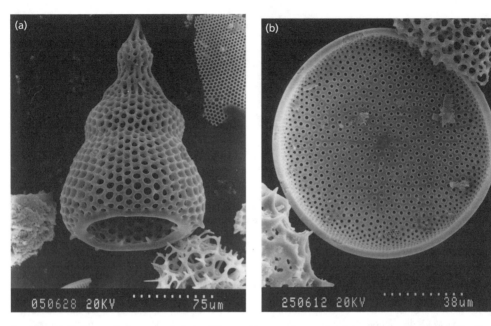

Fig. 6.12 (a) Skeleton of siliceous radiolarian *Theocorythium vetulum*, early Pleistocene, equatorial Pacific. Scale bar = 75 μm. (b) Siliceous diatom *Coscinodiscus radiatus*, early Pleistocene, equatorial Pacific. Scale bar = 38 μm. Photographs courtesy of B. Funnell.

tation rates are high, associated with nutrient-rich upwelling waters and polar seas, particularly around Antarctica (Fig. 6.9). Seawater is undersaturated with respect to silica and it is estimated that 95% of opaline silica dissolves as it sinks through the water column or at the sediment/water interface. Thus, the preservation of opaline silica only occurs where it is buried in rapidly accumulating sediment, beneath the sediment/water interface. Subsequent dissolution of opal in the sediment saturates sediment pore waters with silica. The pore water cannot readily exchange with open seawater and saturation prevents further opal dissolution. High sedimentation rates in the oceans can be caused by high mineral supply rates from the continents, but are usually caused by high production rates of biological particles (Section 6.5.4). In high productivity areas, for example parts of the Southern Ocean bordering Antarctica, diatoms are the common phytoplankton species, and this enhances the importance of these regions as silica sinks. The biological removal of silicon (Si) from seawater is calculated from the opal content of sediments and rates of sedimentation (Table 6.2).

6.4.6 Sulphides

The oxidation of organic matter proceeds by a number of microbially mediated reactions once free oxygen has been used up (see Section 5.5 & Table 4.7). Although small amounts of nitrate (NO_3^-), manganese (Mn) and iron (Fe) are available as electron acceptors in marine sediments, their importance is small in comparison with SO_4^{2-}, which is abundant in seawater (Table 6.1). At seawater pH around 8, sulphate-reducing bacteria metabolize organic matter according to the following simplified equation.

$$2CH_2O_{(s)} + SO_{4(aq)}^{2-} \longrightarrow 2HCO_{3(aq)}^- + HS_{(aq)}^- + H_{(aq)}^+$$

$$\text{eqn. 6.12}$$

This process is widespread in marine sediments but is most important in continental margin sediments, where organic matter accumulation is largest. Sulphate reduction in sediments occurs at depths (varying from a few millimetres to metres below the sediment/water interface) where seawater SO_4^{2-} can readily diffuse, or be pumped by the actions of sediment-dwelling organisms. The reaction yields highly reactive hydrogen sulphide (HS^-), most of which diffuses upward and is reoxidized to SO_4^{2-} by oxygenated seawater in the surface sediment. However, about 10% of the HS^- rapidly precipitates soluble Fe(II) in the reducing sediments to yield iron monosulphide (FeS).

$$Fe_{(aq)}^{2+} + HS_{(aq)}^- \rightarrow FeS_{(s)} + H_{(aq)}^+ \qquad \text{eqn. 6.13}$$

Iron monosulphides then convert to pyrite (FeS_2). At Eh below $-250\,mV$ (see Box 5.4) and at pH around 6, the conversion of FeS to FeS_2 may occur via oxidation of dissolved FeS by hydrogen sulphide, for example:

$$FeS_{(s)} \rightarrow FeS_{(aq)} \qquad \text{eqn. 6.14}$$

$$FeS_{(aq)} + H_2S_{(aq)} \rightarrow FeS_{2(s)} + H_{2(g)} \qquad \text{eqn. 6.15}$$

Alternatively, under less reducing conditions, conversion may involve the addition of sulphur (S) from intermediate sulphur species (e.g. polysulphides, polythionates or thiosulphate ($S_2O_3^{2-}$)), which are the products and reactants in microenvironmental sulphur redox cycling. The reaction involving $S_2O_3^{2-}$ can be summarized as:

$$FeS_{(s)} + S_2O_{3(aq)}^{2-} \rightarrow FeS_{2(s)} + SO_{3(aq)}^{2-} \qquad \text{eqn. 6.16}$$

The sulphite (SO_3^{2-}) is subsequently oxidized to SO_4^{2-}. Sedimentary pyrite, formed as a byproduct of sulphate reduction in marine sediments, is a major sink for seawater SO_4^{2-}. The presence of pyrite in ancient marine sediments shows that SO_4^{2-} reduction has occurred for hundreds of millions of years. On a geological timescale, removal of SO_4^{2-} from seawater by sedimentary pyrite formation is thought to be about equal to that removed by evaporite deposition (Section 6.4.2). Compilations of pyrite abundance and accumulation rates are used to calculate modern SO_4^{2-} removal by this mechanism and to derive the estimate in Table 6.2.

Sulphate reduction (eqn. 6.12) also produces HCO_3^-, and this anion slowly diffuses out of the sediment into seawater, accounting for about 7% of the HCO_3^- flux to the oceans. The slow diffusion also means that HCO_3^- may build up to sufficiently high concentrations in sediment pore waters for the ion activity product of Ca^{2+} (from seawater) and HCO_3^- to exceed the solubility product for

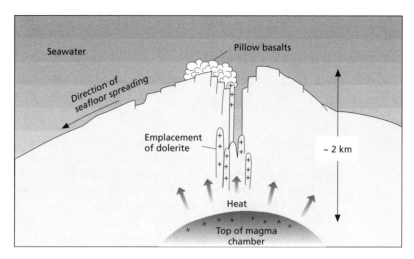

Fig. 6.13 Simplified structure of a fast spreading mid-ocean ridge (e.g. East Pacific Rise).

$CaCO_3$ (eqn. 6.6). This allows $CaCO_3$ to precipitate as nodules (concretions) in the sediment. These are not quantified in the global budget as they are volumetrically small sinks of $CaCO_3$.

6.4.7 Hydrothermal processes

The hydrothermal (hot water) cycling of seawater at mid-ocean ridges has a profound effect on the chemistry and budgets of some major and trace elements in seawater. To understand the chemical changes it is necessary to know a little about geological processes at mid-ocean ridges.

Basaltic ocean crust is emplaced at mid-ocean ridges by crystallization from magma that is sourced from a magma chamber at shallow depth (about 2 km) below the ridge. The magma chamber and newly emplaced crust is a discrete heat source, localized below the ridge (Fig. 6.13). Successive emplacement of new ocean crust gradually pushes the older crust laterally away from the ridge axis at rates of a few millimetres per year. This ageing crust cools and subsides as it travels away from the ridge axis. The resulting thermal structure, i.e. a localized heat source underlying the ridge with cooler flanking areas, encourages seawater to convect through fractures and fissures in the crust. This convection is vigorous close to the ridge axis but more passive on the off-axis flanks (Fig. 6.14).

The deep waters of the oceans are cool (around 2–4°C) and dense relative to overlying seawater. This dense water percolates into fissures in the basaltic crust. Slowly it penetrates deep into the crust, gradually heating up, particularly as it approaches the heat source of the underlying magma chamber. This massive heat

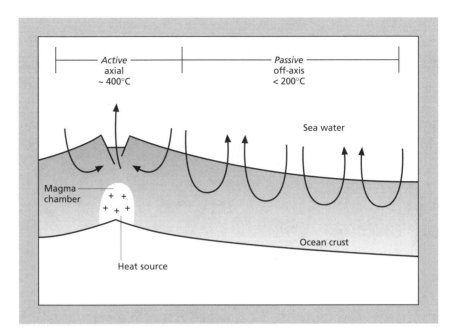

Fig. 6.14 Convection systems at mid-ocean ridges. Active circulation at the ridge axis is driven by magmatic heat. Off-axis circulation is driven by passive cooling of the crust and lithosphere. Modifed from Lister (1982).

source warms the water, causing it to expand and become less dense, forcing it upward again through the crust in a huge convection cell (Fig. 6.15). We can view these convection cells as having a recharge zone and low-temperature 'limb' of subsiding seawater, a hot reaction zone closest to the magma chamber, and a high-temperature rising 'limb' of chemically modified seawater discharging at the seabed (Fig. 6.15). The overall process is called 'hydrothermal' (hot water) convection. It is currently estimated that about $3 \times 10^{13}\,kg\,yr^{-1}$ of seawater is cycled through the mid-ocean ridges of the Earth's crust. It therefore takes about 3.3×10^7 years to cycle the entire volume of seawater through the axial part of the mid-ocean ridges (Table 6.5).

It is not possible to measure directly the maximum temperature to which water becomes heated in the basaltic crust. However, hot springs of hydrothermal water discharge from the seabed at the apex of the convection cell. Temperature measurements taken from ridge axis hot springs range from 200 to 400°C (average around 350°C). This implies that temperatures in the hot reaction zone above the magma chamber (Fig. 6.15) are typically not less than 350–400°C. Owing to chemical reactions between this convecting hot water and the basaltic crust (shaded region on Fig. 6.15), these waters are acidified (typical pH 5–7) and rich in dissolved transition metals leached from the crust. Iron (Fe), manganese (Mn), lead (Pb), zinc (Zn), copper (Cu) and hydrogen sulphides rapidly precipitate a

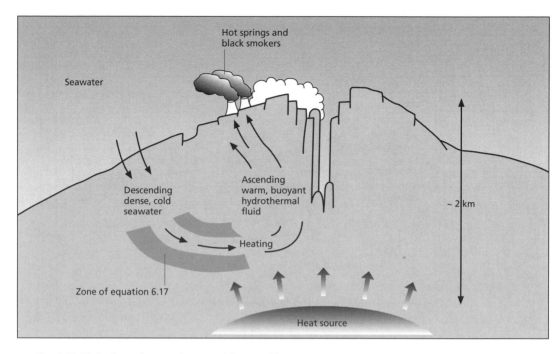

Fig. 6.15 Hydrothermal convection at a mid-ocean ridge.

Table 6.5 Estimates of ocean volume circulation times through hydrothermal systems. Modified from Kado *et al.* (1995).

Axial hydrothermal convection (350°C fluid)	Off-axis convection (20°C fluid)	Recycling through 'black smoke' plume*
3.3×10^7 yr	5.5×10^5 yr	2.8×10^3 yr water 2.4×10^5 yr reactive elements

*Extrapolated from study of the Endevor Ridge. Although the ocean volume is cycled through the plume in 2.8×10^3 years, the rate of reaction of the plume with seawater is much slower. To strip the reactive elements from seawater it is necessary to cycle the ocean volume approximately 100 times through the plume (2.4×10^5 yr).

cloud of iron, zinc, lead and copper sulphides and iron oxides on injection into cold, oxic oceanic bottom waters. This sulphidic plume of particles identifies clearly the location of the hot springs and gives rise to their colloquial name — 'black smokers' (Plate 6.1, facing p. 138).

The high temperatures encountered in hydrothermal circulation cells at mid-ocean ridges increase substantially the rate and extent of chemical reaction

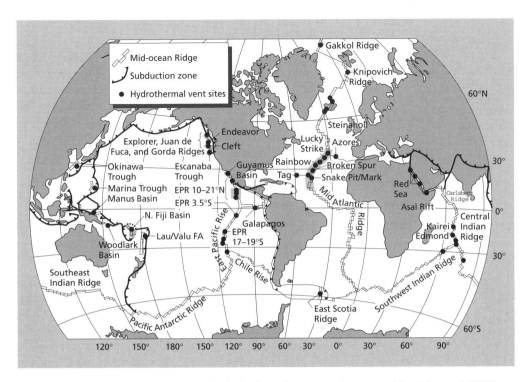

Fig. 6.16 Location of known mid-ocean ridge hydrothermal systems in the oceans. After Baker *et al.* (1995).

between seawater and basaltic ocean crust. However, estimates of the element fluxes involved in these processes are uncertain, mainly because representative sampling in these remote environments is difficult and expensive. The flux estimates are based on a few studies at individual sites on the East Pacific and Mid-Atlantic ridges (Fig. 6.16). Global fluxes have been calculated from these sites, using various geophysical and geochemical approaches. A major problem however, still to be resolved, is a rigorous quantification of the amount of hydrothermal activity occurring at high temperatures near the ridge axis, versus lower temperature circulation on the ridge flanks (Fig. 6.14 & Table 6.5). This is important, because temperature affects the degree, rate and even direction of some chemical reactions. Despite these problems, for some elements the direction of the fluxes agree from site to site allowing construction of tentative global fluxes (Table 6.2). We should, however, note that the magnitudes of the fluxes in Table 6.2 are uncertain.

In the following section we describe the effects of hydrothermal activity on major ions in the oceans: the effects on minor components in the oceans are discussed in Sections 6.5.2 and 6.5.5.

Hydrothermal reactions as major ion sinks

Of the major elements, the case for magnesium removal from seawater during hydrothermal cycling at mid-ocean ridges is most convincing. Experimental work and data from many black smokers (Fig. 6.16) suggests that the hydrothermal fluids exiting from the crust have essentially zero magnesium concentration. This implies that magnesium is removed from seawater by reaction with basalt at high temperature. The precise chemistry of this reaction is not known, but it can be represented as the generalized reaction:

$$11Fe_2SiO_{4(s)} + 2H_2O_{(l)} + 2Mg^{2+}_{(aq)} + 2SO^{2-}_{4(aq)}$$
$$\text{(fayalite)} \qquad\qquad \text{(seawater)}$$

$$\rightarrow Mg_2Si_3O_6(OH)_{4(s)} + 7Fe_3O_{4(s)} + FeS_{2(s)} + 8SiO_{2(aq)}$$
$$\text{(sepiolite)} \qquad \text{(magnetite) (pyrite)} \quad \text{(silica)} \qquad\qquad \text{eqn. 6.17}$$

The basalt, represented here by iron-rich olivine (fayalite), is leached of its iron and hydrated by seawater, whilst Mg^{2+} from seawater is used to form the magnesium clay mineral (sepiolite in eqn. 6.17), which represents altered basalt. The formation of magnesium clay mineral also removes OH^- from water to make the $(OH)_4$ component of the clay mineral in equation 6.17. In laboratory experiments, it is this removal of OH^- from H_2O that leaves the fluid enriched in H^+, explaining the acidity of hydrothermal fluids. The H^+ does not show up in the products of equation 6.17 because the equation is a summary of a number of processes going on over time (see Section 2.4). However, acidity generation is an important feature that, along with complexation by Cl^- anions (Box 6.4), enhances iron solubility. The reaction (eqn. 6.17) also predicts the formation of iron oxide (Fe_3O_4), iron sulphide (FeS_2) and silica, all of which are found at hydrothermal vent sites. Although it is difficult to quantify the amount of Mg^{2+} removed from seawater by this process, it is probably the most important Mg^{2+} sink in the modern ocean. There is also uncertainty about the fate of Mg^{2+} in altered basalt (sepiolite in eqn. 6.17) as it moves away from the ridge axis during seafloor spreading. There is evidence that Mg^{2+} is leached from altered basalt by cold seawater. If large amounts of Mg^{2+} are resupplied to seawater by such low-temperature basalt–seawater interactions, then mid-ocean ridge processes may not cause net Mg^{2+} removal from seawater.

Sodium is by far the most abundant cation in hydrothermal fluids, simply because the fluid is sourced from seawater (Table 6.1). It has long been thought that Na^+ must be removed from seawater at mid-ocean ridges, mainly because the global Na^+ budget does not otherwise balance. The existence of Na^+-enriched basalts (spilites), believed to have formed by reaction with seawater at high temperature, is tangible evidence that Na^+ removal from seawater occurs during mid-ocean ridge hydrothermal activity. The formation of sodium-feldspar (albite) probably accounts for the removal process from the fluid to the altered basalt, but data to quantify fluxes are scant. Hydrothermal activity probably accounts for no more than 20% of the total Na^+ removal from seawater, which

Table 6.6 Changes in seawater major constituents upon reacting with mid-ocean ridge basalt at high temperature. Hydrothermal fluid data are typical ranges from Von Damm (1995).

Constituent	Seawater* (mmol l⁻¹)	Hydrothermal fluids (mmol l⁻¹)	Δ (mmol l⁻¹)
Mg^{2+}	53	0	−53
Ca^{2+}	10	10–100	0–90
K^+	10	15–60	5–50
SO_4^{2-}	28	0–0.6	−28
H_4SiO_4	0.1	5–23	5–23

*Data from Table 6.1.
Δ, difference between typical range in hydrothermal water and seawater.

on a geological timescale is dominated by the formation of evaporites (Section 6.4.2).

Hydrothermal reactions as major ion sources

The chemistry of hydrothermal fluids indicates that basalt–seawater interactions are a source of some elements that have been stripped from ocean crust and injected into seawater. Data from hydrothermal fluids show that both Ca^{2+} and dissolved silica are concentrated in the hydrothermal waters compared with seawater (Table 6.6). Calcium is probably released from calcium feldspars (anorthite) as they are converted to albite by Na^+ uptake, a process called albitization. Silica can be leached from any decomposing silicate in the basalt, including the glassy matrix of the rock. Globally, basalt–seawater interaction seems to provide an additional 35% to the river flux of Ca^{2+} and silica to the oceans.

Hydrothermal reactions involving sulphur

Seawater sulphate is removed from hydrothermal fluids, mainly by the precipitation of anhydrite ($CaSO_4$) as the downward-percolating seawater is heated to temperatures around 150–200°C.

$$Ca^{2+}_{(aq)} + SO^{2-}_{4(aq)} \xrightarrow{\text{heat}} CaSO_4$$

(anhydrite)

eqn. 6.18

(at temperatures >150°C the equilibrium
for this reaction lies well to the right)

This reaction probably consumes all of the seawater-derived Ca^{2+} in the fluid, and about 70% of the SO_4^{2-}; if more calcium is added to the fluid from albitization reactions (see above), even more SO_4^{2-} is consumed. Anhydrite formation limits the amount of SO_4^{2-} entering the higher-temperature (>250°C) parts of the hydrothermal system. At these high temperatures the sulphate is reduced by reaction with FeS compounds in the basalt, and by oxidation of Fe^{2+} compounds, forming hydrogen sulphide (H_2S) or hydrogen bisulphide (HS^-). Anhydrite also forms as part of the black smoker chimney system when the hot vent fluids exit at the seabed. The hot fluids heat the surrounding seawater, which causes anhydrite to form as predicted by equation 6.18.

Most of the $CaSO_4$ formed in the crust (and around vents) probably redissolves in the ocean bottom waters as the crust ages and cools; it thus has little effect on the overall SO_4^{2-} budget of the oceans. It is well known that H_2S precipitates as iron sulphide in venting hydrothermal fluids, giving rise to extensive zones of sulphide mineralization and to the 'black smoke' (Plate 6.1, facing p. 138). However, the total removal of SO_4^{2-} from seawater by this mechanism is again likely to be small, since evaporite and sedimentary sulphide formation adequately removes the river flux of SO_4^{2-} on geological timescales. This suggests that over long timescales much of the hydrothermal sulphide is oxidized on, or just below, the seabed.

6.4.8 The potassium problem: balancing the seawater major ion budget

The major ion budget for seawater (Table 6.2) is quite well balanced (i.e. inputs equal outputs) for all elements except K^+. Laboratory studies predict that K^+ behaviour will change with temperature in hydrothermal fluids. Above 150°C, in the hotter part of hydrothermal systems, K^+ should be leached from basalt (Table 6.6), representing an input to the seawater budget. However, in cooler parts (<70°C) of hydrothermal systems, K^+ adsorption on to altered basalt may be important, resulting in the formation of clay-like minerals such as celadonite (illitic) and phillipsite (a zeolite mineral). As there is no well-documented major removal process for K^+ from seawater, it is generally believed that ridge flank low-temperature hydrothermal activity removes all of the high-temperature hydrothermal K^+ input to seawater and probably some of the river flux also.

A process that might affect the K^+ budget in a small way is K^+ fixation during ion-exchange reactions on clay minerals. Laboratory experiments have shown that degraded micas and illites (see Section 4.5.2), stripped of their K^+ during weathering, but which retain much of their layer charge, are able to fix, irreversibly, K^+ from seawater. The process involves the replacement of hydrated cations for dehydrated K^+ in the interlayer site, fixing the K^+ in its 'mica' site (see Section 4.5.2). Globally, this process might remove another 10–20% of the K^+ river flux to the oceans (Table 6.3).

The imbalance in the K^+ budget and small imbalances in other budgets may be nullified by a number of processes. One possibility is the concept of 'reverse weathering reactions'. In reverse weathering, highly degraded clay minerals react with cations, HCO_3^- and silica in seawater to form complex clay mineral-like silicates. An example reaction addressing the K^+ problem would be:

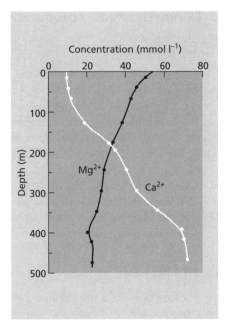

Fig. 6.17 Depth distribution of dissolved Ca^{2+} and Mg^{2+} concentrations in sediment pore waters. After Gieskes and Lawrence (1981).

$$\text{Degraded aluminosilicate}_{(s)} + K^+_{(aq)} + HCO^-_{3(aq)} + H_4SiO_{4(aq)}$$
$$\rightarrow \text{K aluminosilicate}_{(s)} + CO_{2(g)} + H_2O_{(l)} \qquad \qquad \text{eqn. 6.19}$$

This reaction shows that reverse weathering is exactly opposite to continental weathering reactions, which consume CO_2 and liberate HCO^-_3 (see Section 4.4.3). Experimental studies with Amazon delta sediments in the mid 1990s have provided the first evidence that reverse weathering reactions occur naturally. The K^+ sink in Amazon continental shelf sediments alone is calculated to be $6.8 \times 10^{12} \, g \, yr^{-1}$, representing about 10% of the annual global K^+ river flux to the oceans.

Some removal of ions from seawater occurs through permanent burial in sediment pore water. The total removal of major ions by this process is small, less than 2% of the river input for all elements except Na^+ and Cl^-. The burial flux may be significant for Na^+ and Cl^- (20–30% of the river flux), but the data are uncertain.

Seawater buried in marine sediment may react with components of the sediment, particularly fine-grained basaltic volcanic ash. Pore water concentrations of Ca^{2+}, Mg^{2+} and K^+ in deep-sea cores show removal of Mg^{2+} (and, to a lesser extent, K^+) from pore water, mirrored by increases in Ca^{2+} pore water concentration (Fig. 6.17). These results suggest that basaltic ash is converted to Mg^{2+} and K^+ clay minerals, accompanied by the release of Ca^{2+} to pore water. The quantitative importance of this mechanism on a global scale is probably small, but good data are sparse.

6.5 Minor chemical components in seawater

6.5.1 Dissolved gases

Gases dissolve from the atmosphere into the oceans according to the Henry's law constant (see Box 3.4). In the absence of biological processes the ocean surface waters would therefore be saturated with all atmospheric gases. Some of these gases, such as argon (Ar) and helium (He), are chemically inert, while others such as nitrogen gas (N_2) are available as a nutrient source to only a few specialized nitrogen-fixing organisms and are hence effectively inert in seawater, and thus very close to saturation. We will not consider these species further. Other gases, such as carbon dioxide (CO_2) and oxygen (O_2), are intimately involved in biological cycles, similar to those already discussed in freshwater systems (see Section 5.5). We will consider these gases later when discussing the impact of biological cycles on ocean chemistry (Section 6.5.4).

6.5.2 Dissolved ions

Seven major ions dominate the chemistry of seawater, but all of the other elements are also present, albeit often at extremely low concentrations. The major ions in seawater are little affected by biological processes or human activities because seawater is a vast reservoir and the major ions have long residence times. By contrast, complex cycling processes and involvement in biological systems typify the behaviour of dissolved trace elements (components present at $\mu mol\,l^{-1}$ concentrations or less) in seawater. The concentration of some dissolved metals in seawater is very small—typically a few nanomoles per litre ($nmol\,l^{-1}$). Sampling so as to avoid contamination and measuring such tiny concentrations, in the presence of major ions with millimolar concentrations, is difficult. These difficulties prevented routine analysis of trace metals in seawater until the 1970s, although reliable nutrient measurements were available earlier.

Particulate matter concentrations in the deep ocean are low (a few $\mu g\,l^{-1}$), whereas in surface waters particulate matter concentrations are relatively high (generally $10–100\,\mu g\,l^{-1}$) dominated by material produced by biological processes in the euphotic zone (Fig. 6.18). Similarly high values can be encountered within tens or hundreds of metres of the deep ocean floor, caused either by resuspension of deep-sea sediments, or from hydrothermal fluid plumes, sourced from hydrothermal vents (Section 6.4.7 & Fig. 6.18). Apart from this region near the seafloor, particulate matter in the oceans is predominantly of organic origin, generated by primary production in surface seawater. The euphotic zone, where this production occurs, has variable depth, generally around 100 metres in clear open-ocean waters. Since the oceans are on average almost 4000 metres deep, the primary production that drives global biological cycling throughout the oceans occurs in a shallow surface zone.

Dissolved metals in seawater have various sources, for example the dissolution of redox-sensitive metals from reducing ocean floor and mid-ocean ridge hydrothermal sediments. Hydrothermal sediments, for example, are typically

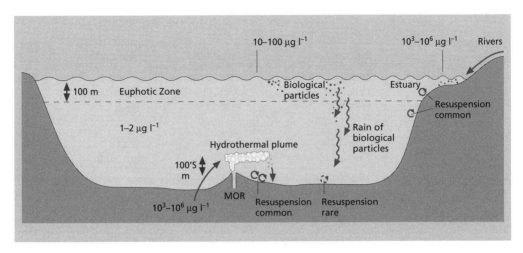

Fig. 6.18 Sketch to show variability in the concentration of particles in seawater. MOR, mid-ocean ridge.

Table 6.7 Source of dissolved manganese to the oceans ($10^9 \, mol \, yr^{-1}$). After Chester (2000).

	Rivers	Atmosphere	Hydrothermal (MOR)*
Source	5	0.5	11–34

*An unknown but potentially significant amount of manganese emitted from mid-ocean ridge (MOR) hydrothermal sources is precipitated near black smokers (i.e. immediately removed), so this estimate is not directly comparable to the other inputs.

iron- and manganese-rich. These sediments derive from FeS particles that precipitate in the 'black smoke' plume as the hot and acidic hydrothermal water is rapidly cooled and mixed with alkali seawater. The particles fall out both close to the vent and some kilometres from it. It is currently difficult to quantify the global metal fluxes from these processes, although estimates have been made. In the case of manganese, mid-ocean ridge hydrothermal sources can be up to hundreds of $mmol \, kg^{-1}$, 10^6 times higher than ambient seawater, and thus may be significant to the global oceanic budget (Table 6.7). Modern atmospheric fluxes of some metals are larger than river inputs (Table 6.8), caused by various combustion processes—coal burning, metal smelting and automobile engines. The shift towards a larger industrial atmospheric source for some metals may increase their concentrations in open-ocean waters, since riverine metal inputs are often removed in estuaries (Section 6.2).

The chemistry of dissolved metals in seawater can be grouped into three classes, which describe the behaviour of the metal during chemical cycling. These classes—conservative, nutrient-like and scavenged—have been recognized by the shapes of concentration profiles when plotted against depth in the oceans.

Table 6.8 Comparison of total atmospheric and riverine inputs to the world's oceans $(10^9 \, mol \, yr^{-1})$. Based on Duce *et al.* (1991).

Element	Riverine*	Atmospheric
Nitrogen (ex N_2)	1500–3570	2140
Cadmium	0.0027	0.02–0.04
Copper	0.16	0.25–0.82
Nickel	0.19	0.37–0.48
Iron	19.7	580
Lead	0.01	0.43
Zinc	0.09	0.67–3.5

Total input—dissolved plus particulate. Estimates are based on data available in early 1990s and so include significant amounts of material mobilized by human activity.
*Dissolved input only; particulate components are assumed to sediment out in estuaries and the coastal zone.

6.5.3 Conservative behaviour

Elements with conservative behaviour are characterized by vertical profiles (similar to major ion profiles) that indicate essentially constant concentrations with depth. These elements behave like the major ions, having long residence times and being well mixed in seawater. These elements are not major components of seawater, simply because their crustal abundances are very low compared with the major ions resulting in correspondingly low input rates. Elements showing this sort of behaviour form either simple cations or anions (e.g. caesium (Cs^+) or bromine (Br^-) with low z/r ratios and little interaction with water (see Section 5.2)), or form complex oxyanions (see Section 5.2), for example molybdenum (Mo) and tungsten (W)—which exist in seawater as MoO_4^{2-} and WO_4^{2-} respectively (Fig. 6.19). Conservative elements have little interaction with biological cycles.

6.5.4 Nutrient-like behaviour

As in continental waters (see Section 5.5.1), NO_3^-, DIP (dissolved inorganic phosphorus) and silicate are usually considered to be the limiting nutrients for biological production, although in some situations it has been suggested that trace elements, particularly iron, may be limiting (Section 6.6). Excepting high-latitude areas the oceans are so large and deep that they are effectively permanently stratified. The production of biological material removes nutrients from surface waters (Box 6.6). After death, this biological material sinks through the water column, decomposing at depth to re-release the nutrients. The nutrients are then slowly returned to surface waters by deep-ocean mixing processes and diffusion. The net result is that the vertical profiles of nutrients are characterized by low concentrations in surface waters (where biological utilization rates exceed supply rates) and deep-water maxima, where decomposition rates exceed uptake rates because of the absence of light (Fig. 6.20). Nitrogen and DIP are cycled

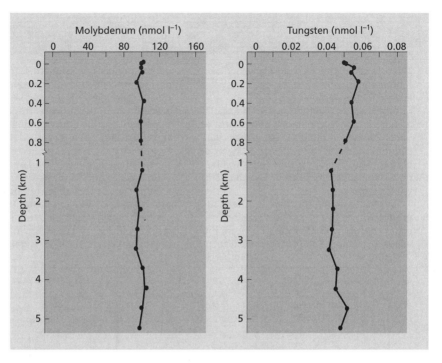

Fig. 6.19 Vertical distribution of dissolved molybdenum and tungsten in the North Pacific. After Sohrin *et al.* (1987), with permission from Elsevier Science.

with the organic tissue of organisms, while silicon and calcium (Ca) are cycled as skeletal material. The decomposition of organic tissue is mainly by bacterial respiration, a rapid and efficient process. By contrast, skeletal material is dissolved slowly (Sections 6.4.4 & 6.4.5). The effect of these different decomposition rates is that the NO_3^- and DIP concentration profiles show rapid increase with depth, implying shallower regeneration of material in the water column than silicon. Nitrate and DIP distributions are therefore closely correlated (Fig. 6.21) with a slope of approximately 16:1. This ratio reflects the relative proportions of nitrate and DIP regeneration and utilization by phytoplankton. This ratio is often referred to as the Redfield Ratio in honour of Alfred Redfield who first described the close linking of these two ions in the ocean.

Biological cycling not only *removes* some ions from surface waters, it also *transforms* them. The stable form of iodine (I) in seawater is iodate (IO_3^-), but biological cycling results in the formation of iodide (I^-) in surface waters, because the production rate of the reduced species is faster than the rate of its oxidation. Biological uptake of IO_3^- in surface water results in a nutrient-like profile, contrasting with the conservative behaviour of most halide ions, for example Cl^- and Br^-. The biological demand for NO_3^- also involves transformation. Phytoplankton take up NO_3^- and reduce it to the –3 oxidation state (see Box 4.3 & Fig. 5.12) for

Box 6.6 Oceanic primary productivity

The rate of growth of phytoplankton (primary productivity) in the oceans is mainly limited by the availability of light and the rate of supply of limiting nutrients (usually accepted to be nitrogen (N), phosphorus (P), silicon (Si) and iron (Fe)). The need for light confines productivity to the upper layers of oceans. Also, in polar waters there will be no phytoplankton growth during the dark winter months.

In temperate oceans there is little winter productivity because cooling of surface waters destroys the thermal stratification and winds mix waters and phytoplankton to depths of hundreds of metres. This deep mixing, which occurs in all temperate and polar oceans, also means that vertical gradients in nutrient concentrations are temporarily eliminated, allowing a corresponding rise in surface-water nutrient concentrations. In spring, surface-water stratification is re-established as the waters warm and winds decrease. Once enough light is available, vigorous plankton growth begins. Nutrient concentrations are high, having built up over the winter, resulting in the 'spring bloom' of phytoplankton. The bloom is generally later

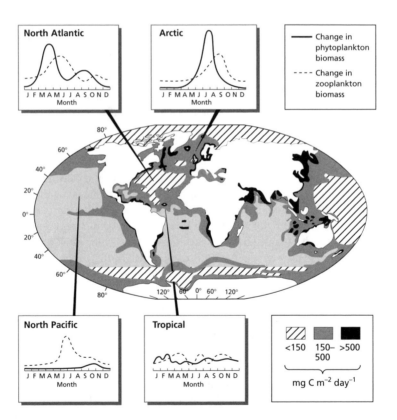

Fig. 1 Seasonal cycles of productivity and global average rates of primary production. Reprinted from Koblentz-Mishke *et al.* (1970), with permission from Scientific Exploration of the South Pacific, Courtesy of the National Academy of Sciences, Washington, DC.

(continued)

in the spring—and larger—moving polewards.

The duration of the spring bloom is limited by nutrient availability and/or grazing by zooplankton. Phytoplankton growth and abundance then decline to lower levels, which are maintained throughout the summer by nutrient recycling within the euphotic zone. In some locations, limited mixing in autumn can stimulate another small bloom, before deep winter mixing returns the system to its winter condition.

In tropical waters, vertical stratification persists throughout the year and production is permanently limited by nutrient supply rates, which are controlled by internal recycling and slow upward diffusion from deep water. Under these conditions, productivity is low throughout the year.

These seasonal cycles of productivity are shown schematically in Fig. 1, along with a map of current estimates of primary production rates. In the north Pacific zooplankton population growth supresses the spring bloom.

Since production rates vary with time and place, the data on the figure are uncertain, but are consistent with satellite-derived maps of chlorophyll concentrations (see Plate 6.3, facing p. 138). These maps show that, on an annual basis, the short, high-production seasons of temperate and polar areas fix more carbon in organic tissue than organisms in tropical waters. There are a few exceptions to this, in so-called upwelling areas, for example the Peruvian, Californian, Namibian and North African coasts and along the line of the equator (Plate 6.3, facing p. 138)& Fig. 1). In upwelling areas, ocean currents bring deep water to the surface, providing a large supply of nutrients in an area with abundant light. Very high rates of primary production ensue and the phytoplankton are the basis of a food chain that supports commercially important fisheries.

The spring bloom and upwelling areas are not simply times and regions of higher productivity; changes in the structure of the whole ecosystem result. For example, the phytoplankton community in areas of higher production is usually rich in diatoms, organisms which, upon death, efficiently export carbon and nutrients to deep waters. This contrasts with the tropics, where the phytoplankton community has adapted to the low-nutrient waters by recycling nutrients very efficiently, with little export to deep waters. Phytoplankton are able to live in low-nutrient tropical waters because they are typically smaller, with larger surface area to volume ratios that increase their efficiency in diffusing nutrients across the cell wall.

utilization in proteins. When phytoplankton die they decompose, releasing the nitrogen as ammonium (NH_4^+) hence N is still in its −3 oxidation state. Similarly, when phytoplankton are eaten by zooplankton, the consumers excrete nitrogen primarily as NH_4^+. This NH_4^+ is then available for reuse by phytoplankton: NH_4^+ is the preferred form of available nitrogen since there is no energy requirement in its uptake and utilization. Alternatively, the ammonium is oxidized via nitrite (NO_2^-) to NO_3^-, the thermodynamically favoured stable species. These rapid recycling processes maintain euphotic zone NH_4^+ at low concentrations. In the deep ocean, the only NH_4^+ source is from the breakdown of organic matter sinking from the surface waters. The amount of NH_4^+ released from this source is small and rapidly oxidized, maintaining very low NH_4^+ concentrations.

Elements showing nutrient-like distribution often have long oceanic residence times, although shorter than conservative elements. The residence times of NO_3^- silicon and DIP have been estimated to be 57 000, 20 000 and 69 000 years respectively (Table 6.9). The vast reservoirs of nutrients in the deep ocean mean that increases in the concentrations of NO_3^- in riverwaters due to human activity (see

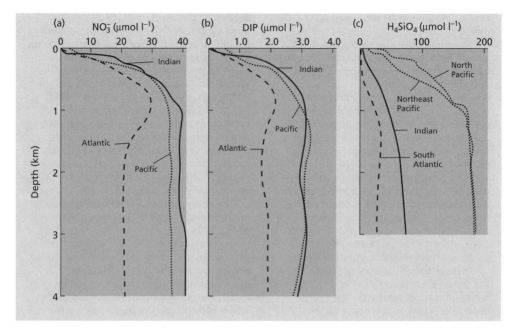

Fig. 6.20 Vertical distribution of dissolved nitrate (a), phosphorus (b) and silicon (c) in the Atlantic, Pacific and Indian Oceans. After Svedrup *et al.* (1941), reprinted by permission of Pearson Education Inc., Upper Saddle River, NJ.

Section 5.5.1) have little effect on oceanic NO_3^- concentration, assuming that NO_3^- is effectively mixed throughout the ocean volume (Section 6.8.3).

In addition to the actual nutrient elements, many other elements show nutrient-like behaviour in the oceans, i.e. low concentrations in surface waters and high concentrations at depth (Fig. 6.22). This distribution implies that biological removal rates from surface waters are rapid, although it does not prove that these elements are limiting, or even essential, to biological processes. In the case of some metals (e.g. zinc (Zn)), a clear biological function has been established. However, for other metals (e.g. cadmium (Cd)), there is less evidence for a biological role; cadmium is usually thought of as a poison, although not at the extremely low concentrations ($<0.1\,nmol\,l^{-1}$) found in seawater. Elements like cadmium probably show nutrient-like behaviour (Fig. 6.22) because they are inadvertently taken up during biological processes, or substitute for other elements when these are in short supply. The Cd^{2+} ion has chemical similarities to Zn^{2+}, thus the nutrient-like cycling of cadmium may reflect inadvertent biological uptake with—or substitution for—zinc.

Finally, we should be aware that even metals with a clear biological role (e.g. Zn^{2+}) can be toxic at sufficiently high concentrations (see Box 5.5). This reminds

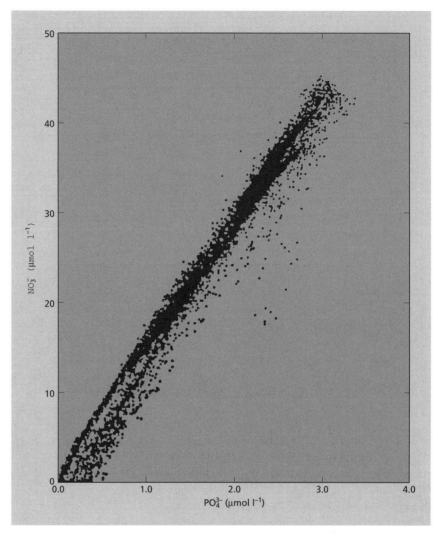

Fig. 6.21 Global relationship between dissolved nitrate and phosphate in seawater based on US Geochemical Ocean Sections (GEOSECS) programme data.

us that all elements are potentially toxic, making terms like *nutrient* impossible to apply in an absolute sense.

6.5.5 Scavenged behaviour

Elements that are highly particle-reactive, characterized by large z/r ratios (see Section 5.2), often have vertical profiles with surface maxima and decreasing concentrations with depth; aluminium (Al)

Table 6.9 Concentrations of nutrients and metals in deep (>3000 m) water in the North Atlantic and North Pacific, together with estimated oceanic residence times.

Component	North Atlantic	North Pacific	Estimated oceanic residence time (years)
Nitrate (μmol l^{-1})	20	40	57 000
Silicon (μmol l^{-1})	25	170	20 000
DIP (μmol l^{-1})	1.3	2.8	69 000
Zinc (nmol l^{-1})	1.7	8.0	4 500
Cadmium (nmol l^{-1})	0.3	0.9	32 000
Aluminium (nmol l^{-1})	20	0.4	50
Manganese (nmol l^{-1})	0.6	0.2	30

DIP, dissolved inorganic phosphorus.

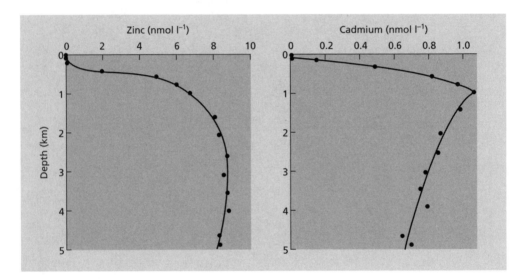

Fig. 6.22 Vertical distribution of dissolved zinc and cadmium in the North Pacific. After Bruland (1980).

is an example (Fig. 6.23). These profiles arise because the inputs of these elements are all in the surface waters, producing concentration maxima there. Poorly understood processes lower these concentrations by removal to particulate phases. The removal processes probably involve adsorption on to particle surfaces, known by the general term *scavenging*. Consequently, oceanic concentrations of scavenged elements are many orders of magnitude below those predicted from simple mineral solubility considerations.

Scavenged species are all metals and their residence times in seawater are estimated to be a few hundred years, short in comparison with nutrient and conservative elements (Table 6.9). These rapid removal rates imply that river inputs are

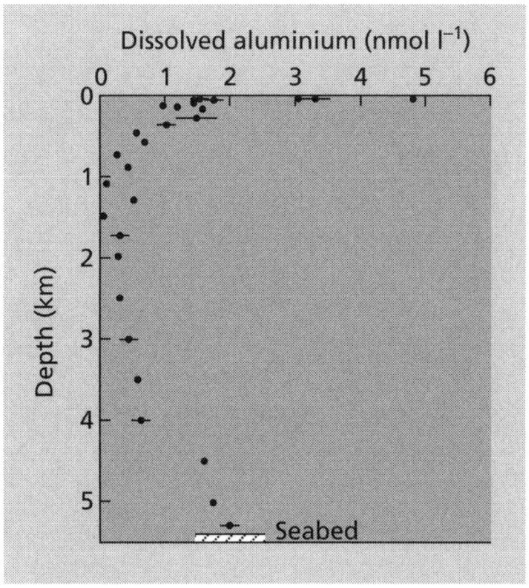

Fig. 6.23 Vertical distribution of aluminium in the North Pacific showing low values at 1–2 km depth, indicative of scavenging. The deepwater increase in concentration arises from aluminium inputs from the sea-bed sediment. After Orians and Bruland (1986), with permission from Elsevier Science.

removed mainly in estuaries, where suspended solid concentrations are high (Section 6.2.1). Consequently, the atmosphere provides the main input of particle-reactive metals to the surface waters of the central ocean. This atmospheric flux has a natural component, the fallout of wind-blown dust particles, which subsequently dissolve in seawater to a small extent (typically a few per cent). Aluminium (Fig. 6.23) and manganese are examples. The second source of particles is human activity: lead (Pb) is an example, entrained into the atmosphere principally from automobile exhaust emissions. Lead use, particularly as a petrol additive, increased rapidly during the 1950s until concern over the possible health effects resulted in a dramatic decline in its use from the late 1970s onward.

We do not have a direct history of dissolved lead concentrations in seawater, but we do have an indirect record from corals. Coral skeletons are made of annual layers of $CaCO_3$, producing growth rings similar to those in trees. These rings can be counted and sampled for lead analysis. The lead ion, Pb^{2+}, is almost the same size and charge as Ca^{2+} and substitutes for it in the $CaCO_3$ coral skeleton, faithfully documenting the history of lead concentrations in surface seawater (Fig. 6.24). The coral data have recently been augmented by ocean water data as we now have the analytical capability to measure Pb^{2+} concentrations in seawater. Data collected in recent years show a continued decline in surface water Pb^{2+} concentrations reflecting declining inputs. Deeper in the water column the Pb^{2+} profiles are more difficult to interpret because decreasing rates of input are fast compared to the residence time of Pb^{2+}. This means that the profiles do not represent a steady-state distribution, rather they record the

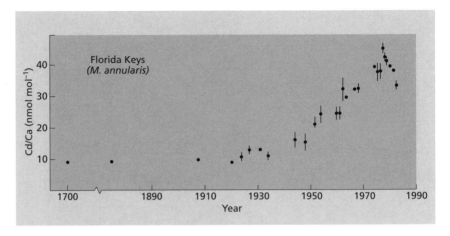

Fig. 6.24 Lead concentrations in dated year bands from a coral collected from the Florida Keys. After Shen and Boyle (1987), with permission from Elsevier Science.

Table 6.10 Removal fluxes of elements by hydrothermal plumes (mol yr^{-1}). Based on data in Elderfield and Schultz (1996), with permission from the *Annual Review of Earth and Planetary Sciences*.

Element	Hydrothermal plume removal flux	River input flux	% removal of river input flux
V	4.3×10^8	5.9×10^8	73
P	1.1×10^{10}	3.3×10^{10}	33
As	1.8×10^8	8.8×10^8	20
Cr	4.8×10^7	6.3×10^8	8

system still adjusting to rapid changes in Pb^{2+} input. Although it is regrettable that humans have caused global-scale lead pollution, chemical oceanographers can at least make use of the information as these transient pollutants act as tracers to allow a better understand of material cycling in the oceans (see also Box 7.1).

In the deep oceans, particle-rich hydrothermal plumes (Plate 6.1, facing p. 138), and the iron- and manganese-rich hydrothermal sediments that fall out from them, cause intense scavenging of some elements from seawater, making mid-ocean ridge environments important sinks for some metals. It is calculated that the entire ocean is stripped of its reactive constituents by hydrothermal plumes in about 2.4×10^5 years (Table 6.5). To do this, the volume of the oceans has to pass through the hydrothermal plumes about 100 times and this means that scavenging in the plume is significant for elements that have oceanic residence times (Box 6.3) in excess of 10 000 years. Scavenging in the plume is a significant removal process for a number of elements, particularly vanadium (V), phosphorus (P), arsenic (As) and chromium (Cr) (Table 6.10). These elements appear to coprecipitate during iron oxyhydroxide formation, for example the oxyanion HPO_4^{2-} is incorporated in particles of $CaHFe(PO_4)_2$; it is assumed that VO_4^{2-}, $HAsO_4^{2-}$ and CrO_4^{2-} form similar compounds with iron oxyhydroxides.

6.6 The role of iron as a nutrient in the oceans

Although abundant in the Earth's crust (see Fig. 1.3), iron is present in seawater at very low concentrations (about $1\,nmol\,l^{-1}$ or less) because the thermodynamically stable Fe(III) species is both insoluble (see Fig. 5.2) and particle reactive, being a highly charged small ion (Section 6.5.5). Despite this, iron is an essential component for a number of life-supporting enzyme systems including those involved in photosynthesis and nitrogen fixation. It may appear surprising that phytoplankton have evolved with an essential requirement for an element present at such low oceanic concentrations. However, the primitive algae that eventually gave rise to modern phytoplankton probably evolved at a time of much lower global oxygen concentrations in a mildly reducing ocean, when soluble Fe(II) was the dominant iron species.

It is only relatively recently that oceanographers have been able to measure accurately dissolved iron concentrations in the oceans; earlier efforts were hampered by a combination of the low dissolved iron concentrations and the ubiquitous sources of iron contamination that compromised sampling and analysis. The distribution of dissolved iron in the oceanic water column is now known to be mostly nutrient-like (Fig. 6.25). Surface waters can have concentrations below $0.1\,nmol\,l^{-1}$ due to phytoplankton uptake, increasing to about $1\,nmol\,l^{-1}$ in deep waters throughout the oceans. These higher deep-water concentrations are maintained by complexation of Fe(III) by strong organic ligands (Box 6.4), which prevent scavenging onto particles. Despite this, the residence time of dissolved iron is thought to be only a few hundred years, similar to that of other scavenged elements. Thus the behaviour of iron in seawater is similar to both nutrient and scavenged elements.

As much of the potential riverwater input of dissolved iron is rapidly stripped out of water during estuarine mixing, the main external source of iron to the open oceans is from the (very limited) dissolution of wind-blown soil and dust. This is material derived primarily from the great Asian, African and Middle Eastern deserts (Plate 6.2, facing p. 138) of the northern hemisphere. Dust has an atmospheric residence time of only a few days, much shorter than hemispheric atmospheric mixing times. This means that the southern hemisphere oceans receive much lower atmospheric dust inputs than those in the northern hemisphere. For example, dust inputs to the North Pacific are 11 times greater than those to the South Pacific.

In some areas of the oceans, it is now clear that the supply of dissolved iron from terrestrial-derived atmospheric dust and from upwelling is inadequate to

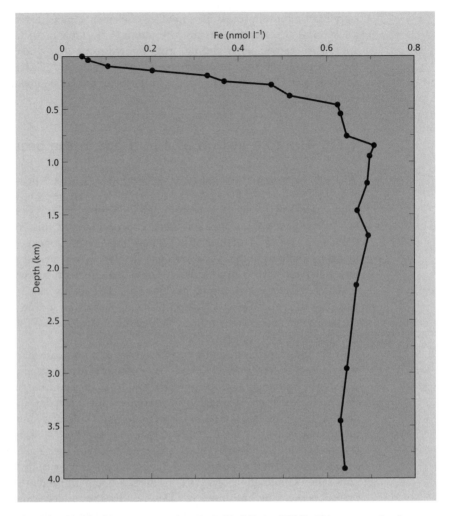

Fig. 6.25 Dissolved iron concentrations in Gulf of Alaska (NE Pacific) seawater showing nutrient-like profile (compare with Fig. 6.20). After Martin *et al*. (1989), with permission from Elsevier Science.

sustain high levels of primary production. In these regions, dissolved iron rather than the macronutrients nitrogen, DIP or silicon becomes the limiting nutrient. This appears to be particularly the case in the Southern Ocean around Antarctica, the ocean area furthest from dust sources. Oceanographers have puzzled for many years over the problem of why dissolved Southern Ocean macronutrient concentrations remain high in summer while phytoplankton growth rates are relatively low. This situation contrasts strongly with other upwelling areas where ocean mixing processes bring nutrient-rich deep water to the surface, promoting very high rates of phytoplankton growth (Plate 6.3, facing p. 138). Iron limitation of the Southern Ocean explains this puzzle. Clear evidence of iron limitation in these areas has come from a recent experiment in which iron was added to a small patch of 'chemically labelled' surface water in the Southern Ocean. The chemical label (an artificial compound called SF_6) was used to trace the movement of the iron-enriched patch of water, and within a few days, phytoplankton abundance had increased markedly (Fig. 6.26).

During the last ice age, approximately 20 000 years ago, dust fluxes to the oceans were higher because the climate was drier and windier. It has been proposed recently that the resulting enhanced supply of dust-derived iron at this time increased ocean productivity directly by increasing photosynthesis. The high dust-derived iron flux may have also promoted enhanced productivity by bacterial nitrogen fixation in the nitrate-deficient central ocean areas, as this enzyme system has a requirement for iron (see above). If increased productivity did take place, it would have also consumed CO_2 because of increased photosynthesis (see eqn. 5.19) lowering global atmospheric CO_2 (see Section 7.2.2) and hence sustaining the glacial climate. The proposed linkage between the biogeochemical cycling of iron and CO_2 is debated hotly between scientists and is the subject of much ongoing research.

6.7 Ocean circulation and its effects on trace element distribution

The preceding discussion of trace elements in seawater has assumed that the oceans have a uniform, warm, nutrient-depleted surface mixed layer and a static deep zone. In fact, at high latitudes surface seawater is cold enough to destroy any density stratification, mixing the oceans to depths of up to 1000 m. This dense surface water sinks and flows slowly into the centre of the oceans as a layer of

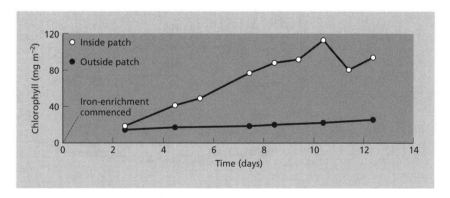

Fig. 6.26 Chlorophyll concentrations as indicators of phytoplankton growth, inside and outside an iron-enriched patch of seawater in the Southern Ocean. After a few days chlorophyll concentrations had increased markedly in the iron-enriched water indicating phytoplankton growth. After Boyd *et al.* (2000), reprinted with permission from *Nature*. Copyright (2000) Macmillan Magazines Limited.

cold, oxygen-rich water, displacing the bottom water in its path. The displaced bottom water is forced to move upwards slowly, setting up an oceanic circulation (Figs 6.27 & 6.28).

The deep mixing at high latitudes only occurs in two locations: in the North Atlantic and around Antarctica. Deep mixing does not occur in the North Pacific, mainly because a physical sill, related to the Aleutian Arc, prevents water exchange between the Arctic and the Pacific (Fig. 6.28). This asymmetry in deep mixing drives a global ocean circulation, in which surface water sinks in the North Atlantic, returns to the surface in the Antarctic and then sinks again and enters the Pacific and Indian Oceans (Fig. 6.27). The deep flow tends to concentrate at the western edge of ocean basins, but allows a slow diffusion of water throughout the ocean interiors. This slow, deep-water flow is compensated by a poleward return flow in surface waters (Fig. 6.29). A 'parcel' of seawater takes hundreds of years to complete this global ocean journey, during the course of which the deep water continually acquires the decay products of sinking organic matter from surface seawater. Waters in the North Pacific have more time to acquire these decay products because they are the 'oldest', in the sense of time elapsed since they were last at the surface and had their nutrients removed by biological processes. As a result the deep waters of the Pacific Ocean have higher macronutrient concentrations than those in the Atlantic Ocean (Fig. 6.20). The waters of the North Pacific also have the lowest dissolved oxygen concentrations and high

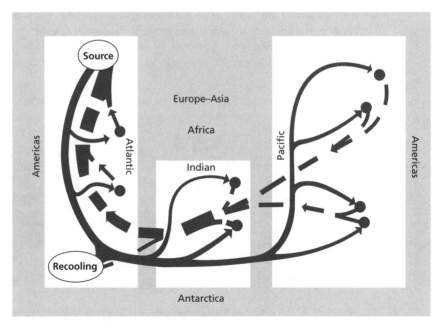

Fig. 6.27 Idealized map of oceanic deep-water flow (solid lines) and surface-water flow (dashed lines). Open circles represent areas of water sinking and dark circles areas of upwelling. After Broecker and Peng (1982).

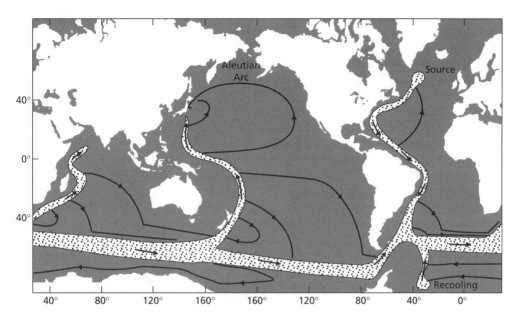

Fig. 6.28 Global oceanic deep-water circulation. Major flow routes are marked by stippled ornament. Deep mixing in the North Pacific is prevented by the topography of the seabed around the Aleutian Arc. Modified from Stommel (1958), with permission from Elsevier Science.

dissolved CO_2 concentrations, since oxygen has been used to oxidize greater amounts of organic matter to CO_2. Overall, the supply of dissolved oxygen to seawater is adequate to oxidize the sinking organic matter and, apart from a few unusual areas in the oceans, oxygen concentrations in the bottom waters are adequate to support animal life. The higher dissolved CO_2 concentration in the Pacific results in a shallower calcite compensation depth (CCD) in the Pacific, relative to the Atlantic Ocean (Section 6.4.4).

The slower regeneration of silicon compared with nitrogen and phosphorus (Section 6.5.4) means that relatively more silicon is regenerated in deep waters, producing steeper interocean concentration gradients (Table 6.9 & Fig. 6.20). Similarly, other elements with nutrient-like behaviour, such as zinc and cadmium, have higher concentrations in the North Pacific compared with other oceans. By contrast, scavenged elements have concentrations that are lower in the deep waters of the Pacific compared with the North Atlantic, because of the longer time available for their removal by adsorption on to sinking particulate matter (Table 6.9).

This pattern of global oceanic circulation has probably existed since the end of the last glaciation, 11 000 years ago. Before this, the circulation pattern is thought to have been different, due to changes in glacial climatic regime and changes in polar ice volume. It is unclear whether changes in ocean circulation provoked climatic change at this time or vice versa. Despite the uncertainty, it is clear that ocean circulation and global climate are intimately linked.

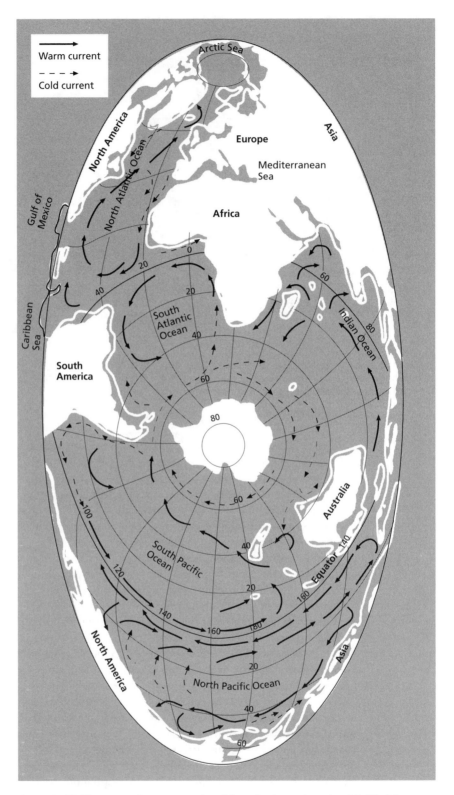

Fig. 6.29 World ocean-surface currents viewed from the Antarctic region. Modified from Spilhaus (1942), with permission from the American Geographical Society.

6.8 Anthropogenic effects on ocean chemistry

The activities of humans have had some impacts on both the major and minor element chemistry of the modern oceans. For example, seawater major ion budgets mostly assume the estimated riverwater input to seawater is that of the pristine (pre-human) system. However, anthropogenic processes have altered some of these fluxes. For example, the riverine Cl^- flux may have increased by more than 40% as a result of human activity and the SO_4^{2-} flux may have doubled, due mainly to fossil fuel combustion and oxidation of pollution-derived H_2S.

As one might predict, the most obvious human impacts are manifest in smaller regional seas that are much less well mixed with respect to the open ocean, or in continental shelf settings close to land. The impacts in these areas is mainly a result of human activities on land that affect the chemical composition of the runoff and/or the mass of suspended load carried by rivers. Deforestation and increased agricultural activity worldwide make land surfaces more susceptible to erosion. The effects on the major ion chemistry of seawater are mainly related to increased input of detrital solids to continental shelves that increase the amount of ion exchange and other solid–seawater interactions (Section 6.4). However, this situation is still changing; the increasing use of dams on rivers is now reducing sediment inputs to the oceans. Less important in terms of global budgets, but important from an ecological viewpoint, increased suspended loads in tropical rivers issuing into coastal waters choke coral reefs with detritus and decrease biological productivity by reducing water clarity.

6.8.1 Human effects on regional seas 1: the Baltic

The Baltic Sea (Fig. 6.30) is a large regional sea, receiving drainage from much of northern and central Europe. The hydrography of the Baltic is complex, consisting of a number of deep basins separated by shallow sills. As a result, the waters of the deep basins can be isolated from exchange with one another—and from the atmosphere—on timescales of years.

There is a long record (almost 100 years) of dissolved phosphorus (P) and oxygen (O_2) concentrations for the waters of the Baltic. The records are 'noisy' due to complex water exchange and deep mixing, but the increasing concentration of dissolved phosphorus over the last 30 years is clear from Fig. 6.30. This increase in nutrient concentration has fuelled primary production and has increased the flux of organic matter to the deep waters. Measurements of dissolved oxygen in deep waters of the Baltic show a steady decline over the last 100 years (Fig. 6.30), consistent with an increase in rates of oxygen consumption due to increasing organic matter inputs—overall, a clear example of eutrophication.

The isolated deep waters of the Baltic have probably always had low oxygen concentrations. However, the declining trend over recent years means that, in some areas, oxygen concentrations have fallen to zero (anoxic). Under anoxic conditions, respiration of organic matter by microbial sulphate (SO_4^{2-}) reduction has produced hydrogen sulphides (HS^-) (plotted as negative oxygen in Fig. 6.30).

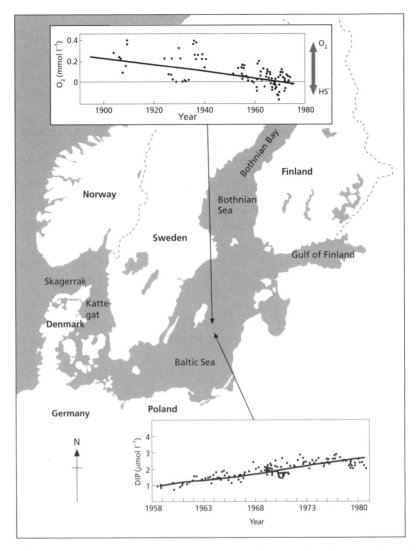

Fig. 6.30 Oxygen and phosphorus concentrations in the Baltic Sea. Dark line through data is a regression line and thin line marks zero O_2 concentration. DIP, dissolved inorganic phosphorus. Data plots after Fonselius (1981) and Nehring (1981), with permission from Elsevier Science.

The Baltic contrasts with the nearby North Sea, where oxygen concentrations rarely fall to low levels, despite large inputs of nutrients. This is because the North Sea is shallow and its waters exchange freely with those of the North Atlantic providing a constant supply of oxygen-rich surface water to the North Sea deeper waters.

6.8.2 Human effects on regional seas 2: the Gulf of Mexico

The Mississippi river system is one of the largest in the world and drains over 40% of the USA, discharging into the Gulf of Mexico through a large and complex delta near New Orleans (Fig. 6.31). The river system drains intensively farmed areas of the USA and nitrate (NO_3^-) concentrations in the river doubled from the 1960s to the 1980s as a result of increased fertilizer use (Fig. 6.31). Since the 1980s NO_3^- concentrations have remained at this high level (see also Section 5.5.1). Increased diatom growth in the riverwater has caused a decrease in silica concentrations (removed to the diatom skeletons) of more than 30% (see also Section 5.5.1).

The Mississippi drains on to the continental shelf of the northern Gulf of Mexico (Fig. 6.31). Here the freshwater flow combines with ocean currents to produce a stratified water column, isolating the shelf bottom water for much of the year. The nutrients from the Mississippi help fuel algal growth in waters off-shore of the delta. After death, some of the algal cells sink into the bottom waters to be degraded by aerobic bacteria, thereby consuming oxygen (see Section 5.5). A 10 000 km² region of low oxygen develops mainly in the spring and summer in these isolated bottom waters (Fig. 6.31).

Records of preserved phytoplankton and organic carbon in dated shelf sediments from this area suggest that increased sedimentation of algal material began the 1960s. The increased agricultural NO_3^- inputs from the Mississippi are very likely responsible—at least in part—for the low oxygen concentrations, although other factors such as wetland loss (Section 6.2.4), changes in river discharge and changes in physical conditions within and around the delta probably also play a part. The discovery of these low oxygen regions in the Gulf of Mexico have led to modification of farming practices throughout the Mississippi drainage region in order to reduce nutrient inputs.

Both the Baltic and Gulf of Mexico examples illustrate that activities taking place hundreds or even thousands of kilometres distant from the oceans can have major impacts on coastal seas. This creates a problem for environmental managers, particularly where inputs in one country impact a neighbour.

6.8.3 Human effects on total ocean minor element budgets

In the preceding sections we showed that riverine nutrient inputs impact regional seas. We also stated in Section 6.5.4 that the huge reservoirs of nutrients in the deep ocean mean that increases in, for example, the concentrations of NO_3^- in riverwaters due to human activity have little effect on oceanic NO_3^- concentration, assuming that NO_3^- is effectively mixed throughout the ocean volume. We can now demonstrate that this is the case using some simple reasoning.

The average NO_3^- concentration in the oceans is 30 µmol l⁻¹. The total oceanic NO_3^- inventory is calculated by multiplying this concentration by the volume of the oceans, i.e.:

$$30 \times 10^{-6} (\text{mol NO}_3^-\text{l}^{-1}) \times 1.37 \times 10^{21} (\text{l}) = 41 \times 10^{15} \text{ mol NO}_3^- \qquad \text{eqn. 6.20}$$

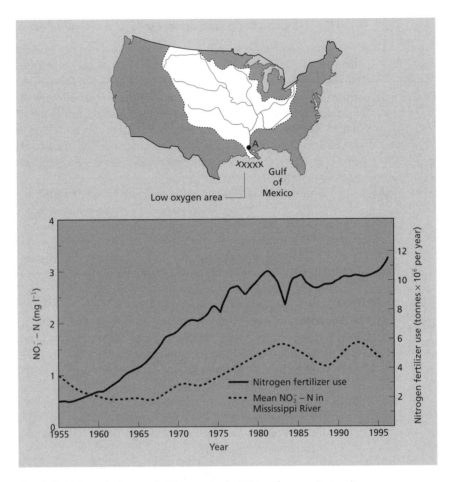

Fig. 6.31 Estimated nitrogen fertilizer use in the USA and mean nitrate-nitrogen concentrations* in the Mississippi River at St Francisville near Baton Rouge, Louisiana (A on inset map) between 1955 and 1995. Inset map shows the Mississippi drainage basin and the region of low oxygen that forms offshore of the Mississippi delta in the Gulf of Mexico. Data courtesy of United States Geological Survey. * Note on units: although we mainly use molar units in this book, in some applications results are reported in mass units. Nitrate is a case in point. Much nitrate data are reported in milligrams of nitrate per litre ($mg\,NO_3^-\,l^{-1}$). However, another common unit is milligrams of nitrogen from the nitrate per litre ($mg\,NO_3^- $-$N\,l^{-1}$) as used in this figure. Conversion of the former to the latter requires division by the molar mass of NO_3^- (62) and multiplication by the atomic mass of nitrogen (14.01).

It is difficult to estimate the natural pre-human NO_3^- concentration in rivers, since most have now been affected by human activities to some extent. Nitrate concentrations in the Amazon are about $20\,\mu mol\,l^{-1}$ and are probably close to natural levels. Multiplying this NO_3^- concentration by the global river flux gives an estimate of the natural nitrate input to the oceans:

$$20 \times 10^{-6}\,(mol\,NO_3^-\,l^{-1}) \times 3.6 \times 10^{16}\,(l\,yr^{-1}) = 0.72 \times 10^{12}\,mol\,NO_3^-\,yr^{-1}$$

eqn. 6.21

The NO_3^- concentrations in some rivers have been increased by human activities to concentrations as high as $500\,\mu mol\,l^{-1}$, although the greatest changes have generally been in small rivers which make a small contribution to the total river flux to the oceans. If, as an extreme case, we suppose that 10% of the world river flow increased to this high concentration, we can calculate how long such a situation would take to double the oceanic NO_3^- concentration, i.e. introduce $41 \times 10^{15}\,mol$. The calculation is simplified by ignoring NO_3^- removal to sediment. In reality, increased NO_3^- inputs would result in increased biological activity and increased NO_3^- removal from the oceans, since the main sink for nitrogen and phosphorus is burial in organic matter.

If 10% of the world's rivers have NO_3^- concentrations increased to $500\,\mu mol\,l^{-1}$, total riverine NO_3^- inputs become:

$$[20 \times 10^{-6}(mol\,NO_3^-\,l^{-1}) \times 3.6 \times 10^{16}(1\,yr^{-1}) \times 0.9]$$
$$+[500 \times 10^{-6}(mol\,NO_3^-\,l^{-1}) \times 3.6 \times 10^{16}(1\,yr^{-1}) \times 0.1]$$
$$= 2.45 \times 10^{12}\,mol\,NO_3^-\,yr^{-1} \qquad\qquad \text{eqn. 6.22}$$

The time needed to double the oceanic nitrate inventory, assuming no removal at all (hence a minimum estimate), is then:

$$\frac{41 \times 10^{15}\,mol\,NO_3^-}{2.45 \times 10^{12}\,mol\,NO_3^-\,yr^{-1}} = 16\,700\,years \qquad\qquad \text{eqn. 6.23}$$

Thus, even drastic perturbations of the freshwater NO_3^- input to the oceans cannot alter the seawater concentration rapidly because of the huge oceanic reservoir of this element. This contrasts strongly with regional seas where NO_3^- concentrations have increased in pace with the enhanced inputs and in some cases have had significant ecological effects.

6.9 Further reading

Berner, K.B. & Berner, R.A. (1987) *The Global Water Cycle.* Prentice Hall, Englewood Cliffs, NJ.

Broeker, W. & Peng, T.-H. (1982) *Tracers in the Sea.* Lamont Doherty Geological Observatory, Palisades, New York.

Chester, R. (2000) *Marine Geochemistry.* Blackwell, Oxford.

Drever, J.I., Li, Y-H. & Maynard, J.B. (1988) Geochemical cycles: the continental crust and oceans. In: *Chemical Cycles in the Evolution of the Earth,* ed. by Gregor, C., Garrels, R.M., Mackenzie, F.T. & Maynard, J.B., pp. 17–53. Wiley, New York.

Falkowski, P.G., Barber, R.T. & Smetacek, V. (1998) Biogeochemical controls and feedbacks on ocean primary production. *Science* 281, 200–206.

Humphris, S.E., Zierenberg, R.A., Mullineaux, L.S. & Thomson, R.E. (1995) *Seafloor Hydrothermal Systems: Physical, Chemical, Biological and Geological Interactions. Geophysical Monograph 91.* American Geophysical Union, Washington, DC.

Libes, S. (1992) *Marine Biogeochemistry.* Wiley, New York.

Turner, B., Clark, W., Kates, R., Richards, J., Mathews, J. & Meyer, W. (1990) *The Earth as Transformed by Human Action.* Cambridge University Press, Cambridge.

6.10 Internet search keywords

marine biogeochemistry
seawater chemistry
estuarine chemistry
ionic strength
halmyrolysis
microbial activity estuaries
phytoplankton estuaries
oxygen estuaries
sewage estuaries
major ions seawater
sea air fluxes
evaporites
calcium carbonate seawater
pyrite marine sediment
hydrothermal processes
black smokers
element distribution oceans
minor components seawater

conservative ions seawater
nutrients seawater
scavenging seawater
phosphate seawater
nitrate seawater
iron seawater
ocean circulation
anthropogenic effects ocean chemistry
anthropogenic effects Baltic chemistry
low oxygen Baltic
low oxygen Gulf Mexico
salinity seawater
residence time seawater
seawater chemistry geological
ion pairs seawater
whitings seawater
ooids seawater
primary production seawater

Global Change 7

7.1 Why study global-scale environmental chemistry?

In previous chapters of this book the chemistry of the atmosphere, oceans and land has been dealt with largely on an individual basis. Using a steady-state model (see Section 3.3), we can envisage each of these environments as a reservoir. In each chapter the cycling of chemicals has been discussed, together with their transformations within the reservoir. Where relevant, some attention has been paid to inputs and outputs into or out of that reservoir from or to adjacent ones. By contrast, the present chapter focuses not on individual reservoirs, but on the ensemble of them that make up an integrated system, of air, water and solids, constituting the near-surface environments of our planet.

As scientists have learnt more about the way chemical constituents of the Earth's surface operate, it has become clear that it is insufficient to consider only individual environmental reservoirs. These reservoirs do not exist in isolation — there are large and continuous flows of chemicals between them. Furthermore, the outflow of material from one reservoir may have little effect on it, but can have a very large impact on the receiving reservoir. For example, the natural flow of reduced sulphur gas from the oceans to the atmosphere has essentially no impact on the chemistry of seawater, and yet has a major role in the acid–base chemistry of the atmosphere, as well as affecting the amount of cloud cover (Section 7.3).

Since integrated systems need to be understood in a holistic way, studies of the global environmental system and natural and human-induced changes to it have become very important. By definition, such studies are on a large scale, generally beyond the resources of most nations, let alone individual scientists. Thus, in recent years several large international programmes have been put in place, the most relevant to environmental chemistry being the International Geosphere–Biosphere Programme (IGBP) (further information can be found at http://www.igbp.kva.se) of the International Council for Science. This has as its aim:

> To describe and understand the interactive physical, chemical and biological processes that regulate the total Earth system, the unique environment that it provides for life, the changes that are occurring in this system, and the manner in which they are influenced by human actions.

This large research agenda is concerned not only with understanding how Earth systems currently operate, but also how they did in the past as well as predicting how they may change in the future as a result of human activities and other factors.

This chapter examines the global cycling of carbon, sulphur and persistent organic pollutants (POPs) at or near the Earth's surface. These examples, concerned with natural or human-induced alterations to new or existing cycles (see Section 1.4), have been selected because they are chemicals that circulate widely in the atmosphere, with potential impacts on large regions of, if not the whole, planet.

7.2 The carbon cycle

The most important component of the carbon cycle is the gas carbon dioxide (CO_2), and almost all of this section will be about this compound and the role it plays in the environment.

7.2.1 The atmospheric record

The best place to start our examination is in the atmosphere, where the observational record is most complete and historical changes are best documented. Figure 7.1 shows values of the atmospheric concentration of CO_2 measured at Mauna Loa (Hawaii) and the South Pole from the late 1950s until the end of the 20th century. It indicates that over this period there was a clear increase in atmospheric CO_2 and that this was a worldwide phenomenon. The rate of increase varied somewhat from year to year, generally ranging from 1 to 2 ppm or about 0.4% per year.

The data in Fig. 7.1 cover the period during which reliable analytical techniques have been in use. In order to extend the record further into the past, resort is made to measurements obtained using cruder techniques (by modern-day standards) in the latter part of the 19th century and the first half of the 20th century. These early data are shown in Fig. 7.2, together with the most complete recent dataset, which is from Mauna Loa. A more reliable way of extending the record backwards has been through the extraction and analysis of bubbles of air trapped in ice cores collected from the polar ice-caps. The principle of this method is that the trapped air bubbles record the atmospheric composition at the time the ice formed. By dating the various layers in the cores from which the air bubbles have

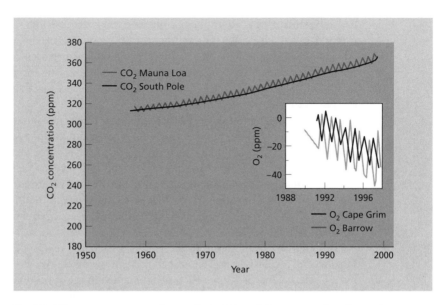

Fig. 7.1 Direct measurements of atmospheric CO_2 and O_2 concentrations for northern (Mauna Loa, Barrow) and southern (South Pole, Cape Grim) hemispheres, showing the changing concentrations and seasonal cycles. After IPCC (2001). With permission of the Intergovernmental Panel on Climate Change.

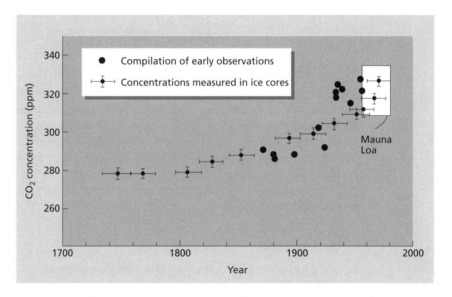

Fig. 7.2 Atmospheric and ice core measurements of CO_2 in air. After Crane and Liss (1985), reprinted with permission of *New Scientist*.

been extracted, the time history of the composition of the atmosphere can be established. The results from measurements made using this approach on an ice core from western Antarctica are also shown in Fig. 7.2. It appears that in the mid-18th century, before major industrialization (and agricultural development) had taken place, the atmosphere contained close to 280 ppm CO_2, and more recent data from ice cores indicate that the 280 ppm level extends back to at least 900 AD. Since about 1850, the CO_2 concentration has increased nearly exponentially, due to humans burning fossil fuels and developing land for agricultural use. In 2001 the level was close to 371 ppm, indicating an increase of almost 33% over the pre-industrial concentration.

Detailed examination of the Mauna Loa data, where measurements are available on a monthly basis (Fig. 7.1), shows a large and regular seasonal pattern of concentration change. Similar seasonalities are found at other sites, although the amplitude of the variation changes with latitude and between hemispheres; note the much smaller seasonal amplitude for the South Pole record in Fig. 7.1. These seasonal effects will be discussed further in connection with biological cycling of CO_2, as will the accompanying changes in atmospheric oxygen shown in the inset to Fig. 7.1 (Section 7.2.2).

Although the trend of atmospheric CO_2 concentrations is clearly upward in Fig. 7.1, the increase is only about half of what would be expected if all the CO_2 from fossil fuel burning since 1958 had remained in the atmosphere, as shown in Fig. 7.8. This indicates that the half that does not appear in the atmospheric record must have been taken up by some other environmental reservoir. This is a simplistic deduction since it assumes that the other reservoirs have themselves not changed in size and that they have not had net exchange with the atmosphere during the relevant period. Despite these simplifications, the calculation forces us to examine the other reservoirs and so stresses the importance of looking at the system as an entity, rather than as disconnected environmental compartments.

7.2.2 Natural and anthropogenic sources and sinks

There are three main sources and sinks for atmospheric CO_2 in near-surface environments: the land biosphere (including freshwaters), the oceans, and anthropogenic emissions from the burning of fossil fuels and other industrial activities. In the natural state the land biosphere and the ocean reservoirs exchange CO_2 with the atmosphere in an essentially balanced two-way transfer. These reservoirs are also sinks for anthropogenic CO_2. Volcanic emissions (see Section 3.4.1) are not considered here since they are thought to be quantitatively unimportant on short timescales.

The land biosphere

In their pristine state the land areas of the Earth are estimated to exchange about 120 GtC (gigatonnes expressed as carbon; $1\,Gt = 10^9$ tonnes $= 10^{15}$ grams) per annum with the atmosphere. This is a balanced two-way flux, with 120 GtC moving from land to air and the same amount going in the opposite direction every year. However, this is a yearly averaged figure and in temperate and polar regions the fluxes are seasonally unequal. For such areas, in spring and summer, when plants are actively extracting CO_2 from the atmosphere in the process of photosynthesis (see Section 5.5), there is a net flux from air to ground. By contrast, in autumn and winter, when the processes of respiration and decomposition of plants remains dominate over photosynthesis (see Section 5.5), the net flux is into the air. Averaged over the whole yearly cycle there is no net flux in either direction. In the tropics, where there is less seasonality in biological processes, the up and down fluxes are in approximate balance throughout the year. However, it should be noted that in the tropics, as at higher latitudes, the fluxes show considerable spatial variability (patchiness).

The seasonal asymmetry in the up and down CO_2 fluxes at middle and high latitudes provides the explanation for the seasonal cycle of atmospheric CO_2 shown in Fig. 7.1. The decreasing values found in spring and summer result from net plant uptake of CO_2 from the air during photosynthesis and the rising limb is due to net release of CO_2 during the rest of the year when respiration and decomposition are dominant. The amplitude of this seasonal pattern varies with latitude, being least at the poles (see CO_2 record for the South Pole in Fig. 7.1) and equator due to lack of biological activity and seasonality respectively. At mid- and sub-polar latitudes the amplitude (peak to peak) is 10–15 ppm, i.e. considerably greater than the average yearly increase (1–2 ppm). The amplitude tends to be greater in the northern compared with the southern hemisphere because of the greater land area in the former compared with the latter. With the uptake and release of CO_2 during photosynthesis and respiration/decomposition there is a concomitant release and absorption of atmospheric oxygen, as indicated in equations 5.19 and 5.20 (see Section 5.5). It has recently become possible to measure these changes in atmospheric oxygen and the data are shown in the inset to Fig.7.1. The oxygen record is much shorter than that for CO_2 due to the very considerable analytical difficulties of measuring the small percentage changes in oxygen compared with CO_2, due to the former being about 550 times more abundant. The seasonality discussed above for CO_2 is observed for oxygen at both Cape Grim in Tasmania (southern hemisphere) and Barrow in Alaska (northern hemisphere), but with the opposite sign (i.e. when atmospheric CO_2 is falling due to plant uptake during photosynthesis, oxygen is rising, and vice versa). It is also clear that the seasonality for atmospheric oxygen is displaced by 6 months between the Barrow and Cape Grim measurement sites, due to their location in the northern and southern hemispheres, respectively.

From the above discussion it is apparent that, while human activities in burning fossil fuel are the primary control on the year-to-year increase in atmospheric CO_2, it is biologically induced exchanges that determine the observed seasonal pattern. Thus, it is clear that the land biota can strongly affect the levels of atmospheric CO_2. This raises the question of whether human activities, for example through change in land use (e.g. clearing of virgin forest), or through enhanced photosynthesis arising from the increasing concentration of atmospheric CO_2, can have produced significant net transfers of carbon into, or out of, the atmosphere.

Turning first to changes in land use, it is clear that when areas formerly storing large amounts of carbon fixed in plant material, for example forests, are converted to urban, industrial or even agricultural use, a large percentage of the fixed carbon is released to the atmosphere as CO_2 quite rapidly. This occurs when the forest is cleared and in part burned, but also by bacterially aided decomposition of dead plant matter, including the soil litter (see Section 4.6.5). None of the new uses for the land store carbon as effectively as the original forest. Even cultivated land, which might appear to be a good store of carbon, contains approximately 20 times less fixed carbon per hectare than a typical mature forest.

Humans have been converting virgin forest and other well-vegetated areas into carbon-poor states for many hundreds of years. This process must therefore be a substantial source of CO_2 to the atmosphere, both in the past and today. It has, however, been difficult to quantify the size of this source. Several attempts have been made to assess how its magnitude has varied over the last century (arguably the period of most rapid change in land use the world has ever experienced). The results from three studies, published in 1983, 1990 and 1993, are shown in Fig. 7.3. There are large discrepancies between the three results and it appears that the earliest attempt overestimated the source compared with the more recent studies. The best estimate of the flux for the 1980s is 1.7 GtC yr^{-1}, with a range from 0.6 to 2.5. The large range confirms the considerable difficulty in trying to quantify CO_2 emissions due to changes in land use.

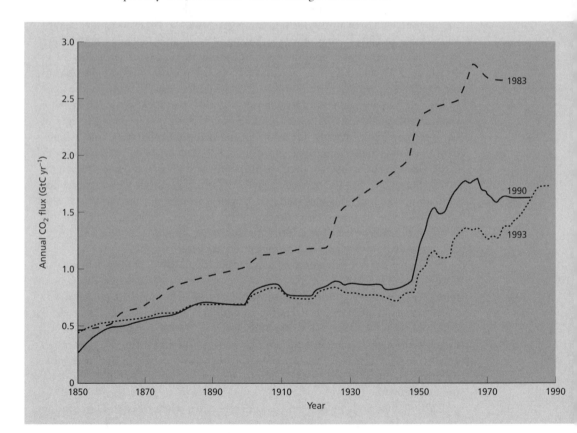

Fig. 7.3 Estimates of CO_2 flux to the atmosphere from land-use changes made in 1983, 1990 and 1993. After Houghton (2000). With kind permission of Cambridge University Press.

It is possible that increasing atmospheric concentrations of CO_2 from fossil-fuel burning and land-use change might cause enhanced growth of plants. This is an important issue, as plant growth could reduce some of the CO_2 emission effects caused by land clearance. Certainly, crops grown in greenhouses under elevated CO_2 regimes produce higher yields. However, extrapolation of such findings to the real environment is problematical. Although CO_2 is fundamental to the process of photosynthesis, in most field situations it is not thought to be the limiting factor for plant growth, availability of water and nutrients such as nitrogen (N) and phosphorus (P) being more important (see Section 5.5.1). It would, however, be wrong to dismiss the possible effect of CO_2 concentration on plant growth, since there may be situations in which the higher CO_2 levels pertaining now, and even more so in the future, may be enough to enhance growth. One suggestion is that elevated CO_2 leads to more efficient use of water by plants, which can then grow in areas previously too dry to sustain them.

The subject of enhanced CO_2 concentration affecting plant growth is being actively researched at present. Studies range from the use of pot-grown plants in controlled (greenhouse) environments, to small-scale field enclosure studies, right through to large-scale field trials, part of the IGBP effort (Section 7.1). In these large-scale experiments, substantial areas ($500\,m^2$) of field crops are exposed to elevated CO_2 concentrations and/or changes in other variables important for growth and the responses monitored over short and long time periods, which can be up to several seasonal cycles. The results of these FACE ('free-air CO_2 enrichment') studies are of considerable interest since, unlike smaller and more confined attempts, they enable the effects of changes in CO_2 and other variables to be studied at as close to real environmental conditions as possible. At the time of writing (2003), over 50 of these experiments have been conducted. In summary, the results indicate that doubled CO_2 can lead to increased plant yield (biomass) by 10–20%, but that factors including changes in temperature, soil moisture and nutrient status, as well as plant species biodiversity, can also affect biomass positively and negatively. Since all these and other factors are operative in the natural environment, prediction of the net effect of such changes is clearly difficult.

In addition to the possibility of enhanced terrestrial take up of CO_2 due to increasing atmospheric levels of the gas, there are other changes that may increase the amount of carbon stored on land. One of these arises from increased deposition from the atmosphere of plant nutrients, such as nitrogen coming from high-temperature combustion sources (automobiles, power stations), in which nitrogen gas from the air is converted into oxides of nitrogen (eqns. 3.22–3.24) which are emitted to the atmosphere. After processing in the atmosphere the nitrogen is deposited on soil where it can fertilize and so potentially enhance plant growth, and thus storage of carbon. A further effect is that of reforestation and regrowth on areas of land previously cleared for agricultural and other purposes. None of these potential enhanced or new sinks for carbon on land is at all straightforward to quantify. The best estimate we have is for the situation in the 1980s when fertilization, whether by elevated CO_2 or nutrients such as nitrogen oxides, together with reforestation and regrowth, were assessed to amount to 1.9 GtC per annum, with a very large range of uncertainty. This once again stresses the great difficulties in trying to estimate changes in carbon uptake and release by the land biosphere resulting directly or indirectly from human actions.

The oceans

As with the land biosphere, the oceans also exchange large amounts of CO_2 with the atmosphere each year. In the unpolluted environment, the air-to-sea and sea-to-air fluxes are globally balanced, with about 90 GtC moving in both directions every year. These up and down fluxes are driven by changes in the temperature of the surface water of the oceans, which alter its ability to dissolve CO_2, as well as by biological consumption and production of the gas resulting from photosynthesis and respiration/decomposition processes in near-surface waters. All of these processes can vary both seasonally and spatially by significant degrees. In general, the tropical oceans are net sources of CO_2 to the atmosphere, whereas at higher and particularly polar latitudes the oceans are a net sink.

Averaged globally and over the yearly cycle, the unpolluted oceans are in approximate steady state with respect to CO_2 uptake/release. This does not mean that over long time periods there is no change in these rates. Indeed, it is thought that the much lower atmospheric CO_2 level which ice core records indicate existed in the past (down to 200 ppm during the most recent glaciations—Fig. 7.10) was due, at least in part, to increased ocean uptake of CO_2 in the cooler waters that existed then, as compared with the present.

The above discussion refers to the ocean/atmosphere system in its pristine state. We know, however, that fossil fuel burning and other human-induced changes have led to substantial additional input of CO_2 into the atmosphere. How much of this extra CO_2 enters the oceans?

Several factors must be taken into account. Firstly, there is the chemistry of seawater itself. Compared with distilled water or even a solution of sodium chloride (NaCl) of equivalent ionic strength (see Box 5.1) to the oceans, seawater has a significantly greater ability to take up excess CO_2. This comes about from the existence in seawater of alkalinity (see Sections 5.3.1, 6.4.4 & Box 5.2) in the form of carbonate ions (CO_3^{2-}), which can react with CO_2 molecules to form bicarbonate ions (HCO_3^-):

$$CO_{3(aq)}^{2-} + CO_{2(g)} + H_2O_{(l)} \rightleftharpoons 2HCO_{3(aq)}^- \qquad \text{eqn. 7.1}$$

This reaction makes seawater about eight times more effective at absorbing CO_2 than a solution of similar ionic strength but not containing CO_3^{2-}.

The discussion above assumes that equilibrium is achieved between the seawater and the air with respect to CO_2. This leads to the second factor which must be taken into account, since the slow mixing time of the oceans means that it takes hundreds, if not thousands, of years for equilibrium to be attained over the whole depth. In general, it is not transfer across the sea surface which is rate-limiting for uptake of CO_2, but mixing of surface water down to the ocean deeps (mean depth 3.8 km, maximum depth 10.9 km). Such mixing is greatly impeded by the existence in most ocean basins of a stable two-layer density structure in the water. At a depth of a few hundred metres there is a region of rapid temperature decrease, the main thermocline. This results in enhanced stability of the water column, which inhibits mixing from above or below. It is only in some polar regions, particularly around Antarctica and in the Greenland and Norwegian Seas in the North Atlantic, where the absence of the thermocline allows direct, and therefore rapid, mixing of surface with deeper waters (see also Section 6.7).

The large, natural, two-way flow of CO_2 across the sea surface makes it very difficult to measure directly the rather small additional flux (about 2% of the gross flux in either direction) resulting from human additions of CO_2 to the atmosphere. Best estimates from this approach (which rely on measurements of pCO_2 (see Box 3.1) across the sea surface) are about $2\,GtC\,yr^{-1}$. In these circumstances resort is often made to mathematical modelling approaches. These models can be of considerable complexity—Box 7.1 shows the principles on which they operate. From modelling studies, the best estimate of the amount of anthropogenic CO_2 being taken up by the oceans is $1.9 \pm 0.6\,GtC\,yr^{-1}$, which is in reasonable agreement with the estimates from direct measurements mentioned above.

As an illustration of how marine biological processes may affect the ability of the oceans to take up CO_2 we now briefly discuss the results of some recent field experiments on the role of iron in controlling photosynthesis in the sea (see also Section 6.6). For many years it had been speculated that in some major ocean areas availability of iron was the limiting factor for phytoplankton growth. However, it was only recently that several direct tests of this idea were carried out by adding iron (in the form of ferrous sulphate) to a small (about $100\,km^2$) area of ocean and observing any resulting effects over periods of days. What was found was that addition of only a very small amount of iron led to a dramatic increase in plankton growth (see Fig. 6.26) with resultant drawdown of CO_2 from the water (and hence potentially from the atmosphere). This is clear in Fig. 7.4 where pCO_2 is lower inside the iron-fertilized patch in relation to the values outside the patch. This is an exciting result in helping to understand what controls marine biological activity. However, its importance for long-term uptake of CO_2 by the oceans is not yet established since it may be that much of the extra carbon incorporated into new phytoplankton growth is rapidly recycled by respiration/decomposition in the near-surface waters. Carbon will only be removed from the atmosphere/surface ocean system for any length of time if dead plankton remains sink into the deep ocean. In order to answer this question it will be necessary to carry out iron-fertilization experiments which last for longer and cover a greater area. This will take a considerable effort at the international level and is currently being planned through programmes such as IGBP (Section 7.1).

Fossil fuel burning

It is relatively easy to quantify the amount of CO_2 that results from the burning of fossil fuel and other industrial activities, such as the manufacture of cement (as part of this process calcium carbonate ($CaCO_3$) is heated to a high temperature and decomposes, yielding CO_2). This source is easier to estimate than those discussed earlier because there is no natural component. All that is required is quan-

Box 7.1 Simple box model for ocean carbon dioxide uptake

In order to calculate how much anthropogenic carbon dioxide (CO_2) the oceans can take up from the atmosphere, it is often necessary to construct a model of the system. The simplest of these models divide the oceans into a series of boxes (numbering from a few to several hundred), with water containing its dissolved carbon (C) flowing between them. The main elements of such models are shown in Fig. 1.

For the relatively well-mixed atmosphere and surface-ocean boxes, the carbon flow between them is assumed to be proportional to their carbon content. Within the deep ocean, where the circulation is much more sluggish, vertical mixing is often modelled as a diffusion process. In addition, the model can include a simple circulation with direct input to the ocean bottom from the surface, balanced by upward water movement throughout the deep ocean, to represent convective processes. The spatial and depth distribution of radioactive substances, such as the isotope ^{14}C (see Section 2.8) (produced both by cosmic rays in the atmosphere and from the detonation of nuclear devices in the

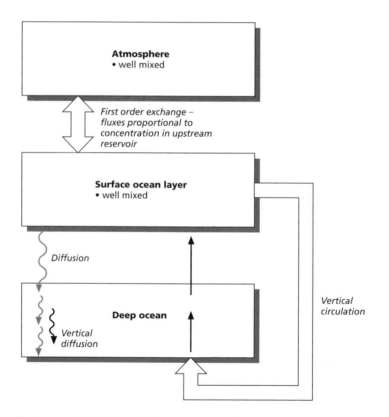

Fig. 1 The main elements of a simple model for ocean CO_2 uptake.

(continued)

1950s and 1960s), can then be used to estimate the rates at which CO_2 is exchanged between the atmosphere and surface ocean, its diffusion into the deep ocean and its transport by vertical circulation.

For the well-mixed reservoirs, a conservation equation is written in which gain of ^{14}C by inflow to the box (atmosphere or surface ocean) is balanced by the outflow to other boxes plus radioactive decay (see Section 2.8) of the tracer during its time in the reservoir. For the deep ocean, conservation is described by a partial differential advection–diffusion equation. The diffusion coefficient is chosen to best fit the measured depth profile of ^{14}C in the oceans.

Using the model, the uptake of fossil fuel CO_2 can be estimated by integrating forward in time from an assumed pre-industrial steady-state value, while adding to the model's atmosphere the estimated year-by-year release of CO_2 from fossil-fuel burning. At each time step, the fluxes of carbon between the various boxes are calculated and the carbon contents and concentration profiles changed accordingly. From such models it is calculated that about 35% of anthropogenic CO_2 is absorbed by the oceans.

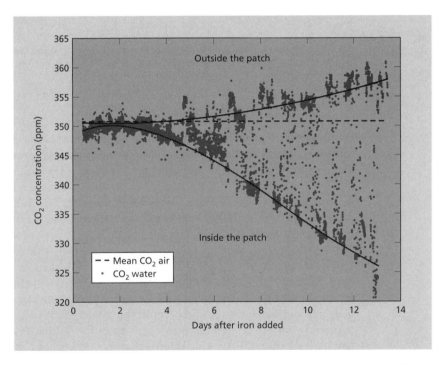

Fig. 7.4 Surface seawater CO_2 concentration during an iron-fertilization experiment. After Watson *et al.* (2000).

tification of the various fuel types burned every year and a knowledge of the amount of CO_2 each produces on combustion. This latter factor, although well known, varies quite a lot between fuels. For example, for each unit of energy produced, coal forms 25% more CO_2 than oil and 70% more than natural gas. This occurs because, in the combustion of gas and oil, a major proportion of the energy comes from conversion of hydrogen atoms (H) in the fuel to water (about 60%

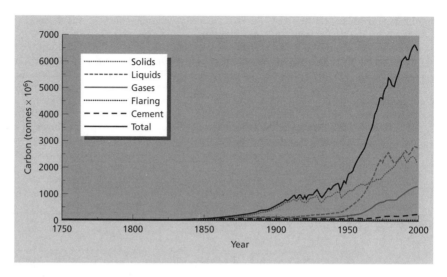

Fig. 7.5 Global CO_2 emissions from fossil-fuel burning (solid, liquid and gaseous fuels), cement production and gas flaring for 1751–1999. After Marland *et al.* (2002).

in the case of gas), rather than from the conversion of carbon to CO_2, which provides 80% of the energy when coal is burned.

Recent data on CO_2 inputs to the atmosphere from fossil fuels and other anthropogenic sources have been compiled by the United Nations in their *Energy Statistics Database*. Earlier data have been obtained from a variety of sources but are more uncertain than the numbers for recent years. The results are presented in Figs 7.5–7.7. In Fig. 7.5 the yearly inputs show a general increase over the period since 1751 when records first become available. The data are plotted on a logarithmic CO_2 emission scale in Fig. 7.6, which shows that the increase has not always been at the same rate. Although for the periods 1860–1910 and 1950–70 the growth rate was close to 4%, during the two world wars, in the great industrial depression of the 1930s and since the 1970s the rate of increase has been closer to 2%. The slackening of emissions in the last 25 years is due to large increases in the price of oil at the beginning of the period, conservation measures generally and economic retrenchment in the 1990s. Wars, like depressions, are apparently times of reduced economic activity. In Fig. 7.7 the data are plotted by latitude for 1980 and 1989, which clearly show how strongly emissions are skewed towards the industrialized mid-latitudes of the northern hemisphere. Over the 1980s there is a clear shift in emissions southwards, as industrialization has become more global. In the last year for which full data are available (1991), emissions from fossil fuel burning, etc. are estimated to be 6.2 GtC yr^{-1}, with an uncertainty of less than 10%. Average annual emissions over the 1980s were 5.4 ± 0.3 GtC yr^{-1}. Finally, in Fig. 7.8 the record of fossil fuel emissions and atmospheric concentration increase at Mauna Loa from 1958 to 2000 is shown. This illustrates the variability of the increase in atmospheric CO_2 year by year (discussed later), and also that only about half of the carbon emitted stays in the atmosphere to produce the yearly increase (as mentioned previously).

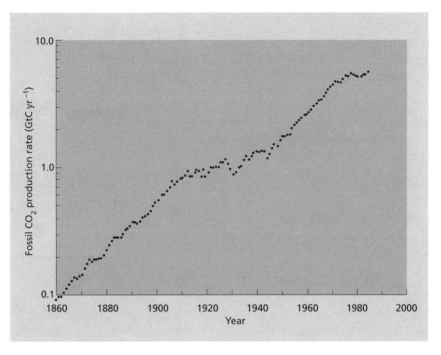

Fig. 7.6 Global annual emissions of CO_2 from fossil-fuel combustion and cement manufacture. After IPCC (1990). With permission of the Intergovernmental Panel on Climate Change.

7.2.3 The global budget of natural and anthropogenic carbon dioxide

We now synthesize much of the knowledge outlined in previous sections on the global budget of CO_2. Firstly, the relative sizes of the natural reservoirs are considered and then the natural flows between them, followed by how anthropogenic CO_2 partitions between the boxes. Finally, likely future levels of atmospheric CO_2 are discussed in terms of possible scenarios of fossil fuel consumption.

Reservoir sizes

A simplified version of the carbon cycle is given in Fig. 7.9. By far the largest reservoir is in marine sediments and sedimentary materials on land (20 000 000 GtC), mainly in the form of $CaCO_3$. However, most of this material is not in contact with the atmosphere and cycles through the solid Earth on geological timescales (see Section 4.1). It therefore plays only a minor role in the short-term cycle of carbon considered here. The next largest reservoir is seawater (about 39 000 GtC), where the carbon is mainly in the dissolved form as HCO_3^- and CO_3^{2-}. However, the deeper parts of the oceans, which contain most of the carbon (38 100 GtC), do not interact with the atmosphere at all rapidly, as discussed in

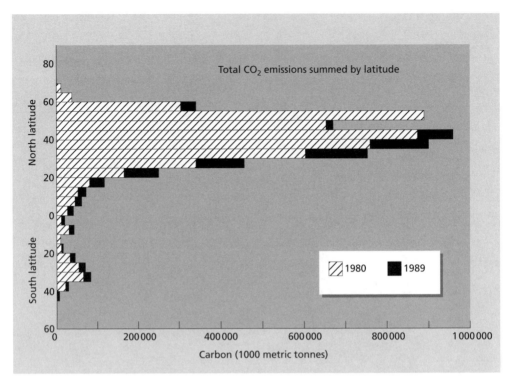

Fig. 7.7 Latitudinal change in CO_2 emissions from 1980 to 1989 as seen in 5° latitude bands. After Andres *et al.* (2000). With kind permission of Cambridge University Press.

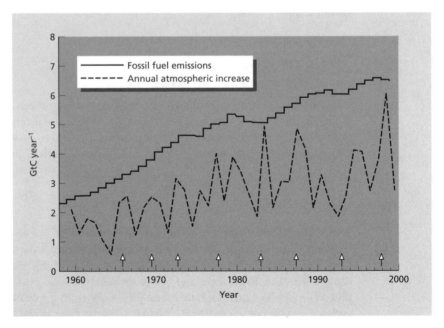

Fig. 7.8 Fossil-fuel emissions and the rate of increase of CO_2 concentrations in the atmosphere. Vertical arrows define El Niño events (see text for discussion). Data from IPCC (2001). With permission of the Intergovernmental Panel on Climate Change.

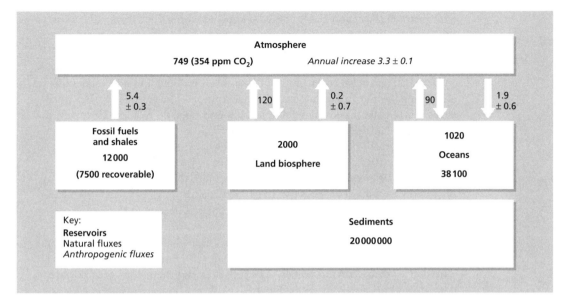

Fig. 7.9 A simplified version of the global carbon cycle for the 1980s. The numbers in boxes indicate the reservoir size in GtC. Arrows represent fluxes and the associated numbers indicate the magnitude of the flux in GtC yr^{-1}. After IPCC (2001). With permission of the Intergovernmental Panel on Climate Change.

Section 7.2.2. The reservoir of carbon in fossil fuels and mudrocks is also substantial and a major portion of the latter is thought to be recoverable and thus available for burning. The smallest reservoirs are the land biosphere (2000 GtC) and the atmosphere (749 GtC, equivalent to an atmospheric concentration of about 354 ppm). It is the small size of the latter which makes it sensitive to even small percentage changes in the other larger reservoirs, where these changes result in emissions to the atmosphere, as, for example, in the burning of fossil fuels.

Natural fluxes

It is often assumed that natural flows between the major reservoirs are balanced two-way fluxes when averaged over the whole year and the total surface of the reservoir. For example, the land biosphere and the oceans exchange approximately 120 and 90 GtC yr^{-1} respectively in both directions with the atmosphere. There is, however, some uncertainty about this assumption over periods of years and it is surely wrong on longer timescales. Evidence for short-term imbalance comes from careful inspection of the atmospheric record (as shown in Fig. 7.8). At the end of the record (in the early 1990s) the rate of increase of atmospheric CO_2 is significantly smaller than for previous years. The explanation for this decrease in the rate of change is very unlikely to be alterations in anthropogenic inputs to the atmosphere, since there is no evidence that fossil fuel burning or land clearance has appreciably altered compared with earlier years. The cause of this decrease in the rate of change appears to be small alterations in the natural fluxes between the atmosphere and land surfaces and the oceans. The large two-way fluxes between these latter reservoirs and the atmosphere mean that only a small imbalance between the up and down fluxes is enough to lead to an observable change in the atmospheric CO_2 concentration. The reasons for such imbalances are largely unknown at present, although possible relationships with, for example, El Niño events (as indicated in Fig. 7.8) are the subject of considerable research effort. El Niño events occur every few years in the tropical Pacific ocean when abnormally warm water near the surface in the eastern part leads to alterations in water circulation and heat exchange with the atmosphere. These cause disruption to fisheries along the coast of Ecuador and Peru, and the altered atmospheric circulation leads to changes in climate throughout the Pacific region and in many other parts of the world.

On the timescale of thousands of years it is clear that changes in the land and ocean reservoirs have led to imbalances in their CO_2 fluxes with the atmosphere. The best evidence for this comes from ice cores and the record of atmospheric composition preserved in them. Figure 7.10 shows how atmospheric CO_2 concentrations and earth-surface temperatures have changed over the last 420 000 years, as recorded in the Vostok ice core from Antarctica. There have clearly been dramatic changes in atmospheric CO_2 levels over this period and the most likely explanation for these shifts is that they arise from (temporary) imbalances between the inter-reservoir fluxes.

While examining Fig. 7.10, it is worth noting that the excursion in atmospheric CO_2 over this 420 000-year period (about 110 ppm) is only marginally greater than that achieved by human activities over the last 200 years (90 ppm), as shown in Figs 7.1 and 7.2. A second point to note from Fig. 7.10 is the close

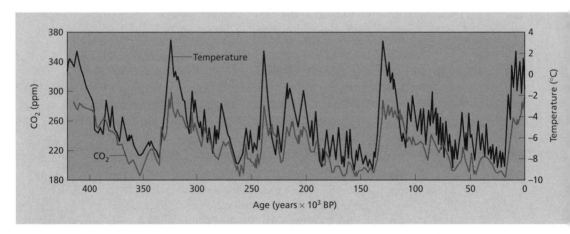

Fig. 7.10 CO_2 concentration in the atmosphere and estimated temperature changes during the past 420 000 years, as determined from the Vostock ice core from Antarctica. BP, before present. Data from Petit *et al.* (1999).

correlation between the CO_2 and the temperature records. This supports the notion of CO_2 as an important greenhouse gas (Section 7.2.4), i.e. when CO_2 is low the temperature is cool (as in glacial periods) and vice versa. Closer inspection of more detailed ice core records indicates that change in CO_2 is apparently not the initiator of the temperature change, which probably arises from alterations in the Earth's orbit and/or changes in the amount of energy coming from the sun. However, orbital or insolation changes cannot account for the magnitude of temperature changes recorded in ice cores, suggesting that CO_2 variations act to amplify the orbital and solar perturbations.

Anthropogenic fluxes

The primary human-induced flux to the atmosphere is that from fossil fuel burning, cement production and so forth, and, as shown in Fig. 7.9, for the 1980s its average value was $5.4 \pm 0.3\,GtC\,yr^{-1}$. Of this input, an amount equivalent to $3.3 \pm 0.1\,GtC\,yr^{-1}$ remains in the atmosphere and leads to the observed year-by-year rise in CO_2 concentrations.

Change in land use arising from human activities leads to an addition to the atmosphere of $1.7 \pm 1.1\,GtC\,yr^{-1}$. On the other hand, fertilization of terrestrial plant growth, reforestation and regrowth are estimated to take up $1.9\,GtC\,yr^{-1}$, giving a small net land sink of $0.2 \pm 0.7\,GtC\,annum^{-1}$.

In Table 7.1 these various flows of anthropogenic CO_2 are given as a budget. Although the budget achieves balance this should not hide the considerable uncertainties over several of the terms, particularly those for exchanges between the land and the atmosphere. For example, the large error associated with the terrestrial biosphere flux term implies that the land could be a small net source, instead of the small net sink shown in Table 7.1, for anthropogenic CO_2.

Emissions and atmospheric carbon dioxide levels in the future

In view of the 'greenhouse' properties of CO_2 (Section 7.2.4) and the fact that atmospheric concentrations of the gas have risen substantially as a result of human activities, considerable effort is currently being devoted to the task of predicting what CO_2 levels will be in the atmosphere over the next century.

In this chapter we have identified the problems that exist in quantitatively accounting for the CO_2 which enters the atmosphere from fossil fuel and other

Table 7.1 Atmospheric sources and sinks of anthropogenic CO_2 for the 1980s. All units are $GtC\,yr^{-1}$. Data from IPCC (2001).

Sources		Sinks	
Fossil fuel burning	5.4 ± 0.3	Atmosphere	3.3 ± 0.1
		Oceans	1.9 ± 0.6
		Land	0.2 ± 0.7
Total	5.4 ± 0.3	Total	5.4 ± 0.6

human activities at the present time. Thus, for any scenario of future anthropogenic CO_2 emissions, there is at least as great an uncertainty over what proportion will remain in the atmosphere as exists for current emissions. In all probability the uncertainty is even greater, since climatological and other global changes, whether human-induced or natural, are likely to alter the rates at which the various environmental reservoirs take up and release CO_2.

Estimating the amount of CO_2 that will be emitted by human activities over the next 100 years is probably even less certain than calculating how it will partition between the air, ocean and land. Although the factors that determine the amounts of anthropogenic emissions can be identified, their quantification can only be guessed at. The size of the human population is a very important factor. We know it is rising and will almost certainly continue to do so (at an unknown rate). Similarly the standard of living of many people from less developed countries is rising and this will lead to greater use of energy in those parts of the world. How this energy is generated will have a profound bearing on how much CO_2 is emitted. The future of CO_2 emissions is crucial for policy decisions. For example, will it be necessary to curb future fossil fuel combustion and, if so, when and by how much, in order to prevent or at least ameliorate undesirable alteration in global climate?

Despite these difficulties some attempts have been made to predict atmospheric CO_2 levels into the next century. The results of one such study are shown in Fig. 7.11. The different curves correspond to different scenarios of population

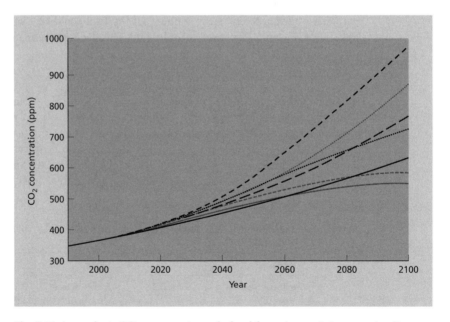

Fig. 7.11 Atmospheric CO_2 concentrations calculated for various emission scenarios. From IPCC (2001). With permission of the Intergovernmental Panel on Climate Change.

growth, energy use and mode of production. All predict a substantial increase in atmospheric CO_2 during the next 100 years, with levels ranging from 500 to more than 900 ppm by 2100. This factor-of-two range does not represent the whole of the uncertainty since other scenarios outside the range used (both higher and lower) are certainly possible. Furthermore, the environmental model used to simulate how much of the emitted CO_2 remains in the atmosphere assumes the environmental system will behave as at present for the whole of the next century.

7.2.4 The effects of elevated carbon dioxide levels on global temperature and other properties

So far we have examined the global cycling of carbon without paying attention to the role CO_2 plays in the Earth's climate. Although CO_2 is a minor component of the atmosphere (see Section 3.2), it plays a vital role in the Earth's radiation balance and hence in controlling the climate. This is illustrated in Fig. 7.12a, which shows the wavelength emission spectrum of the Sun and the Earth, at their effective radiating temperatures of about 5700°C and –23°C respectively.

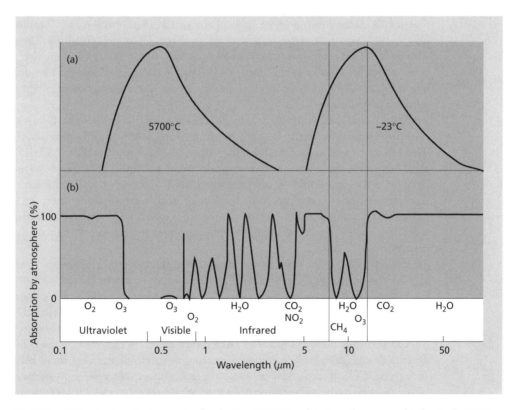

Fig. 7.12 (a) Black body radiation spectra for the Sun (6000 K) and (not on the same scale) the Earth (250 K). (b) Absorption spectrum produced by the principal absorbing gases. After Spedding (1974). Reprinted by permission of Oxford University Press.

Figure 7.12b illustrates how this emitted radiation is absorbed by various atmospheric gases. As Figure 7.12 shows, much of the UV radiation impinging on the atmosphere is absorbed by O_3 molecules in the stratosphere, and this explains the concern that human-induced decreases in stratospheric O_3 may lead to larger amounts of harmful UV radiation reaching the Earth's surface (see Section 3.10). Much of the remainder of the solar energy passes through the atmosphere without major absorption.

Turning now to the Earth's emission spectrum, it is the CO_2 absorption band centred around 15 µm which is particularly important here. This, together with other absorption bands due to water molecules, means that the atmosphere is considerably warmer (mean temperature about 15°C) than the effective emission temperature of the Earth (–23°C; Fig. 7.12a). The combined effect of the atmosphere's transparency to most of the incoming solar radiation, and the absorption of much of the Earth's emitted radiation by water and CO_2 molecules in the atmosphere, is often referred to as the 'greenhouse effect' — by analogy to the role played by the glass of a garden greenhouse.

From the above discussion, it is easy to see why elevated concentrations of CO_2 in the atmosphere resulting from fossil fuel burning are likely to lead to a warmer climate. However, close inspection of Fig. 7.12 indicates that there is sufficient CO_2 in the pre-industrial atmosphere for the 15 µm band to be absorbing almost 100% of the energy in that wavelength range coming from the Earth. Although the CO_2 absorption band will broaden as CO_2 concentrations rise, a major effect is for more of the absorption to occur lower in the atmosphere with less at higher altitudes. The result is that the lower layers warm, whereas higher up there is cooling.

Highly sophisticated mathematical models are used to predict the details of the temperature changes to be expected from rising levels of atmospheric CO_2. The results of one rather straightforward model are shown in Fig. 7.13. The model confirms the simple prediction made above; the lower atmosphere warms by about 3°C for a doubling of atmospheric CO_2 (although the distribution of the increase varies considerably with latitude), with a concomitant decrease in temperatures aloft. Figure 7.11 shows that, for many fossil fuel consumption scenarios, such a doubling might occur some time in the second half of this century.

Although CO_2 is the most important of the anthropogenic greenhouse gases, it is not the only one of significance. Figure 7.14 shows, for the period 1980–90, the relative contributions of various gases to the change in the total greenhouse gas forcing over that decade. Just over half the effect was due to CO_2 but other gases, including methane (CH_4), nitrous oxide (N_2O) and chlorofluorocarbons (CFCs) also contributed substantially to the total effect. In the case of these other gases, although the absolute amounts entering the atmosphere were small compared with CO_2, their contributions to the greenhouse effect were proportionately large due to their absorption of energy being in parts of the Earth's emission spectrum (Fig. 7.12) which are not saturated. To illustrate this we should note that on a molecule-for-molecule basis methane is about 21 times more effective at absorbing energy than CO_2, and CFC-11 is 12 000 times more effective. The

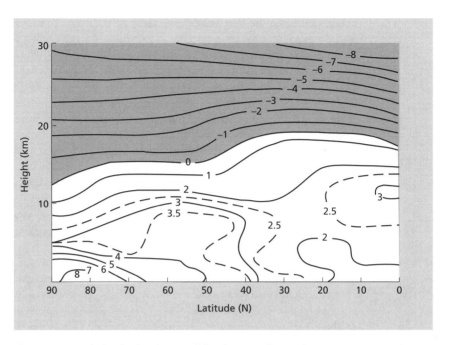

Fig. 7.13 Latitude–height distribution of the change in the zonal mean temperature (K) in response to a doubling of atmospheric CO_2 content. Shaded area identifies decreases in temperature above about 15 km. After Manabe and Wetherald (1980).

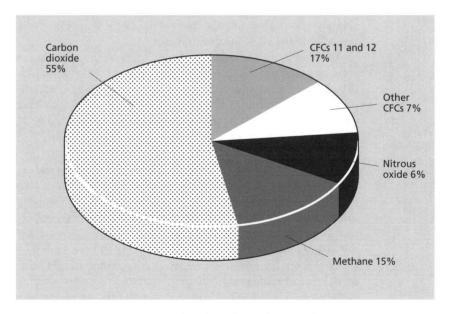

Fig. 7.14 The contribution from each of the anthropogenic greenhouse gases to change in radiative forcing from 1980 to 1990. After IPCC (1990). With permission of the Intergovernmental Panel on Climate Change.

conclusion has to be that an understanding of the cycles of the non-CO_2 greenhouse gases is in total as important as knowledge of the cycling of CO_2.

So far we have concentrated only on greenhouse-gas-induced temperature changes. However, other climatological changes—for example, in the distribution of rainfall—may be more important in a practical sense than temperature increase *per se*. Computer models of likely changes in climate as a result of increase in CO_2 and other gases indicate that global average water vapour, evaporation and rainfall are projected to increase, whereas at the regional scale, both increases and decreases in rainfall are likely. It is the net result of all these changes in rainfall, atmospheric water vapour and evaporation that will determine the agriculturally important property of soil-water content. Water is vital to crop growth and so it is hugely important to be able to predict changes in amounts of soil water, which can be either beneficial or harmful to crop yields. The social, economic and political consequences of such changes and geographical shifts are likely to be considerable.

Another potentially important consequence of global warming would be a global rise in sealevel. This would come about in part due to thermal expansion of seawater and also as a result of melting of glaciers and small ice-caps. Calculations of the magnitude of sealevel rise have considerable uncertainty, but a figure of about half a metre for a doubling of atmospheric CO_2 is the current best estimate. If it occurs, this would have very significant effects in many countries that have centres of population close to the sea or on low-lying land. Further, there is a possibility that warming might eventually lead to the melting of a large mass of grounded ice, for example, the west Antarctic ice sheet. Such an event could produce a more substantial rise in sealevel (several metres), but, even if the temperature rise is great enough to melt the ice, it is estimated that it would take several hundred years for this to occur.

A further impact of rising levels of CO_2 is on the chemistry of the oceans, in particular their pH. Increase in atmospheric CO_2 will lead to a slight lowering of pH of the surface oceans (Section 6.4.4) as a result of extra amounts of the gas crossing from the atmosphere into the oceans. The resulting lowering in pH can be calculated, as can the potential for this increased acidity leading to enhanced calcium carbonate dissolution, for example in corals. Figure. 7.15 shows the pH of surface seawater as a function of temperature calculated for three different concentrations of atmospheric CO_2. A pCO_2 of 280 ppm corresponds to the situation in pre-industrial times with a seawater pH value of just below 8.2. A pCO_2 of 354 ppm, corresponds to 1992 and a seawater pH of just under 8.1 (i.e. a drop in pH of about 0.1 unit from the pre-industrial value). Finally, a pCO_2 of 750 ppm is given as a reasonable estimate of the likely concentration at the close of this century, with a corresponding seawater pH of just under 7.8 (i.e. a drop of 0.4 units from the pre-industrial level). Also shown are pH values derived from measurements made in the Atlantic Ocean surface waters in summer; they cluster in the range 8.0–8.3. What is clear is that by 2100, or whenever the atmospheric pCO_2 reaches double its current value, the pH of surface seawater will be far outside the range of values currently experienced by organisms living in the surface oceans. The effects of such a change on the biology of the oceans and on

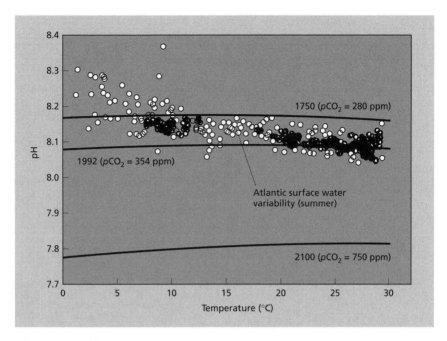

Fig. 7.15 Equilibrium surface seawater pH for various atmospheric pCO$_2$ concentrations assuming no change in alkalinity. Also shown are summertime pH values derived from measurements in the north Atlantic ocean. Courtesy of Doug Wallace, University of Kiel.

associated factors such as CO$_2$ uptake and trace gas emissions (e.g. dimethyl sulphide in Section 7.3) are essentially unknown.

As mentioned above, decreased seawater pH may seriously affect marine organisms that secrete a CaCO$_3$ skeleton (see Section 6.4.4), particularly reef forming corals (e.g. the Australian Great Barrier Reef), since the process is sensitive to the acidity of the seawater. This is illustrated in Fig. 7.16 where the saturation index (Ω) (see Section 6.4.4) for aragonite, the main reef-forming CaCO$_3$ mineral today, is plotted against atmospheric pCO$_2$. The shaded area shows the calculated values of Ω at water temperatures of 25 and 30°C (the range within which reef-building corals live) as a function of the atmospheric pCO$_2$. It is apparent that as pCO$_2$ increases the degree of aragonite supersaturation decreases. Since it is supersaturation that enables coral-forming organisms to synthesize their aragonitic (CaCO$_3$) skeletons, any decrease in Ω will potentially lead to inhibited growth or death of the organisms, and so to die-back of coral reefs. The seriousness of this problem is not well established and is currently a topic of active research.

Given the potentially serious consequences for humans of the climatic changes and other environmental changes discussed above, it is not surprising that emissions of greenhouse gases are now subject to control and regulation through international conventions such as the Kyoto Protocol. Such international action has only come about because the scientific case for action has become compelling. However, as we have seen in this section there still remain many uncertainties in

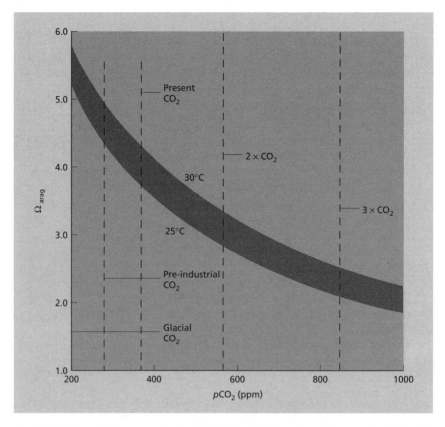

Fig. 7.16 Effect of rising atmospheric pCO_2 (dark stippled band), on the saturation index (Ω) of aragonite ($CaCO_3$). After Buddemeier *et al.* (1998), courtesy of Joan Kleypas.

our understanding of the global carbon cycle and the climatic consequences of changes to it. Predicting such effects into the future is subject to further unknowns, some of these being scientific others being sociological or political. The task of environmental scientists is to reduce the uncertainties in the science so that society can make the serious social changes needed to ameliorate the effects of human-induced climate change in the most cost effective and least socially disruptive way possible.

7.3 The sulphur cycle

7.3.1 The global sulphur cycle and anthropogenic effects

We now turn to the cycling of the element sulphur, outlining the nature of the cycle prior to any major alteration by human industrial and urban activity and examining how these activities have impacted, in a very major way, on the contemporary sulphur cycle.

Comparison of the global sulphur cycle as it is thought to have been prior to any major anthropogenic influence (Fig. 7.17a) with the cycle as it was in the mid 1980s (Fig. 7.17b) reveals some interesting apparent changes in the sizes of some inter-reservoir fluxes. There are also, however, some fluxes for which there is little or no evidence of change, and these are discussed first.

There is no evidence that volcanic emissions of sulphur (mainly as sulphur dioxide, SO_2) have changed significantly during the last 150 years or so (i.e. the time period between parts (a) and (b) of Fig. 7.17) for either terrestrial or marine volcanoes. Similarly, there is no evidence for significant change in the sea-to-air fluxes of either sea-salt sulphate (coming from sea spray arising from wave breaking and bubble bursting at the sea surface) or volatile sulphur, or of emissions of sulphur gases from the terrestrial biosphere. It is important to note that these gaseous fluxes are major components in the cycling of sulphur. The geochemical budget of the element cannot be balanced without them and the total emissions from marine and terrestrial sources is about 70% of the amount of sulphur put into the atmosphere by fossil-fuel burning. The principal component of the marine emissions of volatile sulphur is a gas called dimethyl sulphide (DMS; see also Section 3.4.2 and Fig. 3.4a), produced by phytoplankton (see Fig. 6.10a) and seaweeds that live in the near-surface waters of the oceans. These marine algae also produce lesser amounts of carbonyl sulphide (OCS), carbon disulphide (CS_2) and possibly some hydrogen sulphide (H_2S). Land plants produce a similar suite of gases, but with H_2S playing a major, possibly the dominant, role.

Parts of the sulphur cycle which are thought to have changed significantly as a result of human activities include the following:

1 Aeolian emissions of sulphur-containing soil dust particles are thought to have increased by a factor of about two, from 10 to $20\,Tg\,sulphur\,yr^{-1}$. This is largely as a result of human-induced changes in farming and agricultural practice, particularly through pasturing, ploughing and irrigation.

2 By far the most significant impact on the system has been the input of sulphur (largely as SO_2) direct to the atmosphere from the burning of fossil fuels, metal smelting and other industrial/urban activities. Such emissions have increased approximately 20-fold over the last 120 years. It is not certain that this upward trend will continue indefinitely, since there are ongoing moves in the most advanced industrial nations to restrict emissions by, for example, burning sulphur-poor fuels and removal of SO_2 from power-station stack gases. By contrast, sulphur emissions from the developing nations of the world are likely to increase in the future as they become more industrialized but without the resources to minimize sulphur emitted to the atmosphere. Because of the large magnitude of the fossil-fuel sulphur emissions in relation to other flows in the natural sulphur cycle, this input has substantial impacts on other parts of the cycle, some of which are discussed below.

3 The deposition flux of sulphur from the atmosphere on to the oceans and land surfaces has increased by approximately 25 and 163%, respectively. Although this input has essentially no impact on the chemistry of seawater, due to its buffer capacity and the large amount of sulphate (SO_4^{2-}) it contains (see

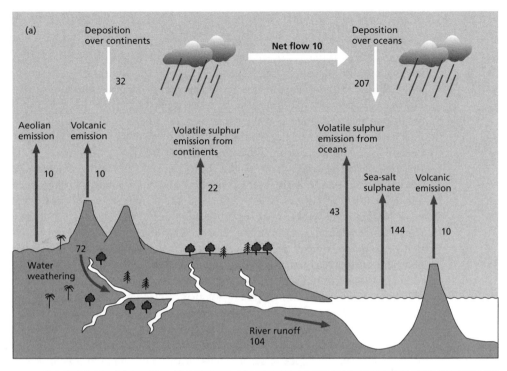

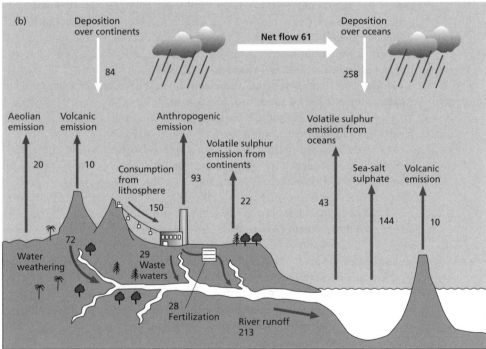

Fig. 7.17 Simplified version of the sulphur cycle (after Brimblecombe *et al.* 1989). (a) Sulphur cycle as it is thought to have been prior to any major anthropogenic influence. (b) Sulphur cycle as it was in the mid-1980s. Units for inter-reservoir flows are in $Tg\,S\,yr^{-1}$ (i.e. $10^{12}\,g\,S\,yr^{-1}$). With permission from the Scientific Community on Problems of the Environment—SCOPE, John Wiley & Sons Ltd.

Table 6.1), it can have a profound impact on poorly buffered soils and fresh waters, as discussed in Section 5.4.1.

4 The amount of sulphur entering the oceans in river runoff has probably more than doubled due to human activities (compare the fluxes in Fig. 7.17a & b). This has been caused in part by sulphur-rich wastewaters and agricultural fertilizers entering river and groundwaters and thence the sea, although another major factor is sulphur deposited directly into surface waters from the atmosphere. The combined (atmospheric and runoff) effects of enhanced sulphur inputs to seawater cause an increase of sulphur (as SO_4^{2-} in the oceans) of only about $10^{-5}\%$ per annum. This estimate is probably an upper limit, since it assumes that removal of seawater sulphur into ocean sediments (see Section 6.4.6) remains as previously and has not increased following the enhanced inputs from the atmosphere and rivers.

5 A final difference highlighted in Fig. 7.17 is in the balance of sulphur flows between the continental and marine atmospheres. In the unperturbed cycle (Fig. 7.17a) there is a small net flow of sulphur from the continental to the marine atmosphere ($10\,Tg\,sulphur\,yr^{-1}$). Today this balance is substantially altered, with about six times greater net flow of sulphur in air flowing seawards ($61\,Tg\,sulphur\,yr^{-1}$) compared with the unperturbed situation.

It is clear from the comparisons above that human activities have substantially changed the cycling of sulphur between the atmosphere, ocean and land surface. This alteration is arguably even greater than that described earlier for human impact on the carbon cycle (Section 7.2.3), and its impact locally and regionally is certainly more apparent, as described below.

7.3.2 The sulphur cycle and atmospheric acidity

If CO_2 were the only atmospheric gas controlling the acidity of rain, then the pH of rainwater would be close to 5.6 (see Box 3.7). However, most pH measurements of rainwater fall below this value, indicating other sources of acidity. Much of this 'extra' acidity arises from the sulphur cycle, as shown in Fig. 7.18. Only two major routes give rise to the sulphur acidity. One is the burning of fossil fuels to produce the acidic gas SO_2. The other is the production of the gas DMS by marine organisms, which then degases to the atmosphere across the air–sea interface (Fig. 7.23). Once in the atmosphere the DMS is oxidized by powerful oxidants, called free radicals (see Section 3.5). The two free radicals important for oxidation of DMS are hydroxyl (OH) and nitrate (NO_3). The products of this oxidation are several, but the two most important are SO_2 and methane sulphonic acid (MSA or CH_3SO_3H). The SO_2 formed in this way is chemically indistinguishable from that coming from the burning of fossil fuels.

The SO_2 from either source exists in the atmosphere either as a gas or dissolved in rain and cloud droplets, whose pH it lowers due to the acidity of the gas. However, within water drops SO_2 can be quite rapidly oxidized to form sulphuric acid (H_2SO_4), which makes them much more acidic since H_2SO_4 is a strong acid (see Box 3.8). MSA formed by oxidation of DMS, via the OH/NO_3 addition route (Fig. 7.18), also contributes to the acidity of atmospheric water. Since this

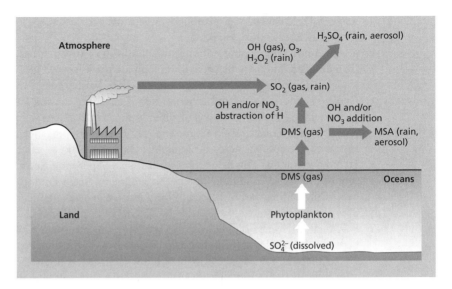

Fig. 7.18 The main natural and anthropogenic routes for atmospheric sulphur dioxide and sulphate.

compound can only be formed from DMS, in contrast to SO_2, it is an unequivocal marker for atmospheric acidity arising from marine biological activity.

The above description is, of course, a considerable simplification of the real situation. For example, rain and cloud droplets contain other dissolved substances important for pH control apart from H_2SO_4—for example, nitric acid (HNO_3) arising from oxides of nitrogen (nitric oxide (NO) and nitrogen dioxide (NO_2)) coming from combustion sources (see Section 3.6.2). Another of these substances, ammonium (NH_4^+, produced by dissolution of ammonia (NH_3) in water), is alkaline and so can partially counteract the acidity arising from the sulphur system. The NH_3 is emitted by soil microbiological reactions (see Section 3.4.2), particularly areas of intensive agriculture, and, according to a recent suggestion, some may come from the oceans (Fig. 7.19) in a cycle somewhat analogous to that of DMS. Another factor is that some of the acid SO_4^{2-} and alkaline NH_4^+ in the atmosphere exist in small aerosol particles (size in the range 10^{-3} to $10\,\mu m$ diameter), which have a chemical composition ranging from 'pure' H_2SO_4 to ammonium sulphate (($NH_4)_2SO_4$), depending on the relative strengths of the sources of SO_4^{2-} and NH_4^+. These particles are formed in part by the drying out of cloud droplets in the atmosphere.

The mass balance for sulphur in Fig. 7.17b represents the various fluxes integrated over the whole globe. Because all the different sulphur compounds shown in Fig. 7.18 have atmospheric residence times (see Section 3.3) of only a few days and so are not well mixed, their distributions in the air are often inhomogeneous. Indeed, for any particular region of the atmosphere, it is likely that one of the major sulphur sources will dominate and thence determine the acidity of rain and aerosols. In general, for remote—particularly marine—areas, the DMS–SO_2–SO_4^{2-} route is likely to control, whereas close to urbanized/industrialized land,

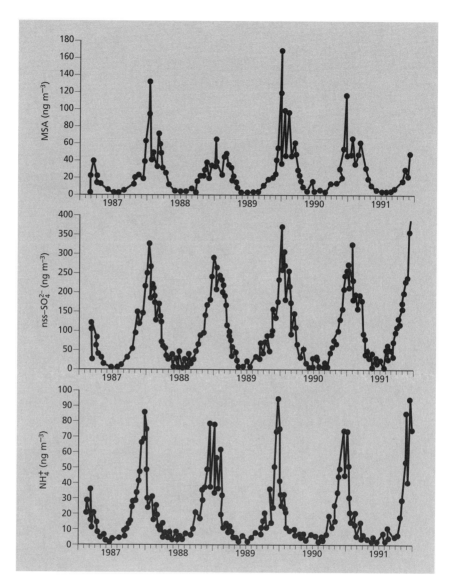

Fig. 7.19 Time series (1987–1991) of weekly average concentrations of methane sulphonic acid (MSA), non-sea-salt-sulphate (nss-SO$_4^{2-}$) and ammonium (NH$_4^+$) measured at Mawson, Antartica. After Savoie *et al.* (1993), with kind permission of Kluwer Academic Publishers.

anthropogenic sources of SO$_2$–SO$_4^{2-}$ will dominate. These contrasting situations are illustrated in Figs 7.19 and 7.20.

In Fig. 7.19 atmospheric measurements of particulate MSA, SO$_4^{2-}$ (after subtraction of the component coming from sea salt, the so-called non-sea-salt-sulphate (nss-SO$_4^{2-}$)) and NH$_4^+$ made in air at Mawson in Antarctica are shown. This site is very remote from human activities and typically receives air which has blown over thousands of kilometres of the Southern Ocean before being

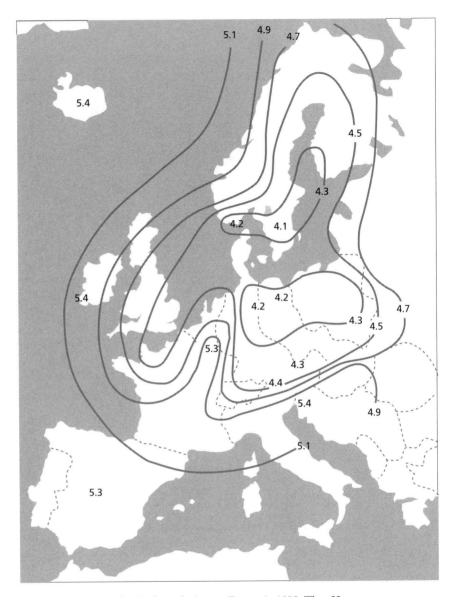

Fig. 7.20 Mean annual pH values of rain over Europe in 1985. The pH contours compare well with rates of acid deposition shown in Fig. 5.7. After Schaug *et al.* (1987). With permission from the Co-operative Programme for Monitoring and Evaluation of the Long Range Transmission of Air Pollutants in Europe (EMEP) and the Norwegian Institute for Air Research (NILU).

sampled. A clear seasonal cycle is apparent, with highest values of MSA and nss-SO_4^{2-} in the austral spring and summer. This is exactly what would be expected if marine biological production of DMS was the dominant source of sulphur, since the phytoplankton are strongly seasonal in their production of DMS. For this site it is well established that marine plankton rather than anthropogenic emissions are the dominant source of sulphur acidity in the air. The same is true for most marine areas of the southern hemisphere. The very similar seasonal cycle seen for NH_4^+ in Fig. 7.19 suggests the possibility of an analogous marine biological source for NH_3 gas also.

Yearly averaged pH values of rain falling over Europe show a very different situation (Fig. 7.20). As might be expected for such a heavily developed area, it is anthropogenic sources which largely control the acidity of the rain. This is shown by the low pH values centred on the most heavily industrialized parts of the region (Germany, eastern Europe, the Low Countries and eastern Britain), with higher (less acidic) pH values to the north, south and far west of the area.

It is not possible to distinguish between SO_2 and SO_4^{2-} coming from fossil fuel burning or marine biogenic (DMS) sources by chemical means. However, recently a differentiation of these two sources has become possible by measuring the ratio of two stable isotopes of sulphur ($^{34}S/^{32}S$, expressed as $\delta^{34}S$; Box 7.2) in rain and aerosol samples. Figure 7.21 illustrates the principle by which the technique works. The $\delta^{34}S$ of sulphur coming from power-station plumes (as SO_2) has a value of between 0 and +5‰ CDT (Canyon Diablo troilite), based on data from power plants in eastern North America and the UK. By contrast, the SO_4^{2-} in seawater, from which phytoplankton make DMS, has a $\delta^{34}S$ value close to +20‰ CDT. This large difference in $\delta^{34}S$ value (between 15 and 20‰ CDT) between the two main sources of atmospheric sulphur is the basis of the method.

If a sample collected in the environment (aerosol, rain, surface water) has a $\delta^{34}S$ value of +20‰ CDT, then it should have its sulphur essentially from the

---

Box 7.2 The delta notation for expressing stable isotope ratio values

Stable isotope (see Box 1.1) abundances cannot at present be determined with sufficient accuracy to be of use in studies of their natural variations. Mass spectrometers can, however, measure the relative abundances of some isotopes very accurately, resulting in stable isotope ratio measurements, for example oxygen $-\,^{18}O/^{16}O$, carbon $-\,^{13}C/^{12}C$ and sulphur $-\,^{34}S/^{32}S$. Stable isotope ratios are reported in delta notation (δ) as parts per thousand (‰ per mil) relative to an international standard, i.e.:

$$\delta = \left(\frac{R_{sample} - R_{standard}}{R_{standard}} \right) \times 1000 \qquad \text{eqn. 1}$$

where R represents a stable isotope ratio and δ expresses the difference between the isotopic ratios of the sample and the standard. δ is positive when the sample has a

larger ratio than the standard, is negative when the reverse is true and is zero when both values are the same. The multiplication by 1000 simply scales up the numbers (which are otherwise very small) to values typically between 0 and ± 100.

For stable sulphur isotopes, the standard is an iron sulphide mineral (troilite) from the Canyon Diablo meteorite. It is known as CDT (Canyon Diablo troilite) and equation 1 becomes:

$$\delta^{34}S = \left(\frac{^{34}S/^{32}S_{sample} - ^{34}S/^{32}S_{standard}}{^{34}S/^{32}S_{standard}} \right) \times 1000$$

$$\text{eqn. 2}$$

Results are reported as $\delta^{34}S$ values relative to the CDT standard, for example, $\delta^{34}S = +20‰$ CDT.

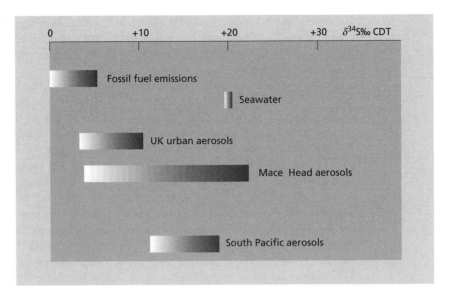

Fig. 7.21 Sulphur isotope ratios ($\delta^{34}S$) for various sources of sulphur and in atmospheric aerosols for several localities. Mace Head (western Ireland) data, courtesy of Nicola McArdle, UEA.

DMS route. On the other hand, if its measured $\delta^{34}S$ is close to the 0 to +5‰ CDT range of fossil fuels, then its contained sulphur is likely to be from this source. Samples with intermediate values will have sulphur from both sources, the ratio being directly calculable by simple mass balance.

There are, of course, several assumptions behind this apparently simple description. One is that the $\delta^{34}S$ signal of all fossil fuel is in the above range. At the moment only a rather small number of samples of power-station flue gases from limited locations have been analysed. A second assumption is that the $\delta^{34}S$ signal of seawater SO_4^{2-} is not altered significantly when DMS crosses the air–sea interface and is oxidized to SO_2 and SO_4^{2-} in the atmosphere. The evidence to date indicates that neither of these assumptions introduces much error, but more work is required to prove this approach.

As might be expected, urban aerosols have a $\delta^{34}S$ signature overlapping to somewhat higher than that from fossil fuels (Fig. 7.21). By contrast, the very few aerosol samples obtained from locations remote from human influence in the South Pacific have a $\delta^{34}S$ value which can approach that of seawater SO_4^{2-}. Results from detailed sampling conducted over a full yearly cycle at Mace Head, a remote site on the west coast of Eire (Fig. 7.21), show almost the whole range of $\delta^{34}S$. Because of the large number of samples collected it has been possible to calculate the percentage of sulphur from the two main sources for different seasons. Thus, in spring and summer approximately 30% of the sulphur in the aerosols at Mace Head comes from DMS (very probably produced by phytoplankton in the northeastern Atlantic, which are only active in any substantial way at these seasons), with the remaining 70% from fossil fuel sources (mainly in Europe, including the UK). In winter essentially all the sulphur is from this latter source. This is a good example of the utility of isotope measurements in environmental sciences, because it is possible to attribute sulphur to its sources without the need to know the strengths of those sources and without recourse to an atmospheric dispersion and deposition model. This avoids the considerable uncertainties associated with estimating these parameters.

7.3.3 The sulphur cycle and climate

In the previous section we examined aerosol particles as sources of acidity in the atmosphere; here we look at their role in controlling climate. First we should note that SO_4^{2-} particles, whether from oxidation of DMS or anthropogenic SO_2, are not the only source of atmospheric aerosols. Other sources include wind-blown dust from soils, smoke from combustion of biomass and industrial processes, and sea-salt particles produced by bubble bursting at the sea surface. However, most study to date has been on SO_4^{2-} aerosols and, for this reason and also because they seem more important in a global context than other types, we will concentrate on them here.

The role of aerosols in climate can be divided into two types: direct and indirect. In the direct effect the particles absorb and scatter energy coming from the sun back to space. This tends to cool the atmosphere since solar radiation, which, in the absence of the aerosols, would warm the air, is now partially absorbed by the particles or reflected upwards out of the atmosphere.

It is difficult to estimate the size of the effect since it depends not only on the total aerosol mass loading in the atmosphere, but also on the chemical composition and size distribution of the particles. However, the effect seems to be significant in terms of climate changes induced by human consumption of fossil fuels. Data in the 2001 report on 'Radiative Forcing of Climate Change' by the Intergovernmental Panel on Climate Change (IPCC) are instructive here. The globally averaged assessment of direct radiative forcing effect of SO_4^{2-} aerosols from fossil fuel burning relative to 1750 (pre-industrial times) is −0.4 (range 0.2 to −0.8) $W m^{-2}$. Similarly the globally averaged figure for biomass burning over the same period is −0.2 (range −0.1 to −0.6) $W m^{-2}$. These numbers can be compared with radiative forcing attributed to greenhouse gas emissions since pre-industrial times of +2.4 (range +2.2 to +2.7) $W m^{-2}$. Four important things should be noted from the comparison. Firstly, the direct effect of aerosols on radiative forcing is smaller globally than that due to greenhouse gases, but is by no means insignificant. Secondly, the sign of the forcing is opposite to that for greenhouse gases, so that the effect of rising aerosol loadings is to reduce to some extent the warming effect of CO_2 and similar gases. Thirdly, the range of the uncertainty on the estimates of the effect of aerosols is very large, for which the Intergovernmental Panel on Climate Change (IPCC 2001, see Section 7.5) classifies the level of scientific understanding as 'low' or 'very low'. Lastly, the spatial distribution of the radiative forcing due to anthropogenic aerosols is very patchy compared with that of the greenhouse gases. This last effect is due to the very different

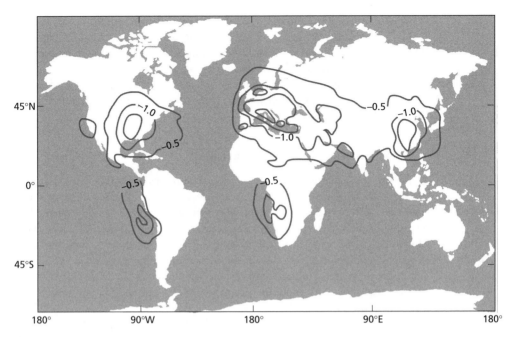

Fig. 7.22 Modelled geographical distribution of annual direct radiative forcing (W m^{-2}) from anthropogenic sulphate aerosols in the troposphere. The negative forcing is largest over, or close to, regions of industrial activity. After IPCC (1995). With permission of the Intergovernmental Panel on Climate Change.

residence times of SO_4^{2-} and other particles in the atmosphere (typically a few days) compared with those of the major greenhouse gases, which remain in the atmosphere for periods of years. An example of this patchiness of the aerosol radiative forcing is shown in Fig. 7.22, which gives the distribution across the globe of the forcing due to anthropogenic SO_4^{2-} aerosols. Not surprisingly in view of where most of the precursor SO_2 is made, coupled with the short residence time of SO_4^{2-} particles, the effect is most pronounced over the continents and especially in regions of high industrial activity.

Indirect effects of aerosols on climate arise from the fact that the particles act as nuclei on which cloud droplets form. In regions distant from land, the number density of SO_4^{2-} particles is an important determinant of the extent and type of clouds. By contrast, over land there are generally plenty of particles for cloud formation from wind-blown soil dust and other sources. Since clouds reflect solar radiation back to space, the potential link to climate is clear. The effect is likely to be most sensitive over the oceans far from land and for snow-covered regions like Antarctica, where land sources of particles have least effect. In such areas a major source of aerosols is the DMS route to SO_4^{2-} particles (Fig. 7.23). Thus, marine phytoplankton are not only the major source of atmospheric acidity but also the main source of cloud condensation nuclei (CCN) and so play an important role in determining cloudiness and hence climate.

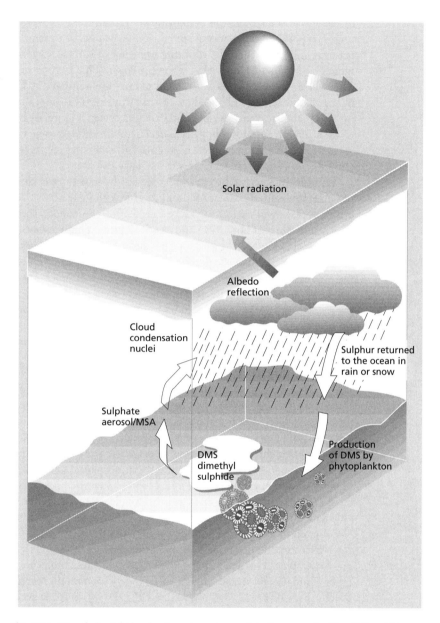

Fig. 7.23 Dimethyl sulphide–cloud condensation nuclei–climate cycle. After Fell and Liss (1993). Reprinted with permission of *New Scientist*.

The concept illustrated in Fig. 7.23 was proposed several years ago, and one group of proponents went further and suggested that the plankton actually played a role in regulating, in contrast to affecting, climate. The idea was that if a change occurred in the temperature of the atmosphere (e.g. due to altered levels of CO_2

or change in solar radiation being received), then the DMS-producing plankton might respond in such a way as to reduce the change. For example, if the air temperature increased then the resulting warming of surface seawater would lead to increased production of DMS by the plankton. This in turn would increase the flux of DMS across the sea surface and so raise the number of CCN in the atmosphere. The resulting enhanced cloudiness would tend to cool the atmosphere, so opposing the warming which initiated the cycle. The process would work in reverse for an initial cooling. If correct, this feedback loop would imply that marine phytoplankton are able to regulate to some degree, at least, the temperature of the atmosphere and thus the Earth's climate.

This idea was tested by examining ice cores from Antarctica for their content of DMS atmospheric oxidation products (MSA and nss-SO_4^{2-}) over the last glacial cycle (as discussed earlier for CO_2; Fig. 7.10). The results, shown in Fig. 7.24, clearly indicate that both MSA and nss-SO_4^{2-} were at higher concentrations during the last glaciation than since its termination about 13 000 years ago. This is the opposite of what would be predicted if planktonic DMS production were reducing any temperature change. Although the results do not support the notion of plankton regulating climate, it is now widely accepted that without CCN formed from DMS the amount of cloudiness, and hence the climate, over large parts of the globe would be significantly different both now and in the past.

7.4 Persistent organic pollutants

Finally, we turn to organic pollutants as examples of exotic chemicals (i.e. those introduced by human manufacture) impacting on the global environment. Organic pollutants are considered persistent when they have a half-life (i.e. the time taken for their concentration to decrease by 50%) of years to decades in a soil or sediment and of several days in the atmosphere. Organic pollutants persist in the environment if they are of low solubility, low volatility or resistant to degradation (see Section 4.10.1 & Box 4.16). Stable aromatic compounds, and highly chlorinated compounds, for example polychlorinated biphenyls (PCBs), polychlorinated dibenzo-*p*-dioxins and furans (PCDD/Fs), 2,2-bis-(*p*-chlorophenyl)-1,1,1-trichloroethane (DDT) and hexachlorocyclohexane (HCH) (Fig. 7.25) are good examples. The deleterious health effects of these molecules on humans and other animals are widely documented being potentially carcinogenic (PCBs, PCDD/Fs, DDT, HCHs), mutagenic (PCDD/Fs) and able to disrupt immune, nervous and reproductive systems.

7.4.1 Persistent organic pollutant mobility in the atmosphere

Many persistent organic pollutants (POPs) are semivolatile organic compounds (SVOCs) having vapour pressures (see Box 4.14) between 10 and 10^{-7} Pa. At these vapour pressures SVOCs can evaporate (volatilize) from soil, water or vegetation into the atmosphere. However, as vapour pressure is temperature dependent (see Box 4.14), it follows that at lower temperatures (lower vapour pressures)

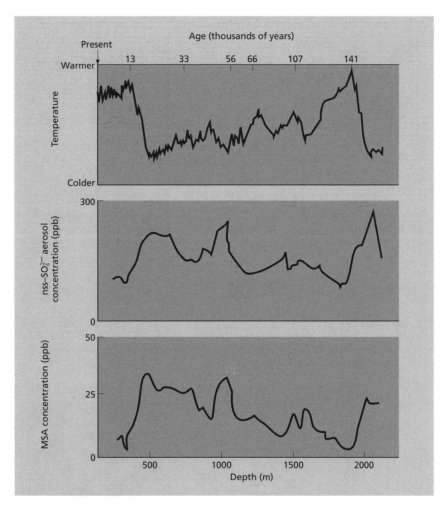

Fig. 7.24 150000-year record of methane sulphonic acid (MSA) concentration, non-sea-salt-sulphate (nss-SO$_4^{2-}$) aerosol concentration and temperature reconstruction (from oxygen isotope data) from an Antarctic ice core. MSA and nss-SO$_4^{2-}$ aerosol concentrations are both high during very cold conditions, e.g. the last glacial period between 18 and 30 thousand years ago (see text for discussion). After Legrand *et al.* (1991).

these SVOCs will condense from the atmosphere back into, or on to, soils, water and vegetation. Given the global poleward movement of air masses (see Section 1.3.2), SVOCs can be transported large distances from their place of manufacture and used in temperate industrial areas to remote polar regions. The low temperatures in polar environments promote condensation of the SVOCs trapping them on the cold land surface and in its vegetation. The overall process has been likened to a global 'distillation system' (Fig. 7.26). Most SVOCs probably require a number of transportation 'hops' to arrive in polar regions, although more volatile organic compounds in rapidly moving air masses may make the journey in a single 'hop' (Fig. 7.26).

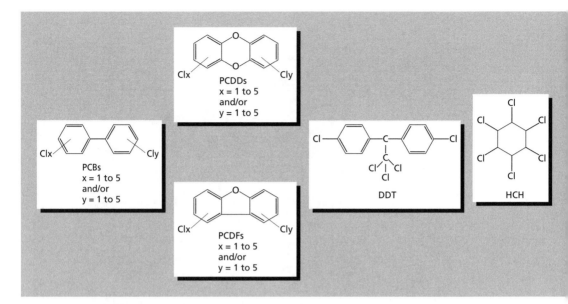

Fig. 7.25 Chemical structures of the contrasting persistent organic pollutants (POPs), polychlorinated biphenyls (PCBs), polychlorinated dibenzo-*p*-dioxins and furans (PCDDs/PCDFs), *p,p′*-dichlorodiphenyl trichloroethane (DDT) and hexachlorocyclohexane (HCH). Symbols x and y indicate the possible number of chlorines attached to the ring structures.

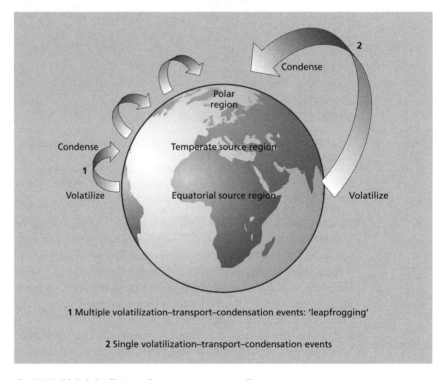

Fig. 7.26 Global distillation of persistent organic pollutants.

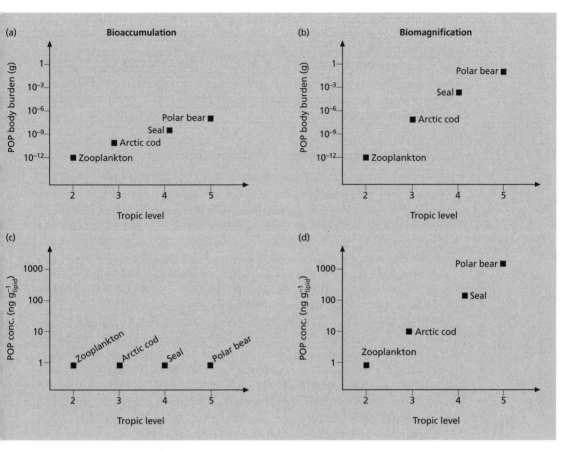

Fig. 7.27 Bioaccumulation (a,c) and biomagnification (b,d) of persistent organic pollutants (POPs) with increasing trophic level (i.e. position occupied by a species in the food chain). Trophic level 2 animals are primary consumers (herbivores) while higher-level animals are carnivores.

Once in the polar environment SVOCs interact with the local ecosystem. Many POPs are hydrophobic (dislike water) and are therefore lipophilic (affinity for fats; see Box 4.14). Consequently, if animals ingest POPs they are partitioned into the organism's fat reserves. This partitioning stores the compound in fats, excluding it from metabolism and excretion. This scenario potentially allows POP concentrations to increase up the trophic level within food chains (Fig. 7.27). We should note here the distinction between *bioaccumulation*, where higher organisms have increased body burdens of POPs on account of the organisms being bigger, and *biomagnification*, where higher organisms not only have increased body burdens of POPs but also exhibit a higher concentration of POPs per unit mass of lipid (Fig. 7.27). It was the bioaccumulation and biomagnification of the insecticide DDT (Fig. 7.25) that was highlighted in *Silent Spring*, Rachel Carson's seminal book of the 1960s. In it, Carson showed how the decline of common North American bird species was due to consumption of food laden with this poison.

The indigenous humans of the arctic rely on local animals, particularly those high up the food chain such as seals and caribou, for their food source. Animals like seals are rich in fat and thus people living in the arctic can have very high exposure to POPs through their diet. For example, a Beluga whale washed up on the St Lawrence seaboard contained PCBs in excess of $50\,\text{mg}\,\text{kg}_{\text{lipid}}^{-1}$. This PCB concentration classified the whale as toxic waste. It is sadly ironic that POPs contaminate arctic environments, ecosystems and peoples who were never involved in their manufacture or use.

7.4.2 Global persistent organic polllutant equilibrium

The manufacture and use of many exotic organic compounds (see Section 1.4) has now been discontinued because of their persistence, potential health effects and global mobility. For example, PCBs were first manufactured and used in the 1930s and their usage increased until the early 1970s. Thereafter PCB usage was banned in many instances or subject to restrictions. However, PCBs did not immediately disappear from the environment. Even today PCBs are present in all of Earth's environmental reservoirs and they are distributed globally. What has changed, following discontinuation of widespread PCB use, is the equilibrium between PCB concentrations in air, water and soil. In the 1970s, when PCB concentrations were at their peak, PCBs in the atmosphere were at much higher concentrations than today. These high atmospheric concentrations caused a net flux of PCBs from the atmosphere to the Earth's surface waters and soils to establish equilibrium between these reservoirs (Fig. 7.28). Today, PCB concentrations in the atmosphere are much lower. This has disturbed the equilibrium between soils and water and the overlying air such that the net flux of PCBs is now out of the soil and water into the atmosphere to re-establish equilibrium (Fig. 7.28). This situation is termed re-emission or secondary emission.

Unequivocal evidence for the re-emission of POPs has been demonstrated in an elegant study of the Great Lakes that border the USA and Canada. This study focused on the POP α-hexachlorocyclohexane (α-HCH), a chiral compound (Box

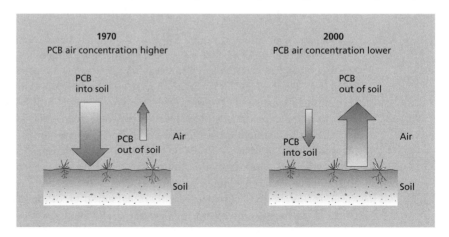

Fig. 7.28 Environmental equilibration of polychlorinated biphenyls (PCBs) between soil and air.

7.3) produced industrially as a racemic mixture of two enantiomers (Fig. 7.29 & Box 7.3). α-HCH is a pesticide that has been widely applied throughout the USA, primarily to control cotton pests. At the time of application both α-HCH enantiomers were present in equal proportions.

α-HCH, is a SVOC, and thus will evaporate following its application as a pesticide, to be transported in the atmosphere and then redeposited by condensation (Section 7.4.1). Interestingly, one of the α-HCH enantiomers is selectively degraded. Thus, as time passes one of the enanitomers decreases in concentration while the other maintains its concentration. This changes the concentration ratio of the enantiomers from an initial ratio of 1 to a progressively lower value with time. The ratio of α-HCH enantiomers in air above the Great Lakes is now 0.85, identifying its source as 'old' α-HCH that has been degraded during storage, but that is now escaping from the lake bed sediments, through the water and re-equilibrating with the atmosphere. Moreover, this re-emission has a seasonal signal, being greatest in the warm summer months when volatilization is encouraged by higher vapour pressures. By measuring not only the concentration of α-HCH above the Great Lakes, but also the concentration ratio of the enantiomers, it is clear that the Great Lakes now behave as a source of α-HCH and not a sink. This case study highlights the complexity of mitigating the effects of pollution by exotic chemicals. Removal of a contaminant from the environment involves far more than just stopping manufacture or use.

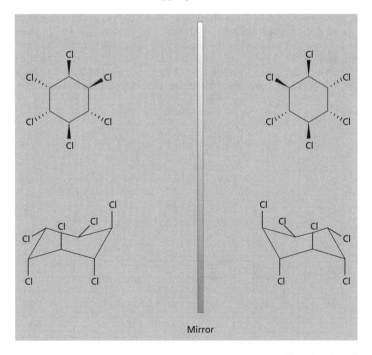

Fig. 7.29 The enantiomers of α-hexachlorocyclohexane (α-HCH). Bold wedge shaped bonds represent bonds rising from the plane of the page toward the viewer, and bonds represented by dashed lines are receding from the plane of the page away from the viewer. See also Box 7.3.

Box 7.3 Chiral compounds

Molecules that cannot be superimposed on their mirror images are said to be *chiral* (from Greek *kheir*—hand). A pair of molecules that fulfil this condition are called enantiomers (from the Greek *enantio*— opposite). Consider a molecule comprising a central carbon atom to which the following groups are attached in a tetrahedral array: $-CH_3$, $-H$, $-Br$, $-COOH$ (Fig. 1.). In this figure the molecule is represented as a tetrahedron. Bonds represented by lines are in the plane of the page; bonds represented by wedged lines are rising from the plane of the page towards the viewer, and bonds represented by dashed lines are receding from the plane of the page away from the viewer. The mirror image of this molecule cannot be superimposed upon the original molecule. Try rotating the mirrored-molecule around the vertical bond; you will find that as the groups rotate round they never arrive in a position that would allow them to be superimposed on the original molecule. Thus, this molecule is chiral and the carbon atom to which the groups are attached is termed the *chiral centre*. By contrast, a molecule comprising a central carbon atom to which the following groups are attached in a tetrahedral array,

$-CH_3$, $-CH_3$, $-Br$, $-COOH$ (Fig. 1.), has a mirror image that can be superimposed upon the original molecule. This molecule is *achiral* and the molecule and its mirror image are not enantiomers.

It is possible to describe enantiomers without drawing the structures by assigning a particular enantiomer the letter 'R' or 'S'. This system ascribes priority to the groups attached to the chiral centre in accordance with their atomic weight. The higher the atomic weight the higher the priority assigned. In a simple case where, for example, the central carbon atom is attached to -Cl, -Br, -I and –H, then priority is given in the order I > Br > Cl > H on account of the atomic weights being 127, 80, 35 and 1, respectively. Where for example a $-CH_3$ group and a $-COOH$ group are attached to the central carbon atom, i.e. the central carbon atom is attached to another carbon atom in both cases, then the groups attached to these carbon atoms need to be considered. In this example priority is given to -COOH as 'O' has a greater atomic weight than 'H'. Having assigned priority to the groups, the group of lowest priority is projected away from the viewer. The viewer then establishes if the

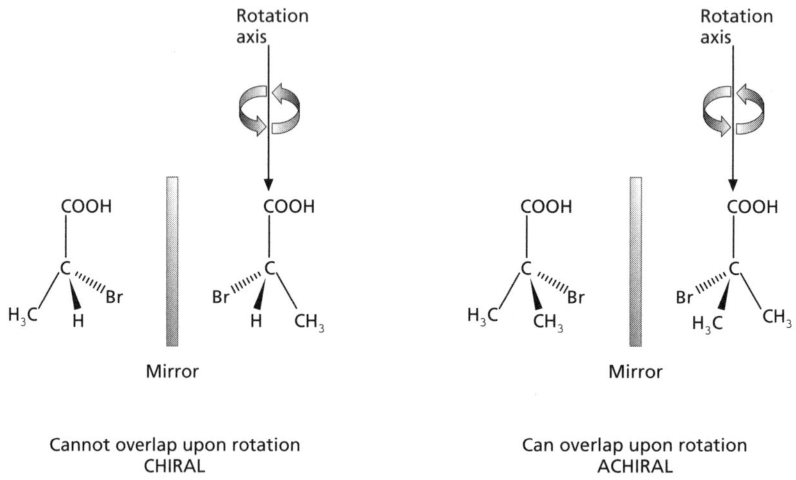

Fig. 1 Contrasting examples of chiral and achiral molecules.

(continued)

route from the highest priority group to the second highest priority group lies in a clockwise or anticlockwise direction. Clockwise is assigned 'R' (from the Latin *rectus*, right) while anticlockwise is assigned 'S' (from the Latin *sinister*, left). In the chiral molecule we began with (Fig. 1), prioritizing the groups yields the order –Br > -COOH > -CH$_3$ > -H. Figure 2 shows the assignment of 'R' and 'S' to the enantiomers. In this figure, as in Fig. 1, the wedge bonds rise from the plane of the page toward the viewer; and the dashed bond recedes from the plane of the page away from the viewer.

A mixture containing equal proportions of two enantiomers is termed a racemic mixture and is indicated by the prefix 'RS-'. During organic synthesis it is usual for a chiral product to be produced as an RS mixture (unless the reaction has been tailored to produce a specific chiral product). Chirality is important because it affects a molecule's activity. For example, the compound thalidomide has one enantiomer that is a

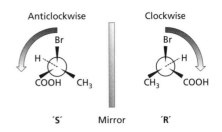

Fig. 2 Assignment of 'R' and 'S' to the enantiomers of a molecule comprising a central carbon atom (central circle) to which –CH$_3$, –H, –Br, –COOH groups are attached.

valuable therapeutic drug (a sedative) while the other enantiomer is highly toxic. Between 1957 and 1960 thalidomide was prescribed in at least 46 countries to alleviate morning sickness associated with pregnancy. As a consequence of the two enantiomers being present thousands of babies were born with birth defects.

7.5 Further reading

Cox, R.A., Hewitt, C.N., Liss, P.S., Lovelock, J.E., Shine, K.R. & Thrush, B.A. (eds) (1997) Atmospheric chemistry of sulphur in relation to aerosols, clouds and climate. *Philosophical Transactions of the Royal Society of London* **352B**, 139–254.

IPCC (2001) *Climate Change 2001: The Scientific Basis. Contribution of Working Group I to the Third Assessment Report, Intergovernmental Panel on Climate Change.* Cambridge University Press, Cambridge.

Jones, K.C and De Voogt, P. (1999) Persistent organic pollutants (POPs): state of the science. *Environmental Pollution* **100**, 209–221.

Schimel, D.S. & Wigley, T.M.L. (eds) (2000) *The Carbon Cycle.* Cambridge University Press, Cambridge.

Turner, B.L., Clark, W.C., Kates, R.W., Richards, J.F., Mathews, J.T. & Meyer, W.B. (eds) (1990) *The Earth as Transformed by Human Action.* Cambridge University Press, Cambridge.

7.6 Internet search keywords

IGBP
global carbon cycle
past global changes
CO$_2$ Mauna Loa
atmospheric CO$_2$

CO$_2$ ice cores
CO$_2$ biosphere
CO$_2$ O$_2$ seasonality
CO$_2$ fossil fuel burning
CO$_2$ deforestation

FACE experiments
IPCC
CO_2 oceans
CO_2 oceans gas exchange
CO_2 iron fertilization
CO_2 climate
CO_2 greenhouse effect
future atmospheric CO_2 predictions
global warming sealevel
global sulphur cycle

dimethyl sulphide
atmospheric acidity
atmospheric aerosols
cloud condensation nuclei DMS
persistent organic pollutants
persistent organic pollutants polar
    bioaccumulation
persistent organic pollutants
    biomagnification

Index